Urban Life and the Ambient in Smart Cities, Learning Cities, and Future Cities

H. Patricia McKenna
AmbientEase, Canada

IGI Global
PUBLISHER of TIMELY KNOWLEDGE

A volume in the Advances in Civil and Industrial Engineering (ACIE) Book Series

Published in the United States of America by
IGI Global
Engineering Science Reference (an imprint of IGI Global)
701 E. Chocolate Avenue
Hershey PA, USA 17033
Tel. 717-533-8845
Fax: 717-533-8661
E-mail: cust@igi-global.com
Web site: http://www.igi-global.com

Library of Congress Cataloging-in-Publication Data

Names: McKenna, H. Patricia, 1954- author.
Title: Urban life and the ambient in smart cities, learning cities, and future cities / by H. Patricia McKenna.
Description: Hershey, PA : Engineering Science Reference, [2023] | Includes bibliographical references and index. | Summary: "This publication is about urban life and the urbanizing process in the context of more aware people interacting with more aware technologies, contributing to understandings of the ambient turn in smart cities, learning cities, and future cities embracing rethinking of more sustainable and livable approaches to urban life in terms of research and practice"-- Provided by publisher.
Identifiers: LCCN 2022025442 (print) | LCCN 2022025443 (ebook) | ISBN 9781668440964 (h/c) | ISBN 9781668440971 (pbk) | ISBN 9781668440988 (eISBN)
Subjects: LCSH: Smart cities. | City and town life. | City planning.
Classification: LCC TD159.4 .M384 2023 (print) | LCC TD159.4 (ebook) | DDC 307.76--dc23/eng/20220725
LC record available at https://lccn.loc.gov/2022025442
LC ebook record available at https://lccn.loc.gov/2022025443

This book is published in the IGI Global book series Advances in Civil and Industrial Engineering (ACIE) (ISSN: 2326-6139; eISSN: 2326-6155)

British Cataloguing in Publication Data
A Cataloguing in Publication record for this book is available from the British Library.

Advances in Civil and Industrial Engineering (ACIE) Book Series

Ioan Constantin Dima
University Valahia of Târgovişte, Romania

ISSN:2326-6139
EISSN:2326-6155

Mission

Private and public sector infrastructures begin to age, or require change in the face of developing technologies, the fields of civil and industrial engineering have become increasingly important as a method to mitigate and manage these changes. As governments and the public at large begin to grapple with climate change and growing populations, civil engineering has become more interdisciplinary and the need for publications that discuss the rapid changes and advancements in the field have become more in-demand. Additionally, private corporations and companies are facing similar changes and challenges, with the pressure for new and innovative methods being placed on those involved in industrial engineering.

The **Advances in Civil and Industrial Engineering (ACIE) Book Series** aims to present research and methodology that will provide solutions and discussions to meet such needs. The latest methodologies, applications, tools, and analysis will be published through the books included in **ACIE** in order to keep the available research in civil and industrial engineering as current and timely as possible.

Coverage

- Coastal Engineering
- Production Planning and Control
- Construction Engineering
- Materials Management
- Quality Engineering
- Ergonomics
- Operations Research
- Productivity
- Hydraulic Engineering
- Optimization Techniques

IGI Global is currently accepting manuscripts for publication within this series. To submit a proposal for a volume in this series, please contact our Acquisition Editors at acquisitions@igi-global.com or visit: https://www.igi-global.com/publish/.

The Advances in Civil and Industrial Engineering (ACIE) Book Series (ISSN 2326-6139) is published by IGI Global, 701 E. Chocolate Avenue, Hershey, PA 17033-1240, USA, www.igi-global.com. This series is composed of titles available for purchase individually; each title is edited to be contextually exclusive from any other title within the series. For pricing and ordering information please visit https://www.igi-global.com/book-series/advances-civil-industrial-engineering/73673. Postmaster: Send all address changes to above address.

Titles in this Series

For a list of additional titles in this series, please visit: www.igi-global.com/book-series/advances-civil-industrial-engineering/73673

Remapping Urban Heat Island Atlases in RegenerativeCities
Hisham Abusaada (Housing and Building National Research Center, Egypt) Abeer Elshater (Ain Shams University, Egypt) and Marwa Khalifa (Ain Shams University, Egypt)
Engineering Science Reference • © 2022 • 267pp • H/C (ISBN: 9781668424629) • US $250.00

Clean Technologies and Sustainable Development in Civil Engineering
Raluca- Andreea Felseghi ("Ştefan cel Mare" University of Suceava, Romania) Nicoleta Cobîrzan (Technical University of Cluj-Napoca, Romania) and Maria Simona Raboaca (National Research and Development Institute for Cryogenic and Isotopic Technologies, Râmnicu Vâlcea, Romania)
Engineering Science Reference • © 2022 • 290pp • H/C (ISBN: 9781799898108) • US $250.00

Health and Well-Being Considerations in the Design of Indoor Environments
Roberto A. González-Lezcano (Universidad CEU San Pablo, Spain)
Engineering Science Reference • © 2021 • 286pp • H/C (ISBN: 9781799872795) • US $240.00

Visibilities and Invisibilities in Smart Cities Emerging Research and Opportunities
H. Patricia McKenna (AmbientEase, Canada)
Engineering Science Reference • © 2021 • 289pp • H/C (ISBN: 9781799838500) • US $215.00

Transforming Urban Nightlife and the Development of Smart Public Spaces
Hisham Abusaada (Housing and Building National Research Center, Egypt) Abeer Elshater (Ain Shams University, Egypt) and Dennis Rodwell (Independent Researcher, UK)
Information Science Reference • © 2021 • 290pp • H/C (ISBN: 9781799870043) • US $215.00

Handbook of Research on Driving Transformational Change in the Digital Built Environment
Jason Underwood (University of Salford, UK) and Mark Shelbourn (Birmingham City University, UK)
Engineering Science Reference • © 2021 • 569pp • H/C (ISBN: 9781799866008) • US $325.00

Structural Dynamics and Static Nonlinear Analysis From Theory to Application
Mourad Belgasmia (Setif 1 University, Algeria)
Engineering Science Reference • © 2021 • 347pp • H/C (ISBN: 9781799843993) • US $215.00

Handbook of Research on Developing Smart Cities Based on Digital Twins
Matteo Del Giudice (Politecnico di Torino, Italy) and Anna Osello (Politecnico di Torino, Italy)
Engineering Science Reference • © 2021 • 674pp • H/C (ISBN: 9781799870913) • US $325.00

IGI Global
PUBLISHER of TIMELY KNOWLEDGE

701 East Chocolate Avenue, Hershey, PA 17033, USA
Tel: 717-533-8845 x100 • Fax: 717-533-8661
E-Mail: cust@igi-global.com • www.igi-global.com

Table of Contents

Preface

OVERVIEW

This work is inspired by the challenges and opportunities enabled by 'the ambient' through the development of technologies with awareness-enabling capabilities in smart environments and the potential for extending and augmenting human capabilities. Building upon the considerable research and writing on the ambient (Marzano, 2006; Rickert, 2013; Streitz et al., 2019) much remains to be done, thus motivating the explorations in this work. This book is intended to encourage us to become more aware and seek to understand more about the ambient in relation to urban life and everyday life in the context of smart cities, learning cities, and smart cities. While 'the ambient' is not a new phenomenon, manifesting as it does in such things as people being able to sense and detect the presence and direction of the wind for example, what is new is the emergent potential of 'the ambient' in relation to technologies in smart environments and in turn, people interacting with such technologies. McCullough (2013) describes the ambient in a variety of ways including "a continuum of awareness" and "that which surrounds but does not distract." Many books are available on the topic of smart cities and this work expands upon, broadens, and extends current thinking by contributing an ambient perspective to explorations of urban life in smart environments, now and going forward. As such, ambient perspectives are offered in this work on urban life in the context of smart cities, learning cities, and future cities, enriching understandings of the experience and interaction potential of people in technology-rich environments characterized by increased awareness, adaptability, and in the moment responsiveness.

In terms of organization, each chapter in this book focuses on a particular dimension of the conceptual framework for urban life and the ambient in smart cities and is intended to be a complete work able to be read on its own or as part of the book. Accordingly, the following elements are included in each chapter: an introduction including the main objectives and research question; a background including definitions for key terms used; a review of the research literature for the dimension and associated factors being explored, identifying key issues, controversies, and problems; formulation of a conceptual framework guiding the proposition to be explored as a reformulation of the research question; the research design; a description of findings followed by a discussion; limitations and future directions for research and practice; a conclusion including insights, ideas, and exercises encouraging further explorations; identification of the target audience; references; additional readings; and key terms and definitions. This book is organized into four sections consisting of twelve chapters where Section 1 includes three chapters and focuses on providing an introduction, ambient perspectives, and the challenges and opportunities of urban engagement and participation. Section 2 includes three chapters and focuses on urban life and the ambient in relation to infrastructures, literacies, and data streams. Section 3 includes two chapters and focuses on urban life and the ambient in relation to emergent developments pertaining to smart technologies in relation to artificial intelligences (AI), ambient learning, smart homes, electric vehicles (EVs), robotics, and extended reality. Section 4 includes four chapters and focuses on theory, approaches, and metrics for urban life and the ambient; governance, livability, and sustainability for enriching urban life; forging a way forward involving human flourishing; and the final chapter provides a summary and conclusion.

A CHAPTER-BY-CHAPTER DESCRIPTION

What follows is a brief, chapter-by-chapter description, highlighting the importance of each while outlining the flow of the book.

Chapter 1 provides an introduction to the book along with a broad conceptual framework for urban life and the ambient while introducing the various dimensions to be addressed in each chapter, based on a review of the research literature for urban life, smart cities, learning cities, and future cities. Patterns and potential relationships are identified along with issues, controversies, and problems. A research question for each chapter is provided along with a reformulation of each question into a proposition to be explored chapter by chapter. Providing an example of the research design used in each chapter, Chapter 1 combines an exploratory case study approach with an explanatory correlational design in conducting an exploration of assessments for key variables identified, based on the research literature. As such, the issue of *funding* is assessed in relation to the variables of *success of smart city projects* and *livability*.

Chapter 2 provides an ambient perspective for urban life and smart cities, learning cities, and future cities with a review of the research literature focusing on culture, economy, and mobility. A conceptual framework is formulated in support of the exploration conducted, focusing on a variety of variables pertaining to the ambient including *attuning to urban spaces* and *heightening urban sensibilities*. For urban culture, economies, and mobilities the variables of *multiple modes of transport* and *walkability* are surfaced for exploration and assessment, among others. This chapter is intended as an aid to understanding the ambient dimension of everyday life involving people and their interactions in urban and everyday spaces, together with awareness-enabling technologies.

Chapter 3 takes a practical approach to understanding urban life and the ambient through explorations of engagement and participation focusing on the elements of collaboration, openness, and privacy. Guided by the literature review, a conceptual framework for urban engagement and participation in smart cities is formulated enabling an exploration of a range of variables surfaced, including *collaborating* and *technology-driven services*; *collaborating* and *privacy*; and *community participation* and *privacy*. This chapter is attentive to emergent forms of the ambient such as ambient collaboration and ambient networking, discussed in relation to virtual conference events and environments.

Chapter 4 provides perspectives on evolving infrastructures in smart cities through a review of the research literature for urban life, experiences, interactions, and augmentations. A conceptual framework for urban life and evolving infrastructures in smart cities is then formulated, guiding an exploration of variables identified, including *invisible infrastructures* and *visible infrastructures*, and the experience of everyday interactions in urban spaces focusing on elements such as *air quality*.

Chapter 5 provides a review of the research literature pertaining to literacies for urban life and the ambient in smart cities focusing on the digital, the algorithmic, play, and inclusion enabling formulation of a conceptual framework to guide explorations. A variety of variables are explored in coming to broader understandings of the evolving nature of urban literacies focusing on *education* and *placemak-*

ing; *interactive public spaces* and *inclusiveness*; and *interactive public spaces* and *ICTs* (Information and Communication Technologies).

Chapter 6 explores the research literature pertaining to data, information, and knowledge in the context of smart cities focusing also on ways of knowing and computational intelligence (CI). A conceptual framework is formulated guiding the exploration of data, information, and knowledge in relation to the variables of *access to public data* and *information*; *information* and *ICTs* (Information and Communication Technologies); and *meaningfulness* and *visual ways to show success in real time*.

Chapter 7 addresses the issue of urban life and artificial intelligence (AI) including machine learning (ML) and deep learning (DL) while also exploring the notion of ambient learning in the context of smart cities, learning cities, and future cities. Aided by a review of the research literature, a conceptual framework for AI in urban life and the ambient in smart cities is formulated, guiding an exploration focusing on the variables of *collaboration* and *creative opportunities,* revealing what is considered to be a very good correlation (Creswell, 2018).

Chapter 8 provides a review of the research literature for smart technologies involving smart homes, electric vehicles (EVs), and robotics in smart cities, incorporating considerations of extending realities in urban life through augmented, mixed, and virtual applications. A conceptual framework is formulated from this review guiding explorations of variables identified such as *connecting* and *trust* (pertaining to awareness-enabling technologies) and *access to public data* (pertaining to success) and *trust* (pertaining to data value).

Chapter 9 explores theories, approaches, and metrics for understanding urban life and the ambient in the context of smart cities, learning cities, and future cities, highlighting the importance of activity theory and ambient theory, among other theories. A conceptual framework is formulated to guide researchers and practitioners in their use of theories, approaches, and metrics for smart cities. The framework is operationalized for use in the chapter guiding an exploration of variables identified such as *collaborating* and *visualizations of data that inform, educate, and inspire in real time.*

Chapter 10 explores governance, livability, and sustainability in smart cities from the perspective of enriching urban life and the ambient involving greater adaptability, awareness, and creativity through a review of the research literature. A conceptual framework is formulated guiding the exploration of variables identified such as *housing affordability* and *useful community functions and purposes.*

Chapter 11 provides an exploration of human flourishing (Ekman and Simon-Thomas, 2021; CGC, 2014) in the context of smart cities and future cities in providing an integration of the chapters in this book while seeking to navigate a way forward. A review of the research literature is conducted focusing on aspects of flourishing including connection, positivity, and resilience. A conceptual framework is formulated guiding the exploration in this chapter of variables such as *connecting*, *technology-driven services*, *collaborating*, and *trust.*

Chapter 12 provides a conclusion to the book by presenting tabular summaries of key aspects of the book chapters. A componential view of urban life and the ambient is developed for understanding and working with smart cities, learning cities, and future cities. The importance of flourishing (Hedlund & Esbjörn-Hargens, 2022; Miller, Douglass, & Rigg, 2020), more broadly, to include all life forms is advanced, extending as well to the planet earth and possibly beyond. Key aspects of flourishing serve as a unifying element for this book in that chapter by chapter, many such aspects emerge for exploration as critical components of smart cities now and going forward.

As such, key terms shedding light, at-a-glance, on the focus of the chapters in this book include the following:

- Air quality
- Artificial intelligence (AI)
- Collaboration
- Community participation
- Computational intelligence (CI)
- Connecting
- Creativity
- Data
- Education
- Engagement
- Flourishing
- Funding
- Housing affordability
- Human flourishing
- ICTs (Information and Communication Technologies)
- Inclusiveness
- Information
- Interactive public spaces
- Knowledge
- Knowing
- Livability
- Meaningfulness
- Multiple modes of transport
- Openness
- Placemaking
- Privacy
- Resilience
- Sharing
- Sustainability
- Technology-driven services
- Trust
- Useful community functions and purposes
- Visualizations of data
- Walkability

IMPACT AND CONTRIBUTION

As an interdisciplinary work, this book is intended to have a broad reach, weaving together as it does key aspects of everyday life in formulation of an integrated conceptual framework for urban life and the ambient in smart cities while being attentive to beneficial outcomes such as improved quality of life (QoL) and quality of urban life (QoUL). This book is intended to be used as a textbook for college and university courses in a variety of domains, including but not limited to data analytics, human-computer interaction, human geography, urban planning, and sociology. Potential exists for this book to be used by educators designing courses for artificial intelligence (AI), software development and design, and technology and society studies, to name a few. Three exercise questions are included at the end of each chapter to encourage and guide further explorations.

The topic of urban life and the ambient in smart cities, learning cities, and future cities is a timely one, fitting as it does in the world today in an interdisciplinary way across many areas of research and practice. While urban life pertains to the practical and complex issues of everyday life in the city, the ambient has relevance to people and their awareness and attention (McCullough, 2013) in urban space (McKenna, 2019), aided and augmented by increasingly pervasive awareness-enabling technologies. As such, the target audience for this work cuts across many domains, traversing boundaries, calling upon everyone to think about and engage with smarter ways of approaching the complex challenges and opportunities afforded by tapping into and augmenting our creativity, awareness, and sensing in smart environments.

This book contributes to the study and practice of urban life and smart cities using an ambient perspectives lens in reinvigorating and enriching thinking on cities, learning cities, and future cities in the face of complex challenges and opportunities intensified by rapidly changing weather patterns, a global pandemic, and other uncertainties affecting many aspects of daily life.

H. Patricia McKenna
AmbientEase, Canada

REFERENCES

CGC. (2014). How can we make cities flourish? In *Challenges of Government Conference.* Blavatnik School of Government, University of Oxford. Retrieved 3 March 2022 from https://www.bsg.ox.ac.uk/news/how-can-we-make-cities-flourish

Creswell, J. W. (2018). *Educational research: Planning, conducting, and evaluating quantitative and qualitative research* (6th ed.). Pearson.

Ekman, E., & Simon-Thomas, E. (2021). Teaching the science of human flourishing, unlocking connection, positivity, and resilience for the greater good. *Global Advances in Health and Medicine: Improving Healthcare Outcomes Worldwide, 10.* Advance online publication. doi:10.1177/21649561211023097 PMID:34123578

Hedlund, N., & Esbjörn-Hargens, S. (Eds.). (2022). *Big Picture Perspectives on Planetary Flourishing: Metatheory for the Anthropocene* (Vol. 1). Routledge. doi:10.4324/9781003140313

Marzano, S. (2006). Ambient culture. In E. Aarts & J. Encarnação (Eds.), *True visions: The emergence of ambient intelligence* (pp. 35–52). Springer. doi:10.1007/978-3-540-28974-6_3

McCullough, M. (2013). *Ambient Commons: Attention in the Age of Embodied Information.* The MIT Press. doi:10.7551/mitpress/8947.001.0001

McKenna, H. P. (2019). *Ambient Urbanities as the Intersection Between the IoT and the IoP in Smart Cities.* IGI Global. doi:10.4018/978-1-5225-7882-6

Miller, M. A., Douglass, M., & Rigg, J. (2020). Governing resilient cities for planetary flourishing in the Asia-Pacific. *Urban Studies (Edinburgh, Scotland)*, *57*(7), 1359–1371. doi:10.1177/0042098020903955

Rickert, T. (2013). *Ambient rhetoric: The attunements of rhetorical being*. University of Pittsburgh Press. doi:10.2307/j.ctt5hjqwx

Streitz, N., Charitos, D., Kaptein, M., & Böhlen, M. (2019). Grand challenges for ambient intelligence and implications for design con- texts and smart societies. Tenth Anniversary Issue. *Journal of Ambient Intelligence and Smart Environments*, *11*(1), 87–107. doi:10.3233/AIS-180507

Acknowledgment

I would like to acknowledge the important contributions from the many individuals who participated in the urban research study forming the basis for this work as well as the many individuals and groups who shared their time, thinking, concerns, and insights in discussions about urban life, the ambient, and smart cities.

A special thanks is extended to IGI Global for the invitation and opportunity to contribute to the smart cities space. Evaluations received through the blind peer review process managed by IGI Global provided important and valuable feedback for the editing and revising of this manuscript. Additionally, the guidance and timely support received from the IGI Global Team throughout this project is very much appreciated.

A very special note of thanks to friends and colleagues always ready to provide supportive and encouraging words. And I am infinitely grateful to Jeanine for her unwavering support in untold ways from the beginning to the end of this undertaking.

Section 1

An Introduction and Background to Urban Life and the Ambient

This section introduces the reader to the notion of urban life and sensing over the continuum of past, present, and future with a focus on smart cities, learning cities, and future cities. The use of an ambient perspective is employed across urban life, touching on many aspects such as culture, economy, and mobility. Challenges and opportunities are addressed focusing on urban engagement and participation and approaches to collaboration, openness, and privacy.

Chapter 1
Urban Life and Smart, Learning, and Future Cities:
Getting a Sense of the City – Past, Present, and Future

ABSTRACT

The purpose of this chapter is to provide an introduction and background to the concept of urban life generally and in relation to evolving understandings of smart cities. A review of the research literature is provided for urban life, smart cities, learning cities, and future cities identifying issues, controversies, and problems. The literature review enables formulation of a conceptual framework for urban life and the ambient in smart cities which is then operationalized for use in responding to the range of issues, controversies, and problems identified in the research literature. Using an exploratory case study approach involving survey and in-depth interviews, a series of research questions are developed for exploration as propositions, chapter-by-chapter in this book. Combining an explanatory correlational design with a case study approach, variables are identified for exploration, pertaining to urban life and smart cities in this chapter and the chapters that follow, informing directions for research and practice.

1. INTRODUCTION

This chapter explores the urban life concept from multiple perspectives and over time in relation to evolving understandings of smart cities, learning cities, and future cities. While urban life has been explored by Psatha, Deffner, and Psycharis (2011) in terms of "quality of urban life (QOUL)" as "a concept able to monitor the multidimensional nature of cities" it is also explored in addressing "the problems that undermine" QOUL such as "social inequalities, urban crime, poor environment, and traffic con[g]estion." Smart cities are described by Townsend (2013) as "places where information technology is combined with infrastructure, architecture, everyday objects and even our bodies, to address social, economic, and environmental problems." Nam and Pardo (2011) note that "a smart city is also a learning city" forming

DOI: 10.4018/978-1-6684-4096-4.ch001

part of the human infrastructure and social capital that make up the human factor component of smart cities, along with creative city, humane city, and knowledge city. Through their work, Nam and Pardo (2011) highlight the need for future research to explore the question: "how do smart technologies change a city?" giving rise to a key ambition of this book. Macionis and Parrillo (2013) provide a global overview of cities and urban life from a sociological perspective, addressing problems encompassing spiraling populations, quality of life (QoL), environment, and shantytowns while also focusing on the future of cities in the context of urban planning and "how current trends give us insight" in terms of growth and development. Schmitt (2019) explores the notion of future cities using "data driven approaches" that are "based on crowdsourcing and sensing" while seeing "the city as the most complex human-made 'organism'." Sumara and Alvermann (2022) draw attention to changing notions of literacy practices, important in this chapter for areas such as funding and what Boggs (2022) refers to as "economic relevance" and the efforts by students and others "to make themselves useful in their communities." As such, this chapter is significant in that it seeks to provide a sense of the city through the continuum of past – present – future, where urban life is the connecting thread and smart cities, learning cities, and future cities are the emerging and evolving concepts giving rise to, and motivating the explorations identified in the following objectives.

Objectives: The key objectives of this chapter are to: a) provide perspectives on urban life while introducing evolving understandings of smart cities; b) explore the concepts of learning cities and future cities in the context of urban life in smart cities; c) develop a conceptual framework for urban life and the ambient in smart cities, learning cities, and future cities; and d) formulate a series of research questions with propositions to be explored in the chapters that follow, based on the example provided in this chapter (e.g., the relationship between *funding* for smart city projects and factors such as *success* and *livability*).

As such, the key research question in this chapter is: *How does support for improving urban life emerge in the context of smart cities?*

2. BACKGROUND

From a sociological perspective, Wirth (1938) highlighted the "three variables" of "number, density of settlement, and degree of heterogeneity" as being able "to explain the characteristics of urban life." Cugurullo (2020) describes the smartness of cities by introducing the notion of "urban artificial intelligence" in the movement from "*automation* to *autonomy*." Concerned with exploring "a sustainable future for human and nature", Zhang and He (2020) provide an introduction to a special issue of smart technologies and urban life from a behavioral and social perspective, in the context of "the explosive growth of sensing technologies, artificial intelligence (AI), and Internet of Things (IoT)" contributing to "a paradigm shift in the ways of experiencing and researching urbanism." It is worth noting that a CitiesX course is being offered in the form of a massive open online course (MOOC), focusing on "the past, present and future of urban life" from an economics perspective enriched with perspectives from anthropology, history, and urban planning, to name a few (EdX, 2021). Batty (2022) speculates on the shape that future cities will take from an advanced spatial analysis perspective, claiming that the city as a concept will disappear in a future that is entirely urban.

2.1 Definitions

Definitions from the research and practice literature are provided in this section for key terms used in this chapter.

- **Future Cities:** Schmitt (2019) develops the notion of future cities in terms of "the importance of information and information architecture" along with simulation to "understand, shape, plan, design, build, manage, and continually adapt a city."
- **Learning Cities:** The learning city is described by UNESCO (2019) as "one that effectively mobilizes resources in every sector to promote inclusive learning, from basic to higher education."
- **Smart Cities:** The smart city is defined by Augusto (2020) as one "where technology is deployed and effectively used to improve the life experience of its inhabitants" and where technology is "a necessary but not sufficient component."
- **Urban Life:** Wirth (1938) employs a theory of urbanism involving three variables – number, density, and degree of heterogeneity – "to explain the characteristics of urban life."

2.2 Overview

What follows is a review of the research and practice literature for urban life and smart cities, learning cities, and future cities identifying issues, controversies, and problems. A conceptual framework for urban life and the ambient in smart cities is then developed, based on the literature review, and is operationalized for use in this chapter. A series of research questions are identified and reformulated as propositions to be explored chapter-by-chapter in this work. The research design is described underlying the explorations and the first of eleven research questions and propositions is undertaken in this chapter, providing an example for explorations in subsequent chapters. Results of the exploration and a discussion of findings are provided, followed by the identification of limitations and future directions. The chapter concludes with an overview of outcomes including the identification of key insights, ideas, and exercise questions encouraging further exploration. The target audience for this chapter is also identified.

3. PERSPECTIVES: URBAN LIFE AND SMART, LEARNING, AND FUTURE CITIES

This chapter provides a review of the research literature for urban life in relation to smart cities, learning cities, and future cities from multiple perspectives.

3.1 Urban Life

Wirth (1938) describes urbanism in terms of "a characteristic mode of life" approachable from the "three interrelated perspectives" of the ecological; social organization; and urban personality and collective behavior. Exploring the issue of global urbanization, through the lens of "a unified theory of urban living", Bettencourt and West (2010) note that "the pace of urban life systematically increases with each expansion of population size" whereby "diseases spread faster, businesses are born and die more often and people even walk faster in larger cities" along with "negative metrics" pertaining to "crime, traffic congestion" and the like. Böhlen and Frei (2010) look to "pervasive technologies as a means to invigorate urban life" giving rise possibly to potentials for more awareness-enabling technologies. Marans and Stimson (2011) provide an overview of the evolving of approaches to the study of the quality of life (QOL) and the quality of urban life (QOUL) in terms of objective indicators, subjective indicators, and the use of geographic information systems (GIS) to integrate the two. Garau and Pavan (2018) consider indicators and assessment tools for the evaluation of the quality of urban life in the context of smart sustainable cities. For Bharne (2013), "dense activity networks and spaces" provide insight into "how cities truly function and 'metabolize' and how urban life is practiced and supported." Nuvolati (2014) defines the quality of urban life, taking into consideration "living conditions in urban areas and mainly in the cities" noting the focus on "urban size and density" by many researchers. Okulicz-Kozaryn and Mazelis (2018) undertake "a test of Wirth's theory of urban life" finding in the United States that "people are less happy in cities, in part because of population size and density, the defining features of the city" while "controlling for urban problems." Conway, Salon, da Silva, and Mirtich (2020) explore how the global COVID-19 pandemic will affect the future of urban life in terms of "long-term, massive societal changes" as "changed behaviors and preferences", given that "people are being compelled to experience new ways of interacting, working, learning, shopping, traveling, and eating meals." It is worth noting that a six-organization collaboration has produced a "manual to define cities, towns and rural areas for international comparisons" (Eurostat, 2021) through application of "the degree of urbanisation." Mateo (2022) identifies key trends giving form to urban life, namely, the office as a community, zero-energy district; everything as a service; distrust of algorithms; and synthetic, permanently antagonistic realities. Mateo (2022) reminds us that while "our built environments – both in the physical and digital sense – do shape us and our future" we live in these spaces as creative beings and as such "we can help evolve them in a way that is a service to our communities and guides society towards a more sustainable future."

3.2 Smart Cities

Marrone and Hammerle (2018) provide a review and analysis of the literature on smart city stakeholders, finding that "citizens tend to be under represented in discussions" while highlighting the need to involve citizens more in discussions and collaborations to improve understanding and the "responsible advancement" of developments going forward. Visvizi and Lytras (2019) address issues and challenges pertaining to smart cities through a mapping of political, social, and economic risks and threats. Additionally, Visvizi and Lytras (2019) identify challenges associated with "contentious issues" associated with "surveillance and monitoring" in order "to ensure safety" in urban spaces. According to Zandbergen and Uitermark (2020), "citizens negotiate urban life by means of data and sensing technologies" and while smart city depictions may be dichotomous, as in "bottom-up participatory" or top down "control" it is nevertheless argued that both forms enable people to "feed into appealing visions of urban life." Moradi (2020) conducted a scientometrics of literature on smart cities to provide an analysis of research trends over a 48-year period from 1970 to 2018 involving a review of the research literature and content analysis. The study by Moradi (2020) was guided by 8 research questions pertaining to elements such as formats of studies, finding mostly conference papers; thematic trends – finding IT infrastructure, smart energy, governance, economy, transportation, and citizen; citation status – finding 63.9% uncited, 35.5% cited 1-9 times, and 0.6% cited more than 10 times. Other elements included historical trends, funding pioneers, publishing supporters, and a geographic atlas of smart city studies. It is worth noting that Moradi (2020) encourages authors "to acknowledge funders in their research" as it seems that an abundance of unfunded research is occurring, giving rise possibly to some other interesting avenues of exploration pertaining to funding and the lack of funding associated with smart city activities. For example, following the guidance by Rinne (2021) to "see what's invisible", in terms of unfunded work, it is noteworthy that Burns and Andrucki (2020) utilize a discourse analysis approach to make visible unpaid, low-paid, and volunteer labor, in the form of "digital care worker" that is "invisibilized in current research and public discourse" about smart cities.

Zheng, Yuan, Zhu, Zhang, and Shao (2020) provide a scientometric analysis of the smart city literature from 1990 to 2019 that is "visual, quantitative, and longitudinal" finding that "smart sustainable cities and sustainable smart cities are the two emerging trends." Zheng et al. (2020) also highlight other findings such as the importance of ICT (information and communication technology) and urban planning for smart city development; people in relation to ICT applications; governance; and smart cities as a "foundation for future city expansion." James, Astoria, Castor, Hudspeth, Olstinske, and Ward (2021) describe key fundamental characteristics of smart cites, chief of which is that "a smart city is based on humans and human interaction for a common purpose." Additionally, James, et al. (2021) identify four main research themes of smart cities as research technology (29%), the nature of smart cities (17%), models and frameworks (13%), and policy and strategy (8%). According to James et al. (2021) "many cities remain focused on technology" while "others rely on a governance model" and some "focus on citizen wellbeing and engagement" and "still others focus on sustainability." In broad terms, James, et al. (2021) provide a definition of a smart city as "an urban environment where technology allows for an efficient relationship between data and its applications" said "to provide an environment that is responsive, resilient, and healthy" where "functional aspects of this relationship" include "responsive, predictive, adaptive, and capable of learning." Kotak and O'Neill (2021) argue that "the pandemic has strengthened the appetite for developments in connectivity, healthcare provision, smart and clean energy production, and innovative and responsive government services" giving rise to "a demand for newly defined occupations." Konstantinou (2022) argues that smart cities, in support of "the need to enhance citizen's quality of life", require "an electric grid" as "the energy backbone" that is "immune to adverse events, both cyber and physical risks" drawing on "renewable sources" highlighting "security and resilience."

3.3 Learning Cities

Learning cities are identified by Nam and Pardo (2011) as a dimension of smart cities. UNESCO (2015) provides a framework of the key features of learning cities with three areas of focus – individual empowerment and social cohesion; economic development and cultural prosperity; and sustainable development. The framework (UNESCO, 2015) identifies six key building blocks for a learning city as – inclusive learning in the education system; revitalized learning in families and communities; effective learning for and in the workplace; extended use of modern learning technologies; enhanced quality in learning; and a vibrant culture of learning throughout life. The framework (UNESCO, 2015) also includes three fundamental conditions for building a learning city – strong political will and commitment; governance and participation of all stakeholders; and mobilization and utilization of resources. Laitinen and Stenvall (2016) focus on the learning dimension of smart cities, and "new, innovative models for collaboration" involving "many different partners/stakeholders." McKenna (2016) identifies several types of relationships important for innovating learning in smart cities such as funding to value citizen engagement and the activities of independent researchers and practitioners. Other relationships identified by McKenna (2016) include data, governance, professional development, to name a few. In a book chapter on learning cities, Nichols (2018) notes that "learning happens in places and spaces of our lives which includes places where we dwell, work, recreate and pass through" adding that "how to live in these places and spaces is learned, as we observe, sense and practice modes of participation and as we reflect on these experiences." Levy, Shafriri, and Alef (2019) describe smart learning with local places while Chauncey and Simpson (2020) advance the notion of learning city 'smart teams' in support of extending the community school model. Lister (2020) advances the notion of "smart pedagogy" in the context of smart cities, exploring smart learning on a community level focusing on digital skills and literacies. The notion of "just in time learning" is advanced by Lister (2021) in that it "forms part of the ambient and pervasive interactions 'ubiquitous computing' landscape of digitally connected learning cities." A recent Conference on Learning Cities (UNESCO, 2022) highlights a pledge by "learning cities around the world to placing lifelong learning for health and the development of resilience at the centre of their agendas" where sub-themes include but are not limited to health, well-being, and climate action.

3.4 Future Cities

Marzano (2006) describes the use of a "strategic futures" process "to 'imagineer' or create scenarios or hypotheses about what everyday life may be like in the future." Claiming that the smart city is "in an early stage of development", Dainow (2017) argues for the need to "place the smart city within a future studies framework" since "the smart city is not today's city with some digital tech laid over the top of it" but "a fundamentally different type of environment from any we have previously seen." It is again worth noting that a future cities massive open online course (MOOC) was developed (Schmitt, 2019) with the aim of providing "the basis to understand, shape, plan, design, build, manage, and continually adapt a city." Dawson (2019) addresses issues associated with future cities and urban life where cities are said to drive and be driven by climate change. Augusto (2020) highlights the importance of "understanding how to evolve the concept" of smart city with an "emphasis on the future." Robinson and Attuyer (2020) address the contested space of density, focusing on "the 'nexus' of height, targets and finance in London's new developments" revealing concerns with livability, quality of life, and access in the context of designing and building future cities. Conway et al. (2020) explore the effects of the COVID-19 global pandemic on "the future of urban life" finding that "a sizeable increase in telework and decreases in both business air travel and restaurant patronage are likely here to stay" with people indicating they "are planning to bike and walk more than they used to." Conway et al. (2020) also found among respondents that "many appreciate some aspects of the pandemic-induced changes to our way of life." Using an evidence synthesis approach, Bibri (2021) identifies problems and challenges associated with sustainability and urbanism in the context of explorations of data-driven smart sustainable cities while noting controversies associated with potentials for technocratic governance in cities of the future. It is worth noting that Heacox (2021) makes a case for a Department of the Future in support of a vision to realign societal priorities. And for a number of years now, the Institute For The Future (IFTF, 2021a) has been developing a Cities Futures Research Collection and is now offering a Futures Thinking massive open online course (MOOC) (IFTF, 2021b). It is perhaps worth noting that Rinne (2021) advances the notion of a "flex mindset" where, among the "superpowers for thriving in constant change" are the two components – "let go of the future" and "see what's invisible." In an urban event on cities (WRLDCTY, 2021), of note is the number of sessions addressing issues pertaining to the future in one way or another, from the future city to the future of the central business district (CBD), mobility, urban events, entertainment, and conferences, to name a few. The need for cities to become more sustainable and adapt to rising waters and the like is highlighted perhaps in a question posed by Palfrey (2021) – "Are floating cites our future?" Lister (2021) speaks in terms of "ad hoc 'smart enough' cities" as an "unfolding future learning terrain" where design of learning environments for "learning effectiveness" is critical. Kahachi (2022) argues for livable future cities theory as "the most appropriate theory for the cities worldwide" because "it directly links the city with people/community needs" providing an opportunity perhaps for formal development and testing of this theory that so far seems to be at the early conceptual stages.

3.5 Literature Review Overview

An overview by author and date is provided in this section for the research and practice literature pertaining to urban life and smart cities, learning cities, and future cities.

Table 1 provides an overview by author and date of the research and practice literature for urban life spanning the timeframe of 2010 to 2022 while also including a reference to earlier work by Wirth in 1938 on urbanism from a sociological perspective focusing on the importance of size, density, and heterogeneity. How population size affects the pace of urban life in cities is addressed by Bettencourt and West (2010) while Böhlen & Frei (2010) look to the potential of pervasive technologies for the invigorating of urban life. Marans & Stimson (2011) as well as Psatha et al. (2011) are concerned with the quality of urban life in terms of types of indicators, particularly for the multi-dimensionality of cities and associated challenges. For Bharne (2013), "dense activity networks" reveal insight into the practice of urban life and how cities function and metabolize. Macionis and Parrillo (2013) provide a sociological perspective on population trends and the concern with living conditions is addressed by Nuvolati (2014). Okulicz-Kozaryn & Mazelis (2018) test fundamental features of cities identified by Wirth (1938), finding that people are less happy in cities in the United States, in part, due to population size and density. Conway et al. (2020) consider the future of urban life in the context of global pandemics such as COVID-19 and the potential for changed behaviors and preferences. From a behavioral and social perspective, Zhang and He (2020) address the impact on urbanism of sensing, AI, and IoT technologies in terms of experience and research. An economics perspective on urban life, enriched with perspectives from many other disciplines, is provided in a MOOC on cities (EdX, 2021) while Eurostat (2021) seeks to enable urbanisation comparisons through the collaborative development of indicators. Mateo (2022) identifies key trends shaping urban life in terms of work, energy, services, distrust of algorithms, and the realities we create.

Table 2 provides an overview by author and date of the research and practice literature for smart cities spanning the timeframe of 2013 to 2022.

For Townsend (2013), smart cities enable the combining of information technology with infrastructure, architecture, everyday objects, and the human body as a way of addressing a range of problems from the social to the economic to the environmental. Marrone and Hammerle (2018) argue for the importance of citizens as stakeholders in the responsible development of smart cities while Visvizi & Lytras (2019) identify challenges pertaining to surveillance and monitoring and the importance of ensuring safety. Augusto (2020) defines smart cities in relation to technology, acknowledging that it is a necessary component yet insufficient alone in understanding the full meaning, purpose, and potential, giving way possibly to a componential approach to the understanding of smart cities. Burns & Andrucki (2020) make visible the invisible in the form of unpaid or under paid 'digital care workers' in smart cities research and discourse. Zandbergen & Uitermark (2020) claim that sensing technologies are used in the negotiating of urban life, bridging perhaps, top down and bottom up approaches while the work of Zheng et al. (2020)

Table 1. Overview of research literature for urban life

Author(s)	Year	Urban Life
Wirth	1938	Urbanism, sociological perspective – size, density, heterogeneity
Bettencourt & West	2010	Pace of life and population size
Böhlen & Frei	2010	Pervasive technologies
Marans & Stimson	2011	Quality, indicators – objective, subjective & GIS for integration
Psatha et al.	2011	Quality of
Bharne	2013	Density – insight into functioning, metabolizing, and practice
Macionis and Parrillo	2013	Global view of cities from a sociological perspective
Nuvolati	2014	Living conditions, size, and density
Okulicz-Kozaryn & Mazelis	2018	Challenging Wirth (1938): size, density & happiness in cities
Conway et al.	2020	Global pandemic & changed behaviors and preferences
Zhang & He	2020	Smart technologies and AI, social perspective
EdX	2021	Past, present, and future – MOOC as CitiesX
Eurostat	2021	Degree of urbanisation for international comparisons
Mateo	2022	Key trends – work, energy, services, distrust/algorithms, realities

points to the trends of smart sustainable cities and sustainable smart cities. Moradi (2020) encourages researchers to acknowledge funders for smart cities research while James et al. (2020) describe fundamental characteristics of smart cities involving human interaction along with research themes such as technology, models and frameworks, well-being and engagement, sustainability, and policy and strategy. Kotak and O'Neill (2021) describe the pandemic-related appetite for connectivity, healthcare provision, smart and clean energy, and innovative and responsive government services contributing in turn to the defining of new occupations. Rinne (2021) speaks of unfunded work and the need for exploring this type of invisibility. Konstantinou (2022) advances an electric grid based on renewable energy sources attentive to security and resilience in support of the quality of life (QoL) of citizens.

Table 3 provides an overview by author and date of the research and practice literature for learning cities spanning the timeframe of 2011 to 2022.

Nam and Pardo (2011) identify learning cities as a dimension of smart cities while UNESCO (2015) provides a framework for learning with key features focusing on individual empowerment and social cohesion; economic development and cultural prosperity; and sustainable development. Laitinen and Stenvall (2016) explore smart learning environments involving new forms of innovative partner/stakeholder collaboration. McKenna (2016) identifies relationships that are important for innovating learning including funding, valuing citizen engagement, and the activities of independent researchers and practitioners. Nichols (2018) notes that "learning happens in our lives" as we observe, sense, and practice while Levy et al. (2019) demonstrate smart learning with local places. Chauncey and Simpson (2020) seek to extend the community school model with learning city smart teams. Lister (2020) describes smart pedagogy, exploring smart learning in the community in terms of digital skills and literacies and also advances the "just in time learning" notion as "ambient and pervasive interactions" in "digitally connected learning cities" (Lister, 2021). The UNESCO (2022) Institute for Lifelong Learning (ILL)

Table 2. Overview of research literature for smart cities

Author(s)	Year	Smart Cities
Townsend	2013	Approach to social, economic, and environmental problems
Marrone & Hammerle	2018	Citizens found to be underrepresented as stakeholders
Visvizi & Lytras	2019	Contentious issues of surveillance and monitoring; safety
Augusto	2020	Technology as "a necessary but not sufficient component"
Burns & Andrucki	2020	Invisible 'digital care workers' in SC research and discourse
Moradi	2020	Research and funding sources
Zandbergen & Uitermark	2020	Data and sensing technologies in negotiating urban life
Zheng et al.	2020	Trend toward smart sustainable cities and sustainable SCs
James et al.	2021	Human interaction; Governance; Sustainability; Well-being
Kotak & O'Neill	2021	Connectivity, healthcare, energy, government services – new occupations
Rinne	2021	The invisibility of unfunded research
Konstantinou	2022	Electric grid, QoL – renewable energy – security, resilience

highlights commitments at a recent learning cities conference, to health and development of resilience through lifelong learning.

Table 3. Overview of research literature for learning cities

Author(s)	Year	Learning Cities
Nam & Pardo	2011	Dimension of smart cities
UNESCO	2015	Empowerment, social cohesion, economic, cultural, sustainable
Laitinen & Stenvall	2016	Smart learning environments involving collaboration
McKenna	2016	Relationships for innovating learning – funding, engagement, etc.
Nichols	2018	Learning in places & spaces of our lives - observe, sense, practice
Levy et al.	2019	Smart learning with local places
Chauncey & Simpson	2020	Smart teams – extending the community school model
Lister	2020	Smart pedagogy – smart learning & digital skills & literacies
Lister	2021	Just in time learning as ambient and pervasive interactions
UNESCO	2022	Health and development of resilience through lifelong learning

Table 4 provides an overview by author and date of the research and practice literature for future cities spanning the timeframe of 2016 to 2022.

Marzano (2016) speaks in terms of creating strategic futures through imagining scenarios for everyday life. Dainow (2017) highlights the need for a future studies framework in coming to an understanding of the "functional characteristics" of smart cities. Schmitt (2019) developed a massive open online course (MOOC) on future cities incorporating multiple aspects such as capabilities to adapt, build, design, plan, manage, and shape and more recently, a course was offered on livability in future cities. Dawson

Table 4. Overview of research literature for future cities

Author(s)	Year	Future Cities
Marzano	2016	Strategic futures and imagining
Dainow	2017	Future studies framework
Dawson	2019	Climate change
Schmitt	2019	Data driven approaches, crowdsourcing, and sensing; Livability
Augusto	2020	Evolving the smart cities concept for the future
Conway et al.	2020	Pandemic-induced changes to urban life – work and travel
Robinson and Attuyer	2020	Density and livability
Bibri	2021	Sustainability and urbanism in data-driven spaces
IFTF	2021a	Cities Futures Research
IFTF	2021b	Futures Thinking course
Heacox	2021	Department of the Future vision to realign societal priorities
Lister	2021	Future learning terrain
Palfrey	2021	Cities and water
Rinne	2021	Flexibility mindset for change
WRLDCTY	2021	Imagining cities into the future
Batty	2022	City concept disappears in a fully urban future
Kahachi	2022	Liveable future cities theory – people/communities

(2019) focuses on cities and climate change while Augusto (2020) points to the importance of evolving the smart cities concept with the future in mind. Conway et al. (2020) explore the future of urban life in relation to pandemic-induced changes affecting things such as work and travel practices. Robinson and Attuyer (2020) highlight the importance of density and livability in the building of future cities. Bibri (2021) focuses on sustainability in the context of data-driven environments, attentive to technocratic challenges. The IFTF (2021a) has developed a Cities Futures Research collection and more recently is offering a Futures Thinking MOOC (IFTF, 2021b). Heacox (2021) argues for a Department of the Future to rethink the priorities of society while Lister (2021) speaks of an emergent future learning terrain in ad hoc, smart enough cities. Palfrey (2021) questions whether cities will become floating ones as waters rise and Rinne (2021) focuses on a flexibility mindset to accommodate constant change. At a WRLDCTY global forum (2021), of note are the number and range of sessions pertaining to the future, from mobility to events, to the central business district (CBC). Kahachi (2022) advocates for liveable future cities theory, based as it is on the needs of people and communities. Batty (2022) speculates that the city concept will disappear in a fully urban future.

In terms of patterns and potential relationships, what emerges in Table 5 from this review of the research and practice literature is an overview for urban life, smart cities, learning cities, and future cities.

Elements that emerge in the urban life column of Table 5, to the left, continue to reappear in the other three columns to the right for smart cities, learning cities, and future cities such as density, pandemic effects, the future, technologies, quality of life and livability, the social, and urbanism. Learning in the form of massive open online courses (MOOCs) emerges for both smart cities and future cities, as well

Table 5. Patterns and potential relationships

Urban Life	Smart Cities	Learning Cities	Future Cities
AI	Citizens underrepresented	Collaboration	Cities and Water
Appealing visions	Collaborations	Cultural	Climate change
Collective behavior	Economic issues	Economic	Data-driven approaches
Density	Environmental issues	Empowerment	Density
Ecological	Governance	Governance	Dept of the Future/Gov
GIS	Human interaction	Innovating learning	Flexibility mindset
Pace of life	Invisibility	- engagement, funding	Future learning terrain
Pandemic effects	-underfunded digital work	Interactions: just-in-time	Futures thinking course
Past-Present-Future	-unfunded research	- ambient	Imagining scenarios
Pervasive technologies	MOOC	- pervasive	Livability
Population	Responsive, resilient	Lifelong learning	MOOC
Quality of life	Social issues	Participation	Pandemic effects
Quality of urban life	Sustainability, QoL	Resilience	- travel, work
Size	Technologies	Smart pedagogy/literacy	Quality of life
Smart technologies	-ICT, data, sensing, safety	Smart teams	Research
Social, Sociological	-necessary/not sufficient	Social cohesion	Strategic
Urbanism	-surveillance/monitoring	Sustainable	Sustainable, Adaptive
Urbanization	Well-being, Engagement	Well-being	Urbanism

as sustainability, benefits and challenges associated with technologies, social and economic issues, government and governance, and so on.

These key elements emerging from the literature review inform the research questions and propositions developed later in this chapter. Of note are the unfunded aspects of smart city activities associated with invisible work such as digital work and scholarly research activities. And elements such as collaboration, engagement, governance, resilience, sustainability, and well-being appear for smart cities and learning cities.

3.6 Issues, Controversies, Problems

Based on this review of the research and practice literature, Table 6 provides an overview of issues, controversies, and problems associated with urban life and the ambient in smart cities, learning cities, and future cities.

For urban life, issues, controversies, and problems pertain to the need for indicators and assessment tools, figuring out how to invigorate urban life and the importance of pervasive technologies, the need for theory as a tool to assist in understanding urban life, the pace of life, the global pandemic, population density, the environment and the shaping of behavior, sustainability, spiraling populations, quality of life, the environment, and shantytowns.

For smart cities, issues, controversies, and problems pertain to funding, political-social-economic risks/threats, privacy and ICT technologies, smart sustainability, sustainability, urban security and surveillance/monitoring, citizens being underrepresented, the responsiveness of government, safety in public spaces, and security and resilience.

For learning cities, issues, controversies, and problems pertain to climate action, designing digital learning activities, literacies, data relationships, funding to value citizen engagement and participation, lifelong learning, ambient or just in time learning, funding data initiatives for learning, and resilience.

Table 6. Issues, controversies, and problems

	Issues	Controversies	Problems
Urban Life	• Indicators/Assessment tools • Pervasive tech to invigorate • Need for theory to understand • Pace of life	• Global pandemic • Population density • Shaping of behavior • Sustainability	• Spiraling populations • Quality of life • Environment • Shantytowns
Smart Cities	• Funding • Political-social-economic risks/ threats • Privacy & ICT technologies	• Smart sustainability • Sustainability • Urban security & surveillance/ monitoring	• Citizens underrepresented • Responsive government • Safety in public spaces • Security, resilience
Learning Cities	• Climate action • Designing digital learning activities • Literacies	• Data relationships • Funding to value citizen engagement • Lifelong learning	• Ambient learning • Funding data initiatives for learning • Resilience
Future Cities	• Climate change • Design for learning • Future learning terrain • Sustainability	• Density • Livability • Quality of life • Technocratic governance	• Adaptability • Data driven smart cities • Futures thinking • Sustainability & urbanism

For future cities, issues, controversies, and problems pertain to climate change, design for learning effectiveness, future learning terrain, sustainability, density, livability, quality of life, technocratic governance, adaptability, data driven smart cities, futures thinking, and sustainability and urbanism.

It is worth noting that funding emerges in the issues and problems categories across both smart cities and learning cities while density gives rise to controversies in both urban life and future cities. Sustainability emerges as both an issue and a problem for future cities while also emerging for urban life and smart cities. Issues, controversies, and problems emerging from the literature review give rise to areas in need of further exploration and as such, provide the basis for the formulation of research questions that will be reformulated and addressed as propositions in this work, chapter by chapter.

3.7 Research Questions for the Exploration of Urban Life and Smart Cities

Research questions formulated from the review of the research literature for the exploration of urban life, smart cities, learning cities, and future cities are as follows:

Research Question (Q1): How does support for improving urban life emerge in the context of smart cities?
Research Question (Q2): How does the ambient emerge in urban life in the context of smart cities?
Research Question (Q3): Why do engagement and participation matter in urban life in the context of smart cities?
Research Question (Q4): How are urban infrastructures evolving in smart cities?
Research Question (Q5): Why are literacies for urban life in smart cities needed?
Research Question (Q6): How are data, information, and knowledge related in urban life in smart cities?
Research Question (Q7): How and why does artificial intelligence, including machine learning and deep learning, manifest in urban life and the ambient in smart cities?
Research Question (Q8): How do smart homes, electric vehicles, extended realities, and robotics contribute to urban life and the ambient in smart cities?

Figure 1. Urban life and the ambient in smart cities conceptual framework

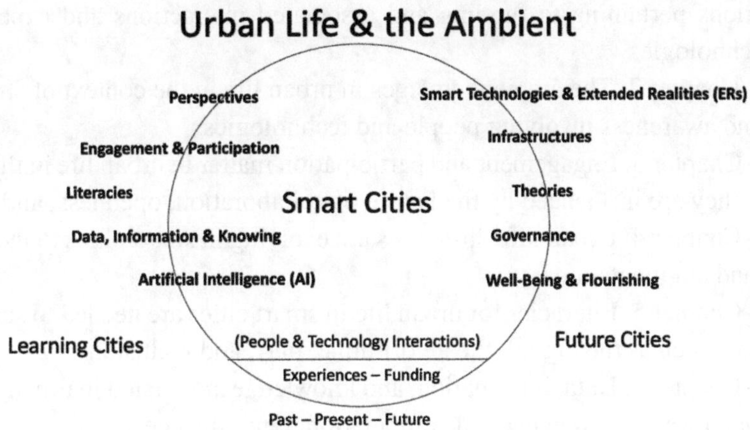

Research Question (Q9): Why are tools such as theory needed for understanding urban life and the ambient in smart, learning, and future cities?

Research Question (Q10): How do governance, livability, and sustainability contribute to the enriching of urban life and the ambient in smart, learning, and future cities?

Research Question (Q11): Why are connection, positivity, and resilience important for urban life and the ambient in smart, learning, and future cities?

4. CONCEPTUAL FRAMEWORK FOR URBAN LIFE AND THE AMBIENT

In response to the issues, controversies, and problems highlighted in this chapter, a conceptual framework for urban life and the ambient in smart cities, learning cities, and future cities is formulated based on the review of the research literature in Section 3. As depicted in Fig. 1, urban life and the ambient, on a past – present – future continuum, is conceptualized in this work in the context of smart cities, involving the notion of learning cities while encompassing that of future cities. In the chapters that follow, perspectives are provided in this work on urban life and the ambient; engagement and participation is addressed as well as literacies; data, information, and knowing; artificial intelligence (AI); smart technologies and extended realities (ERs); infrastructures; theories; governance; and well-being and flourishing. As such, this conceptual framework for urban life and the ambient in smart cities features a componential approach and is operationalized for use in this work using a case study approach combined with an explanatory correlational design. According to Yin (2018), a case study approach is particularly appropriate for the exploration of contemporary phenomena, in this instance, the exploration of urban life in relation to evolving understandings of smart cities, learning cities, and future cities focusing on people and technology interactions where experiences with funding emerge, among a range of other issues, controversies, and problems.

In response to the issues, controversies, and problems identified through the literature review, the eleven research questions are reformulated here as propositions to be explored in this work, chapter by chapter, as follows.

Proposition (P1) – Chapter 1: Support for improving urban life in the context of smart cities emerges through questions pertaining to funding and associated interactions and experiences involving people and technologies

Proposition (P2) – Chapter 2: The ambient emerges in urban life in the context of smart cities through interactions and awareness involving people and technologies

Proposition (P3) – Chapter 3: Engagement and participation matter in urban life in the context of smart cities because they are influenced by the nature of collaboration, openness, and privacy

Proposition (P4) – Chapter 4: Urban infrastructures are evolving in smart cities to include experiences, interactions, and augmentations

Proposition (P5) – Chapter 5: Literacies for urban life in smart cities are needed to support understandings of elements such as the digital, the algorithmic, play, and inclusion

Proposition (P6) – Chapter 6: Data, information, and knowledge are related in urban life in smart cities as shown through ways of knowing and computational intelligence

Proposition (P7) – Chapter 7: Artificial intelligence, including machine learning and deep learning, manifest in urban life and the ambient in smart cities in increasingly pervasive ways in support of ambient learning, mobility, and many other in-the-moment, real time interactivities

Proposition (P8) – Chapter 8: Smart homes, electric vehicles, extended realities, and robotics contribute to urban life and the ambient in smart cities in support of sustainability and livability on the planet

Proposition (P9) – Chapter 9: Tools such as theory are needed for understanding urban life and the ambient in smart, learning, and future cities to identify variables important for the selection of approaches and metrics for researchers and practitioners

Proposition (P10) – Chapter 10: Governance, livability, and sustainability contribute to the enriching of urban life and the ambient in smart, learning, and future cities influencing health and well-being, quality of life, and resilience.

Proposition (P11) – Chapter 11: Connection, positivity, and resilience are important for urban life and the ambient in smart, learning, and future cities, in support of human flourishing generally, and more specifically, creativity, health and well-being, and quality of life.

What follows is a description of the methodology used in exploring the proposition under study in this chapter.

5. METHODOLOGY

The methodology for this chapter is described in sections 5.1 to 5.3 in terms of the research design, data collection, and data analysis.

5.1 Research Design

The research design for this chapter involves an exploratory case study approach combined with an explanatory correlational design. Yin (2018) notes the "complementarity of case study and statistical research" indicating that case studies "have been needed to examine the underlying processes that might explain a correlation" and in this work, the explanatory correlational design is intended to assist with understanding more about the nature of the relationships explored. An online space is used to describe the study and invite people to participate, and interest emerged from individuals in a wide variety of cities in multiple countries (Canada, United States, Greece, Ireland, Israel, etc.) across several continents.

5.2 Data Collection

During sign up for the study, demographic data were gathered including age range, gender, and city, while people could self-identify in one or more categories (e.g., business, community member, student, educator, etc.). A pre-tested survey instrument with closed and open-ended questions was used in this study with questions pertaining to urban life, smart cities, learning cities, and future cities. Survey respondents were also invited to participate in an in-depth interview based on a pre-tested interview protocol.

In parallel with this study, data were also gathered systematically through group and individual discussions conducted with diverse individuals in many sectors (e.g., students, faculty, researchers, and staff in higher education, business, community members and leaders, and city staff and leaders) in several Canadian cities (e.g., Toronto, Vancouver, and Victoria and some surrounding municipalities).

5.3 Data Analysis

Analysis of data in this chapter was guided by the research question posed in Section 1 of this chapter and then reformulated as a proposition for exploration, as follows.

P1: *Support for improving urban life in the context of smart cities emerges through questions pertaining to funding and associated interactions and experiences involving people and technologies.*

Descriptive statistics enabled the analysis of quantitative data that were gathered from survey questions pertaining to funding in relation to success and then in relation to livability. The correlation feature of the Real Statistics Resource Pack containing an-add in for Microsoft Excel (Zaiontz, 2022) enabled the conducting of a Spearman Correlation Coefficient for ordinal data in exploring the nature of relationships between items pertaining to funding.

Content analysis was used for qualitative data gathered through open-ended survey questions, interviews, and group and individual discussions. This inductive analysis was combined with deductive analysis using terminology from the research literature.

Overall, an analysis was conducted for n=79 consisting of 42% females and 58% males for people ranging in age from their 20s to 70s.

6. RESULTS AND DISCUSSION

As shown in Table 7, assessments were conducted for funding in relation to the success of smart city projects and livability. For example, when asked to what extent funding contributes to the success of a smart city project, on a 7-point Likert type scale, where 1= Not at all and 7=Absolutely, 25% of survey respondents selected 7, 50% responded toward the upper end of the scale at 6 (Sure), and 25% responded at position 5 (Sort of). Probing the funding factor further, another question posed pertained to the extent to which funding contributes to the livability of a smart city. In response, 17% selected position 5 on the scale, 17% selected position 6 toward the upper end of the scale, and 66% selected position 7 at the upper

Table 7. Assessments for funding in relation to success and livability

Items	Assessments for Funding	Correlation
Success of SC projects	25% (5); 50% (6); 25% (7)	.81
Livability	17% (5); 17% (6); 66% (7)	

end of the scale. Although broad and general in nature, these two questions and the responses that emerge show that funding is important, appearing as it does at various points on the upper portion of the scale.

Further, when funding as a factor for success in smart city projects is correlated with livability using the Real Statistics Resource Pack add-in for Microsoft Excel (Zaiontz, 2022), a Spearman correlation coefficient for ordinal data of .81 emerges. According to Creswell (2018), correlations in the .66 – .85 range are said to be "very good." As such, this chapter identifies opportunities going forward for explorations, for both practice and research, into the importance of funding for improving upon the quality of urban life in the context of smart cities, learning cities, and future cities based on variables such as success of smart city projects and livability and extending to other possible factors such as that of literacies as discussed by Lister (2020) in relation to smart learning and by Boggs (2022) in relation to making money.

7. LIMITATIONS AND FUTURE DIRECTIONS

This section identifies possible limitations of this chapter while highlighting directions going forward for research and practice opportunities.

Limitations

Possible limitations of this chapter include the scope of the literature review, bounded as it is by a focus on urban life generally and in relation to smart cities, learning cities, and future cities. Additionally, the small sample size in this work calls for caution in the interpretation of correlation findings from the assessment of funding in relation to success of smart city projects and livability in smart cities while opening the way to further, larger sample size explorations involving more cities and countries.

Future Research Opportunities

Opportunities going forward for research include the potential for pursuing additional studies on the variables identified in this work relevant to funding (e.g., success and livability). Additionally, opportunities exist for the identification of other variables relevant to funding that enable further learning, as in smart learning (Lister, 2020) and literacies (Sumara and Alvermann, 2022) in the context of smart cities, learning cities, and future cities. Research opportunities also present in terms of further testing and validating the conceptual framework for urban life and the ambient in smart cities (Fig. 1) advanced in this chapter.

Future Practice Opportunities

Opportunities going forward for practice include the potential for pursuing additional exploration of the variables identified in this work for livability and for the funding of smart city projects and initiatives in terms of success. Also, opportunities exist for the identification of other variables relevant to livability in terms of how people sense, navigate, and experience the city in support of smart, learning, and future cities and other variable relevant to funding involving collaborations across government, business, academia, and the community.

8. CONCLUSION

This chapter contributes to the research literature in terms of providing an overview of urban life from multiple perspectives and in relation to smart cities, learning cities, and future cities. Additionally, this chapter contributes to the study of smart cities by identifying issues, controversies, and problems emerging from the research literature enabling formulation of a series of research questions, reformulated as propositions to be explored, chapter by chapter in this book. And finally, this chapter contributes to the smart cities domain, through formulation of a conceptual framework for urban life and the ambient in smart cities that is then operationalized for use through an exploratory case study approach combined with an explanatory correlational design. The proposition under study in this chapter focuses on the issue of funding where a very good correlation is found based on an assessment of the factors, *livability* and the *success of smart city projects*. However, among the limitations of the study underlying this chapter is that of a small sample size, and as such, caution is urged in the interpretation of findings and further research is encouraged. Nevertheless, a pathway to guide research and practice activities going forward is provided calling for further validating of the variables explored in this chapter while also encouraging activities to identify other, additional factors relevant to urban life in relation to funding in smart, learning, and future cities. This work also suggests the need for more funding for research pertaining to smart cities and more particularly, urban life and the ambient in smart, learning, and future cities.

Insights emerging from this chapter include the following:

1. The importance of funding as a factor contributing to the potential for the success of smart city projects
2. The importance of funding as a factor contributing to the potential for improving livability in smart cities
3. The potential for many more factors to emerge for understanding smart cities, learning cities, and future cities when an emphasis is placed on the experience of the urban in everyday life

Ideas emerging from this chapter include the following:

1. The opportunities potential for exploring funding from different perspectives in the context of urban life and the ambient in smart cities research
2. The opportunities potential for urban initiatives involving more creative funding arrangements
3. The use of a componential approach for the study of urban life and the ambient in smart, learning, and future cities

Exercises for the reader to consider, encouraging further exploration include the following:

1. Identify one or more factors in your city that are contributing to livability
2. Identify one or more ways in which funding would enhance the quality of urban life in your city
3. Describe one or more elements that contribute to making your city a learning city

This work will be of interest to a broad audience including but not limited to community members, educators, the business community, researchers, urban planners, data scientists, and anyone concerned with urban life and associated practice and research opportunities and challenges for funding in support of improving livability in smart cities, learning cities, and future cities.

REFERENCES

Augusto, J. C. (2021). Smart cities: State of the art and future challenges. In J. C. Augusto (Ed.), *Handbook of Smart Cities*. Springer. doi:10.1007/978-3-030-15145-4_95-2

Batty, M. (2022). *The shape of future cities: Three speculations*. Transactions in Urban Data, Science, and Technology. doi:10.1177/27541231221113945

Bettencourt, L., & West, G. (2010). A unified theory of urban living. *Nature*, *467*(7318), 912–913. doi:10.1038/467912a PMID:20962823

Bharne, V. (2013). *The emerging Asian city: Concomitant urbanities and urbanisms*. Routledge.

Bibri, S. E. (2021). Data-driven smart sustainable cities of the future: An evidence synthesis approach to a comprehensive state-of-the-art literature review. *Sustainable Futures*, *3*, 100047. doi:10.1016/j.sftr.2021.100047

Boggs, G. (2022). When you goin' teach us how to make that money? In D. Sumara & D. E. Alvermann (Eds.), *Ideas That Changed Literacy Practices: First Person Accounts from Leading Voices*. Myers Education Press.

Böhlen, M., & Frei, H. (2010). Ambient intelligence in the city: Overview and new perspectives. In H. Nakashima, H. Aghajan, & J. C. Augusto (Eds.), *Handbook of ambient intelligence and smart environments* (pp. 911–938). Springer. doi:10.1007/978-0-387-93808-0_34

Burns, R., & Andrucki, M. (2021). Smart cities: Who cares? *Environment and Planning A. Economy and Space*, *53*(1), 12–30. doi:10.1177/0308518X20941516

Chauncey, S. A., & Simpson, G. I. (2020). The role of learning city "smart teams" in promoting, supporting, and extending the community school model. In. Lecture Notes in Computer Science: Vol. 12425. *HCI International 2020 – Late Breaking Papers: Cognition, Learning and Games. HCII 2020*. Springer. doi:10.1007/978-3-030-60128-7_25

Conway, M. W., Salon, D., da Silva, D. C., & Mirtich, L. (2020). How will the COVID-19 pandemic affect the future of urban life? Early evidence from highly-educated respondents in the United States. *Urban Science*, *4*(4), 50. doi:10.3390/urbansci4040050

Creswell, J. W. (2018). *Educational research: Planning, conducting, and evaluating quantitative and qualitative research* (6th ed.). Pearson.

Cugurullo, F. (2020). Urban artificial intelligence: From automation to autonomy in the smart city. *Frontiers in Sustainable Cities*, 2. Retrieved 22 October 2021 from https://www.frontiersin.org/articles/10.3389/frsc.2020.00038/full

Dainow, B. (2017). Smart city transcendent: Understanding the smart city by transcending ontology. *Orbit (Amsterdam, Netherlands)*, *1*, 1–15.

Dawson, A. (2017). *Extreme cities: The peril and promise of urban life in the age of climate change*. Verso.

EdX. (2021). *CitiesX: The Past, Present, and Future of Urban Life*. Harvard University. Retrieved September 7, 2021, from https://www.edx.org/course/citiesx-the-past-present-and-future-of-urban-life

Eurostat. (2021). *Applying the degree of urbanisation – A methodological manual to define cities, towns and rural areas for international comparisons – 2021 edition*. EC, FAO, UN-Habitat, ILO, OECD, and The World Bank. Retrieved 2 October 2021 from https://ec.europa.eu/eurostat/en/web/products-manuals-and-guidelines/-/ks-02-20-499

Garau, C., & Pavan, V. M. (2018). Evaluating urban quality: Indicators and assessment tools for smart sustainable cities. *Sustainability*, *10*(3), 575. doi:10.3390u10030575

Heacox, K. (2021). Why America needs a Department of the Future. *The Guardian*. Retrieved August 27, 2021 from https://www.theguardian.com/commentisfree/2021/aug/27/why-we-need-a-department-of-the-future

IFTF. (2021). *Cities Futures Research Collection*. Retrieved 7 October 2021 from https://www.iftf.org/our-work/global-landscape/cities-futures/

IFTF. (2021). *Futures thinking*. Coursera. Retrieved September 13, 2021 from https://www.coursera.org/specializations/futures-thinking?utm_source=recommendations&utm_medium=email&utm_campaign=9507&sfmc_id=2491390&sfmc_key=0031U00001OvTmnQAF

James, P., Astoria, R., Castor, T., Hudspeth, C., Olstinske, D., & Ward, J. (2021). Smart cities: Fundamental concepts. In J. C. Augusto (Ed.), *Handbook of Smart Cities*. Springer. doi:10.1007/978-3-030-69698-6_2

Kahachi, H. A. H. (2022). Future cities and reality: Analytical preview of the different theories from information cities, smart cities, to liveable cities. In M. O. Karkush & D. Choudhury (Eds.), *Geotechnical Engineering and Sustainable Construction*. Springer. doi:10.1007/978-981-16-6277-5_65

Konstantinou, C. (2022). Toward a secure and resilient all-Renewable energy grid for smart cities. *IEEE Consumer Electronics Magazine*, *11*(1), 33–41. doi:10.1109/MCE.2021.3055492

Kotak, A., & O'Neill, K. (2021). *Work in Progress: Emerging Smart City Occupations*. Information and Communications Technology Council (ICTC). Retrieved 3 February 2022 from https://www.ictc-ctic.ca/wp-content/uploads/2021/05/Final-smart-city-occupations-May-19.pdf

Laitinen, I., & Stenvall, J. (2016). *From smart cities to smart learning environments: A comparative study*. University of Glasgow. Retrieved 5 April 2022 from http://cradall.org/content/smart-cities-smart-learning-environments-comparative-study

Levy, D., Shafriri, Y., & Alef, Y. (2019). A place to discover, imagine, and change: Smart learning with local places. In M. Antona & C. Stephanidis (Eds.), Lecture Notes in Computer Science: Vol. 11572. *Universal Access in Human-Computer Interaction. Theory, Methods and Tools. HCII 2019*. Springer. doi:10.1007/978-3-030-23560-4_34

Lister, P. (2020). Smart learning in the community: Supporting citizen digital skills and literacies. In N. Streitz & S. Konomi (Eds.), Lecture Notes in Computer Science: Vol. 12203. *Distributed, Ambient and Pervasive Interactions. HCII 2020*. Springer. doi:10.1007/978-3-030-50344-4_38

Lister, P. (2021). What are we supposed to be learning? Motivation and autonomy in smart learning environments. In N. Streitz & S. Konomi (Eds.), Lecture Notes in Computer Science: Vol. 12782. *Distributed, Ambient and Pervasive Interactions. HCII 2021*. Springer. doi:10.1007/978-3-030-77015-0_17

Macionis, J. J., & Parrillo, V. N. (2017). *Cities and urban life* (7th ed.). Pearson.

Marans, R. W., & Stimson, R. (2011). An overview of quality of urban life. In R. Marans & R. Stimson (Eds.), Social Indicators Research Series: Vol. 45. *Investigating Quality of Urban Life*. Springer. doi:10.1007/978-94-007-1742-8_1

Marrone, M., & Hammerle, M. (2018). Smart cities: A review and analysis of stakeholders' literature. *Business & Information Systems Engineering*, *60*(3), 197–213. doi:10.100712599-018-0535-3

Marzano, S. (2006). Ambient culture. In True Visions. Springer. doi:10.1007/978-3-540-28974-6_3

Mateo, C. (2022). Our urban life in 2022: Cristina Mateo explores five trends which will shape urban life this year. *IE Insights*. https://www.ie.edu/insights/articles/our-urban-life-in-2022/

McKenna, H. P. (2016). Innovating relationships for learning in 21st century smart cities. *ICERI2016 Proceedings*, 4695-4704.

Moradi, S. (2020). The scientometrics of literature on smart cities. *Library Hi Tech*, *38*(2), 385–398. doi:10.1108/LHT-12-2018-0203

Nam, T., & Pardo, T. A. (2011). Conceptualizing smart city with dimensions of technology, people, and institutions. In *Proceedings of the 12th Annual International Conference on Digital Government Research* (pp. 282-291). ACM. 10.1145/2037556.2037602

Nichols, S. (2018). Introducing learning cities. In S. Nichols & S. Dobson (Eds.), *Learning Cities. Cultural Studies and Transdisciplinarity in Education* (Vol. 8). Springer. doi:10.1007/978-981-10-8100-2_1

Nuvolati, G. (2014). Urban life, quality of. In A. C. Michalos (Ed.), *Encyclopedia of Quality of Life and Well-Being Research*. Springer. doi:10.1007/978-94-007-0753-5_2383

Okulicz-Kozaryn, A., & Mazelis, J. M. (2018). Urbanism and happiness: A test of Wirth's theory of urban life. *Urban Studies (Edinburgh, Scotland)*, *55*(2), 349–364. doi:10.1177/0042098016645470

Palfrey, J. (2021). Are floating cities our future? *The Guardian*. Retrieved 4 October 2021 from https://www.bbc.com/travel/article/20211003-are-floating-cities-our-future

Psatha, E., Deffner, A., & Yannis Psycharis, Y. (2011). *Defining the quality of urban life: Which factors should be considered? ERSA conference papers ersa11p785*. European Regional Science Association.

Rinne, A. (2021). *Flux: 8 superpowers for thriving in constant change*. Penguin and Random House.

Robinson, J., & Attuyer, K. (2020). Contesting density: Beyond nimby-ism and usual suspects in governing the future city. *Urban Geography*, *41*(10), 1294–1301. doi:10.1080/02723638.2020.1860623

Schmitt, G. (2019). *Future Cities*. ETHzurich. Retrieved 11 December 2017 from https://online.ethz.ch/courses/course-v1:ETH+FC-01x+2019_T1/about

Sumara, D., & Alvermann, D. E. (Eds.). (2022). *Ideas That Changed Literacy Practices: First Person Accounts from Leading Voices*. Stylus Publishing.

Townsend, A. M. (2013). *Smart cities: Big data, civic hackers and the quest for a new utopia*. W. W. Norton & Company.

UNESCO. (2015). *UNESCO global network of learning cities: Guiding document*. UNESCO Institute for Lifelong Learning.

UNESCO. (2019). *Learning cities: Drivers of inclusion and sustainability*. UNESCO Institute for Lifelong Learning. Retrieved September 9, 2021 from https://uil.unesco.org/lifelong-learning/learning-cities/learning-cities-drivers-inclusion-and-sustainability

UNESCO. (2022). From emergency to resilience: Building healthy and resilient cities through learning. Conference Report. In *Fifth International Conference on Learning Cities*. UNESCO, Institute for Lifelong Learning. Retrieved 11 August 2022 from https://unesdoc.unesco.org/ark:/48223/pf0000380657

Visvizi, A., & Lytras, M. D. (Eds.). (2019). *Smart cities: Issues and challenges – Mapping political, social, and economic risks and threats*. Elsevier.

Wirth, L. (1938). Urbanism as a way of life. *American Journal of Sociology*, *44*(1), 1–24. doi:10.1086/217913

WRLDCTY. (2021). *About: Change for good*. Retrieved September 29 from https://www.wrldcty.com/about/

Yin, R. (2018). *Case study research and applications: Design and methods* (6th ed.). Sage Publications.

Zaiontz, C. (2022). *Real statistics using excel*. Retrieved from www.real-statistics.com

Zandbergen, D., & Uitermark, J. (2020). In search of the smart citizen: Republican and cybernetic citizenship in the smart city. *Urban Studies (Edinburgh, Scotland)*, *57*(8), 1733–1748. doi:10.1177/0042098019847410

Zhang, J., & He, S. (2020). Smart technologies and urban life: A behavioral and social perspective. *Sustainable Cities and Society*, *63*, 102460. doi:10.1016/j.scs.2020.102460

Zheng, C., Yuan, J., Zhu, L., Zhang, Y., & Shao, Q. (2020). From digital to sustainable: A scientometric review of smart city literature between 1990 and 2019. *Journal of Cleaner Production*, *258*, 120689. doi:10.1016/j.jclepro.2020.120689

ADDITIONAL READING

Florida, R., Glaeser, E., Sharif, M. M., Bedi, K., Campanella, T. J., Chee, C. H., Doctoroff, D., Katz, B., Katz, R., Kotkin, J., Muggah, R., & Sadik-Khan, J. (2021). How life in our cities will look after the coronavirus pandemic. *Foreign Policy*. Retrieved September 9 from https://foreignpolicy.com/2020/05/01/future-of-cities-urban-life-after-coronavirus-pandemic/

Galloway, L. (2021). The world's five safest cities post-pandemic. *BBC Travel*. Retrieved 7 October 2021 from https://www.bbc.com/travel/article/20211006-the-worlds-five-safest-cities-post-pandemic

Rinne, A. (2021). A futurist's guide to preparing your company for constant change. *Harvard Business Review*, (September), 22.

UNESCO. (2015). *Key features of learning cities*. UNESCO Institute for Lifelong Learning. Retrieved 7 October 2021 from https://uil.unesco.org/lifelong-learning/learning-cities/key-features-learning-cities

KEY TERMS AND DEFINITIONS

Ambient Perspective: An ambient perspective refers to views that take into consideration the dynamic awareness capabilities of people and the dynamic aware-enabling capabilities of technologies and the interactive potentials of the two.

Funding Literacies: Funding literacies, in the context of urban life and smart cities, refer to an awareness of what constitutes "economic relevance" (Boggs, 2022), thus generating income from usefulness contributed to communities.

Literacies: Literacies refer to the multiple skills required to successfully navigate one or more environments such as the physical and digital or a combination or mixing of the two in the context of contemporary urban spaces.

Livability: Livability refers to the quality of life afforded by a particular environment, taking into consideration a range of factors such as cultural, social, economic, environmental, political, and so on.

Quality of Life: Quality of life refers to the extent that people are capable of enjoying and engaging with life-based factors such as health, well-being, and the like.

Quality of Urban Life: Quality of urban life refers to the extent that people are able to enjoy and engage with urban life-based factors that are acceptable and being continually improved upon (e.g., air-quality, traffic congestion, noise, and the like).

Success: Success refers to the achieving of a desired outcome or the completing of one or more tasks in fulfillment of an objective.

Chapter 2
Exploring Urban Life From an Ambient Perspective:
From Culture to Economy to Mobility

ABSTRACT

The purpose of this chapter is to provide an introduction and background to the concept of the ambient in relation to urban life and to evolving understandings of smart, learning, and future cities. A review of the research literature is provided for the ambient and urban life, in the context of smart cities, focusing on culture, economies, and mobility. The literature review surfaces issues, controversies, and problems while also enabling formulation of a conceptual framework for urban life and the ambient in smart cities which is then operationalized for use in this chapter. Using an exploratory case study approach involving survey and in-depth interviews, combined with an explanatory correlational design, how the ambient emerges in urban life in the context of smart cities is explored focusing on a variety of variables such as multiple modes of transport and walkability for mobility. Directions for future practice and research are also discussed.

1. INTRODUCTION

The ambient is described by Stoop (2003) in the context of the ambient intelligence vision as a space "where people will be empowered in their daily activities by technology that surrounds them, responds to them and anticipates their needs, yet remains invisible" as in, "unobtrusiveness" making way for other features such as that of "mobility" involving "relevant and meaningful technology on the move." Bohn, Coroamă, Langheinrich, Mattern, and Rohs (2005) describe the ambient economy as accommodating invisible technologies that are embedded in products in support of smart products and tracking, while acknowledging issues associated with personal privacy, reliability, and social acceptability. The anything, anywhere business model for the ambient economy (Bohn et al., 2005) gives rise to a space for ambient mobility, as described perhaps by Stoop (2003). Marzano (2006) argues that "we are rapidly developing a global civilized culture based on sustainability" contributing to "an appreciation that the world and its

DOI: 10.4018/978-1-6684-4096-4.ch002

inhabitants are ultimately interdependent" resulting in "a culture that goes beyond environmental issues to include questions of social responsibility and economic viability" opening the way possibly for the notion of ambient economies. The ambient has been described by Rauterburg (2007) in terms of the notion of ambient culture in the context of computing as "open systems that understand and support the rituals of our living" and that "adapt themselves to people through time and space." McCullough (2013) describes the ambient as "a continuum of awareness and an awareness of continuum" while Patorniti and Stevens (2017) explore ambience at the street level in urban spaces. This chapter is significant in that it builds upon the work of these researchers in further developing perspectives on the ambient involving people and technologies in smart urban environments, including the suggestion possibly of a mutual augmenting of awareness giving rise to the motivation to provide an exploration as identified in the primary objectives of this chapter, formulated as follows.

Objectives: The objectives of this chapter are to: a) describe the notion of an ambient perspective in the context of smart cities, learning cities, and future cities using the examples of ambient culture, ambient economy, and ambient mobility; b) formulate a conceptual framework for perspectives on urban life and the ambient in smart cities; and c) explore the relationship between multiple modes of transport for smart cities and walkability as factors contributing insight into ambient mobility. As such, the main research question under exploration in this chapter is – *How does the ambient emerge in urban life in the context of smart cities?*

2. BACKGROUND

The ambient is not a new phenomenon in that information of an ambient nature is generally present in the form of information sensed by people through well-worn paths, the presence and direction of the wind, and other such examples. Stoop (2003) argues that "unobtrusive technology is not actually about hiding the technology, but about making it appropriate, comprehensive, more human" such that "the 'real estate' of the body provides endless opportunities" whereby "wearables will adapt to the user, rather than the user to the system." For Stoop (2003) mobility is important, affording as it does, "freedom of both body and mind" supported by technologies for people that enable them "to be productive, to express themselves, feel safe, explore and connect with others" while "on the move" with "access to solutions that meet the needs of their immediate personal context." Said to "trigger the imagination" (Marzano, 2006), ambient intelligence (AmI) is also referred to as a term that gives rise to many other questions (Marzano, 2006) where quality of life is important. Papastergiadis, Hannon, McQuire, Wyatt, and Carter (2020) identify the need to critically examine ambient public culture using "a new methodological tool-kit" through experimenting with "an exercise in understanding ambient culture" involving "a collage of some of the different views and perceptions" of a public square.

2.1 Definitions

Definitions are provided for key terms used in this chapter based on the research and practice literature.

- **Ambient:** McCullough (2013) describes the ambient in various ways, one of which is, as "that which surrounds but does not distract."

- **Ambient Culture:** Marzano (2006) describes ambient culture as one where systems are educated, have "good, social behavior" and "the ability to learn."
- **Ambient Economy:** According to McKenna (2019) ambient economies "refer to more dynamic, adaptive, and evolving understandings of economies enabled by more aware people in combination with more aware technologies."
- **Ambient Intelligence:** According to Bauer (2015), "ambient intelligence refers to electronic environments that are sensitive and responsive to the presence of people" with capabilities "able to sense what the human inhabitant needs and deliver it without being requested to do so."
- **Ambient Mobility:** The Ambient Mobility Lab (Fraunhofer IAO, 2021) describes ambient mobility as "an exploration of future-facing innovation in mobility" that "operates in tomorrow's new urban paradigm."

2.2 Overview

What follows is a review of the research and practice literature for perspectives on urban life and the ambient focusing on culture, economies, and mobility while identifying issues, controversies, and problems. A conceptual framework for perspectives on urban life and the ambient in smart cities is formulated based on the literature review and is operationalized for use in this chapter. The research question identified in Section 1 is reformulated as a proposition for exploration in this chapter. The methodology for the underlying study in this chapter is described followed by the presentation of results and a discussion of findings. Limitations and future directions are identified followed by the conclusion with an overview of outcomes; the identification of insights and ideas; and the formulation of exercise questions encouraging further exploration.

3. PERSPECTIVES ON URBAN LIFE AND THE AMBIENT

This chapter provides a review of the research and practice literature for understandings and interpretations of urban life in relation to the ambient, focusing on culture, economies, and mobility while keeping in mind the broader context of smart cities, learning cities, and future cities.

3.1 Ambient Culture

Marzano (2006) describes the development of ambient culture through explorations of ambient intelligence by Philips Design and "specialist institutes" finding the key principles of "relevant, meaningful, understandable" in the contexts of "the home, on the person, at work, in public places, and on the move" while highlighting the importance of "imagination and creativity" in identifying opportunities. In shedding light on ambient intelligence, Marzano (2006) claims that it "is about creating new types of relationships, not only between people and their objects, but also among those objects themselves." Ambient intelligence is, according to Bauer (2015) "the amalgamation of neural networks, big data, IoT, wearables, and device user interfaces into services that can automate processes and make recommendations to improve the users' quality of life." McKenna (2015a) introduces the ambient culture concept "in the context of 21st century urban educational spaces in further developing the learning dimension of smart cities" focusing on the constructs of awareness, learning, openness to experience, and civic engagement. McKenna (2015b) describes a minimally viable environment in the form of an application – AboutTheCity – as an example of an enabler of ambient culture. The Institute for the Future (IFTF, 2016) presents a vision of the future of ambient communications and the "ambient experience" as pervasive media environments, as in, "embedded, illuminated, anticipatory, multisensory, and programmable." McKenna (2017a) focuses on awareness, learning, and engagement in exploring "mechanisms for realization of the Internet of Cultural Things (IoCT) and Applications in public spaces in the context of learning cities and smart cities" and the possibility for "animating cultural heritage elements or fragments" in contributing to "discussion and debate spaces for ambient culture and ambient heritage." McKenna (2019) explores ambient culture in the context of a conceptual framework for the ambient turn and smart cities where meaningful participation is featured. Fanea-Ivanovici and Pană (2020) explore "how digital transformations can improve citizens' well-being" while "highlighting the main benefits and challenges of developing smart culture" with such a focus. Papastergiadis et al. (2020) note the importance of "attention" and reflection while highlighting that "as culture moves out, beyond the black boxes and white walls of institutional interiors, it is timely to reconsider whether we have the right tools to account for emerging ambient and open experiences of culture." Papastergiadis et al. (2020) refer to the notion of "ambient participation" and "ambient culture" as explored through the experience of a public square where it is found that multiple modes of experience are supported including formality, informality, deliberative, incidental, and the serendipitous. Cubitt, Lury, McQuire, Papastergiadis, Palmer, Pfefferkorn, and Sunde (2021) articulate the notion of "ambient images" where, enabled through online spaces and other technologies, the sheer "increasing number" of images "accentuates how they are stitched together to form new atmospheres, assemblages, iterations" giving way to the potential "to outsource our literacy" through "algorithmic mediation" and the like "to cope with the sensorial complexity." Wu (2022) describes affective computing as "calculative technologies that purportedly recognize emotion based on data detailing a wide range of human behaviors" where "such technologies include 'like' metrics."

3.2 Ambient Economies

Bohn et al. (2005) speak of "the ambient economy" where features such as, being both "reliable" and "socially acceptable" are highlighted as important. Marzano (2006) argues that the construction of an ambient culture must involve "ideas that are technologically feasible, economically viable, and in line with environmental and social sustainability." Regarding sustainability, Zygiaris (2013) speaks in terms of "green economies" and the interconnection layer as one of several layers of a smart city. Indeed, Gray (2015) speaks of the notion of an ambient workforce, described in dynamic terms as "distributed, always-on, at-the-ready." McKenna (2017b) points to the "possibilities and opportunities for emerging notions of ambient creativity and ambient economies" based on "highly dynamic, evolving, adaptive, and complex smart city environments" when re-conceptualizing jobs, work, and labour. As such, McKenna (2019) provides an understanding of ambient economies as more dynamic, adaptive, and evolving while highlighting the importance of awareness in the case of both people and systems. From a practitioners' perspective in emerging economies, Roux (2019) explores ambient out-of-home (OOH) media which, based on work by various researchers, is "also referred to as ambient advertising, creative media advertising, street marketing, non-traditional, unconventional, or alternative media." According to Roux (2019) "the medium itself conveys the message by employing innovative and unexpected elements in the environment, such as public fountains" designed to engage. McKenna (2021) explores the innovation dimension of smart cities noting that "rethinking economies emerges" in relation to variables such as "livability, privacy and trust for digital platforms, and the value of creativity." Eliot (2022) describes ambient economy as "a secondary economy constructed of goods that are not scarce" but that "must be made artificially scarce in order to enter the market economy", giving rise to "turmoil associated with the collection, use, and privatization of data" in that "current economic and policy understandings of economic and property rights used to mediate policy are incompatible with the economic realities of data" in an age of "ambient computing."

3.3 Ambient Mobility

In the context of ambient intelligence, Stoop (2003) speaks of mobility, affording "relevant and meaningful technology on the move." Doherty, du Bousquet, Campos, El Atifi, Falquet, Massink, and Santoro (2006) note that mobility gives rise to "a number of new problems" for designers "by changing the context of interaction with the system" contributing "a new dimension to the design problem" and as such, interaction that is ambient "where systems are attempting to intelligently and unobtrusively help the user" then give way to "understanding the user's intentions and providing relevant and adequate responses" as "a new challenge." Encarnação (2008) describes ambient mobility in relation to "the convergence of mobility, ubiquity, multimediality/multimodality" contributing to the emergence of "a new information & communication technology paradigm" where "interaction evolves from human-computer interaction (HCI) to a human environment interaction (HEI)" and where computers become part of "the infrastructure of our everyday life" in "supporting us in efficiently managing our personal environment." Freudendal-Pedersen, Kesselring, and Servou (2019) address the notion of automation and mobility in the context of the sustainability of smart cities and future cities. The Ambient Mobility Lab (Fraunhofer IAO, 2021) positions ambient mobility at the intersection of trends such as the "pervasive deployment of sensors, networked technology, and the big data they generate" along with "self-driving technology" and electric vehicles and "ride and vehicle-sharing." As such, ambient mobility may contribute to "a reduction of circulating vehicles, travel times, emissions, and parking infrastructure." Looking to the future, Olgen, Mehrzad, Rahimi, and El Khatib (2022) turn to science fiction for a glimpse of transportation modes involving self-driving drones and the like, highlighting challenges associated with "affordability, social cohesion, equity, and climate change" as well as design and planning. Qiao, Hiuang, and Yeh (2022) describe the Mobility as a Service (MaaS) notion involving transport-flow, information-flow, and computing-flow based on "intelligent transportation concepts, such as high-speed communication networks and autonomous fleets."

3.4 Urban Life and the Ambient

Kaiying, Plewe, and Röcker (2015) explore ambient systems in an Asian context, applied to environmental and air quality problems where technologies are employed for detection and regulation. Dainow (2017) claims that "the presence of ubiquitous" information and communication technologies enable "the capacity of the environment to obtain data and to respond to human action" and it is this that "represents a fundamentally new type of built environment for human living." Drawing on the work of Lynch (1960), Iaconesi and Persico (2017) describe how people moving through the city are "transforming the ambient around themselves" as in "a massive, emergent, real-time, dissonant, and randomly harmonic work of time-based art with millions of authors that change all the time" giving rise possibly to some form

Table 1. Overview of research for ambient culture in smart cities

Author(s)	Year	Ambient Culture
Marzano	2006	Relevant, meaningful, understandable; @home, on the person, @work, public space, on the move; Interdependence; social responsibility; sustainability
Rauterburg	2007	Open adaptive systems in support of the rituals of everyday life
Bauer	2015	Quality of life
McKenna	2015a	Urban educational spaces – awareness, learning, openness, civic engagement
McKenna	2015b	Extended framework for ambient culture
IFTF	2016	Vision of ambient communications, ambient experience, pervasive media
McKenna	2017a	IoCT and Applications for ambient culture & heritage
McKenna	2019	Ambient turn and smart cities – meaningful participation
Fanea-Ivanovici & Pană	2020	Digital transformations and well-being
Papastergiadis et al.	2020	Ambient public culture, ambient participation – attention and reflection
Cubitt et al.	2021	Ambient images, algorithmic mediation, outsourcing of literacy
Wu	2022	Affective computing and ambient politics

of connection between the ambient and serendipity. Using a systems approach, Patorniti and Stevens (2017) address ways to achieve street ambience through elements such as "greenery, illumination, aesthetic interest, surfaces for sitting, surfaces for play, places to meet and wait, street-level food, drink and entertainment, and human activity." As such, street ambience research is said to bridge knowledge from the siloed areas of architecture, urban design, engineering, transport planning and economics (Patorniti and Stevens, 2017). From a practice perspective, Kordts, Gerlach, and Schrader (2021) refer to "ambient applications" as "applications that are used within smart environments." Sharma, Aittala, Schechner, Torralba, Wornell, Freeman, and Durand (2021) remove or "subtract" the element of "ambient light" in evolving technologies to reveal information in a space that is otherwise not visible. Considered to be a key component of smart cities, the notion of ambient interfaces is described by Hopkins, Bae, Uhr, Banić, Zheng, and Do (2021) as "a distributed network of sensors and actuators, aimed towards recognizing humans" while being attentive to their needs, citing the work of David, Zhou, Xu, and Chalon (2011) and referring to "context-aware systems", drawing on the work of Kjær (2007). Hopkins et al. (2021) describe ambient interfaces associated with a wide range of urban services and activities from mobility to play to surveillance (in the form of drones), to healthcare. McKenna (2022) explores the usefulness and usability of ambient theory for smart cities focusing on variables such as mixed-use spaces and urbanizing (e.g., adapting for urban uses).

3.5 Literature Review Overview

An overview by author and date is provided in this section for the research and practice literature for ambient culture, ambient economies, ambient mobility, and urban life and the ambient.

Table 1 provides an overview by author and date of the research and practice literature for ambient culture spanning the timeframe of 2006 to 2022.

Marzano (2006) speaks of ambient culture in terms of the interdependence of life, involving not just concerns with the environment but also with social responsibility, sustainability, and the contexts of home,

Table 2. Overview of research for ambient economies in smart cities

Author(s)	Year	Ambient Economies
Bohn et al.	2005	Anything, anywhere business model, reliable, socially acceptable
Marzano	2006	Economic viability; quality of life
Zygiaris	2013	Green economies; interconnection layer
Gray	2015	Ambient workforce – distributed, always on, at-the-ready
McKenna	2017b	Re-conceptualizing jobs, work, and labour
McKenna	2019	More dynamic, adaptive, and evolving understandings of economies
Roux	2019	Ambient out-of-home (OOH) media
McKenna	2021	Rethinking economies in smart environments – value of creativity
Eliot	2022	Data collection, use & privatization in an age of ambient computing

work, and public spaces. Rauterburg (2007) describes an ambient culture of open adaptive systems that support the rituals of everyday living. Bauer (2015) describes ambient intelligence as systems enabling services in support of recommendations for improving quality of life. McKenna (2015a) introduces the ambient culture concept to urban educational spaces focusing on awareness, learning, openness, and civic engagement and also proposes an extended framework for ambient culture (McKenna, 2015b). IFTF (2016) advances a vision of ambient communications with a pervasive media experience in support of an ambient experience. McKenna (2017a) explores potentials for the Internet of Cultural Things (IoCT) and applications in public spaces influencing discussion spaces for ambient culture and ambient heritage. Ambient culture forms part of a conceptual framework for the ambient turn and smart cities formulated by McKenna (2019), featuring meaningful participation. Fanea-Ivanovici and Pană (2020) consider the benefits and challenges of smart culture with a focus on digital transformations in support of well-being. Papastergiadis et al. (2020) highlight the need for attention and reflection on the experience and perception of emergent ambient public culture and ambient participation. Cubitt et al. (2021) describe "ambient images" and the volume and speed of image generation, giving way to the potential "to outsource our literacy" through algorithmic mediation with this emergent complexity. Wu (2022) explores affective computing in relation to ambient politics in the context of culture.

Table 2 provides an overview by author and date of the research and practice literature for ambient economies spanning the timeframe of 2005 to 2022.

Bohn et al. (2005) speak of the ambient economy in relation to the anything, anywhere business model and the importance of reliability and social acceptability while Marzano (2006) advances the importance of economic viability and quality of life. Zygiaris (2013) articulates the notion of green economies and the interconnection layer of smart cities. Gray (2015) describes the workforce as ambient as in, distributed, always on, and at-the-ready. McKenna (2017b) points to potential and opportunities for ambient economies in re-conceptualizing jobs, work, and labour in smart environments. McKenna (2019) refers to ambient economies as more dynamic, adaptive, and evolving with awareness as key in the case of both people and systems. Roux (2019) explores ambient economies from the perspective of practitioners in emerging economies focusing on creative instances of ambient out-of-home (OOH) media. McKenna (2021) highlights the importance of rethinking economies in smart environments and the

Table 3. Overview of research for ambient mobility in smart cities

Author(s)	Year	Ambient Mobility
Stoop	2003	Supportive, responsive, relevant, meaningful technologies on the move
Doherty et al.	2006	Ambience & mobility, design challenges, ambient interaction contexts
Marzano	2006	On-the-move
Encarnaç*ão*	2008	Mobility, ubiquity, multimediality/multimodality in support of HEI
David et al.	2011	Mobile user interfaces
Freudendal-Pedersen et al.	2019	Automation and mobilities, sustainable future cities, techno-policies
Fraunhofer IAO	2021	Sensors, networks and data to improve travel, emissions, and parking
Olgen et al.	2022	Sci-fi glimpse of transportation modes and associated challenges
Qiao et al.	2022	Mobility as a Service (MaaS) – transport, information, computing

value of creativity. Eliot (2022) describes policy issues associated with the ambient economy pertaining to "the collection, use, and privatization of data" in an age of ambient computing.

Table 3 provides an overview by author and date of the research and practice literature for ambient mobility spanning the timeframe of 2003 to 2022.

Stoop (2003) speaks of ambient mobility in terms of technologies that are relevant and meaningful as people are on the move. Doherty et al. (2006) address the design challenges associated with the changing context of ambient interactions. Marzano (2006) also speaks of the ambient in the context of people on-the-move while Encarnação (2008) describes ambient mobility as the convergence of mobility, ubiquity, multimediality/multimodality in support of human environment interaction (HEI). David et al. (2011) describe mobile human interfaces while Freudendal-Pedersen et al. (2019) are concerned with techno-policy discourse and the sustainability of mobilities in smart cities. The Ambient Mobility Lab at Fraunhofer IAO (2021) is concerned with the use of more aware technologies to improve travel times, emissions, and parking infrastructures. Olgen et al. (2022) consider future transportation modes drawing on science fiction while highlighting associated challenges. Qiao et al. (2022) describe the notion of Mobililty as a Service (MaaS) focusing on transport, information, and computing.

Table 4 provides an overview by author and date of the research and practice literature for urban life and the ambient spanning the timeframe of 2007 to 2022.

Kjær (2007) articulates the notion of context-aware systems while Kaiying et al. (2015) explore ambient systems in an Asian context for detecting environmental and air quality problems. Dainow (2017) points to a new type of built environment for human living involving ICTs and the use of data generated in such environments in responding to human action. Iaconesi and Persico (2017) describe how people transform the ambient around them in urban spaces as time-based art. Patorniti and Stevens (2017) consider approaches to achieving street ambience through greenery, illumination, aesthetic interests, surfaces for sitting or play, places to meet or wait or entertain, and more generally, human activity. Kordts et al. (2021) articulate the notion of ambient applications for smart environments while Hopkins et al. (2021) describe ambient interfaces consisting of a network of sensors and actuators, responsive to people and their needs. Sharma et al. (2021) utilized ambient light in evolving technologies in revealing otherwise invisible information. McKenna (2022) addresses the usefulness and usability of ambient theory for smart cities using the variables of mixed-use spaces and urbanizing.

Table 4. Overview of research for urban life and the ambient in smart cities

Author(s)	Year	Urban Life and the Ambient
Kjær	2007	Context-aware systems
Kaiying et al.	2015	Ambient systems for detecting environmental & air quality problems
Dainow	2017	ICTs and human action – built environment for human living
Iaconesi & Persico	2017	People transforming the ambient around them as time-based art
Patorniti & Stevens	2017	Street ambience – greenery, illumination, aesthetics, surfaces, places
Kordts et al.	2021	Ambient applications
Hopkins et al.	2021	Ambient interfaces – sensors & actuators – attentive to people
Sharma et al.	2021	Ambient light – addition or subtraction of, to reveal information
McKenna	2022	Usefulness and usability of ambient theory for smart cities

In terms of patterns and potential relationships, what emerges from this review of the research and practice literature is a comparative overview for ambient culture, ambient economies, ambient mobility, and urban life and the ambient as presented in Table 5.

Table 5. Patterns and potential relationships

Ambient Culture	Ambient Economies	Ambient Mobility	Urban Life and the Ambient
Adaptive systems Ambient communications Ambient experiences Ambient participation Ambient public culture Attention Digital transformation Interdependence Pervasive media Open systems Quality of life Reflection Rituals of everyday life Social responsibility Sustainability Well-being	Adaptive Always on, at-the-ready Ambient workforce Anything, anywhere BM- Complex Distributed Dynamic Economic viability Evolving Green economies Interconnection layer Out-of-home media Policy & data Quality of life Reliability Socially acceptable	Automation Context of interaction Design challenges Mobile user interfaces On the move Techno-policies Technologies - meaningful - relevant - responsive - supportive Travel improvements - data - networks - sensors Sustainable future cities	Air quality Ambient applications Ambient interfaces - actuators - attentive to people - sensors Ambient light - making invisible visible Ambient systems Context-aware systems Environmental quality People & time-based art Street ambience - greenery, places, surfaces Theory - mixed-use, urbanizing

Elements that emerge across columns of Table 5, include, but are not limited to, adaptive, various aspects of the ambient (e.g., applications, communications, experiences, interconnection, interdependence, interfaces, light, participation, public culture, systems, and workforce), openness, people, quality of life, technologies, social aspects, and sustainability.

3.6 Issues, Controversies, Problems

This review of the research literature surfaces a range of issues, controversies, and problems as identified in Table 6 for ambient culture, economies, mobility, and urban life and the ambient.

Table 6. Issues, controversies, and problems

	Issues	Controversies	Problems
Ambient Culture	• Interdependence • Meaningful • Relevant • Understandable	• Multiple modes/experience • New types of relationships • Quality of life (QoL) • Pervasive media environ	• Attention & Reflection • Open adaptive systems • Outsourcing our literacy • Sustainability
Ambient Economies	• Adaptive • Dynamic • Reliable • Socially acceptable	• Ambient workforce Always-on, At-the-ready • Privacy • Tracking	• Economically viable ideas • Livability • Social sustainability • Trust
Ambient Mobility	• Affordability • Environmental • Interaction context • Social cohesion	• Climate change • Equity • Human-environ interaction • Relevant meaningful tech	• Design challenges • Designing & planning • Interactions • Sustainability
Urban Life & the Ambient	• Environment data • Healthcare • Quality of life • Street ambience	• Awareness • Interfaces • Mixed-use spaces • Surveillance	• Air quality • Environmental detection • Environmental regulation • Urbanizing

For ambient culture, issues, controversies, and problems pertain to interdependencies; technologies that are meaningful, relevant, and understandable; multiple modes of experience; new types of relationships; quality of life (QoL); pervasive media environments; attention and reflection; open adaptive systems; the outsourcing of our literacy; and sustainability.

For ambient economies, issues, controversies, and problems pertain to adaptive, dynamic, reliable, and socially acceptable; an ambient workforce that is always on and at-the-ready; privacy; tracking; economically viable ideas; livability; social sustainability; and trust.

For ambient mobility, issues, controversies, and problems pertain to affordability; the environment; the changing interaction context; social cohesion; climate change; equity; human-environment interaction (HEI); relevant and meaningful technologies; design challenges; designing and planning; interactions; and sustainability.

For urban life and the ambient, issues, controversies, and problems pertain to the collecting of data from the built environment; healthcare; quality of life; street ambience; awareness; interfaces; mixed-use spaces; surveillance; air quality; environmental detection and regulation; and urbanizing (e.g., adapting for urban uses).

Of note is the issue of the built environmental and the emergence of associated problems such as air quality where the ambient, in the form of information and data analysis, may be of assistance. As such, issues, controversies, and problems emerging from the literature review give rise to areas in need of further exploration, providing additional support for the research question in this chapter as stated in Section 1.

4. CONCEPTUALIZING URBAN LIFE AND THE AMBIENT IN SMART CITIES

In response to the issues, controversies, and problems emerging from the literature review in the preceding section, a conceptual framework for perspectives on urban life and the ambient in the context of smart cities is presented in Fig. 1. This framework builds upon and enhances the ambient perspectives component of the urban life and the ambient conceptual framework for smart cities presented in Chapter

Figure 1. Perspectives on urban life and the ambient in smart cities conceptual framework

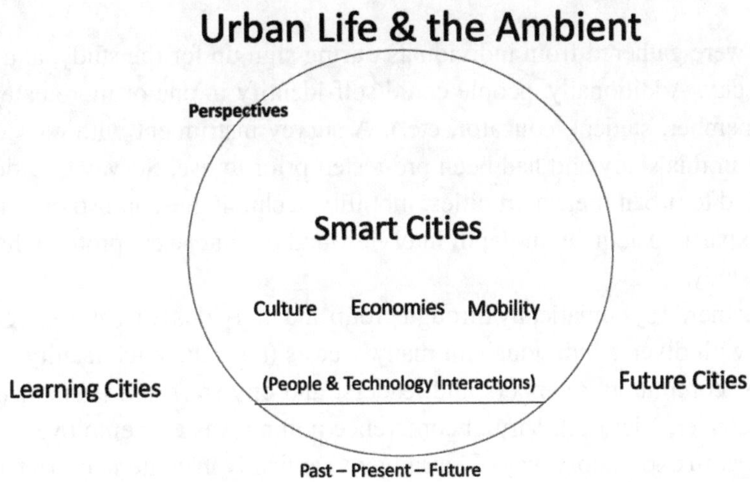

1 (Fig. 1). While being attentive to learning cities and future cities in this chapter, Fig. 1 focuses on the ambient in terms of culture, economies, and mobility, as emergent examples involving people and technology interactions, operating on an underlying awareness continuum attentive to the past-present-future.

The research question identified in Section 1 is reformulated as a proposition for exploration in this chapter, as follows:

Proposition (P2): *The ambient emerges in urban life in the context of smart cities through interactions and awareness involving people and technologies*

The methodology used in exploring the proposition under study in this chapter is described in the following section.

5. METHODOLOGY

The research design for this chapter is described in sections 5.1 to 5.3 in terms of the process used, sources of evidence for data collection, and data analysis.

5.1 Process

The research study underlying this chapter involved an exploratory case study approach combined with an explanatory correlational design since the two are said to be complementary (Yin, 2018) in that the former may be "needed to examine the underlying processes that might explain a correlation." Using an online space to describe the study and invite people to participate, interest emerged from individuals in a variety of cities in many countries (Canada, United States, multiple countries in Europe, Israel, etc.) enabling voices to be heard from several continents.

5.2 Sources of Evidence

Demographic data were gathered from individuals during sign up for the study, and this included age range, gender, and city. Additionally, people could self-identify in one or more categories (e.g., business, community member, student, educator, etc.). A survey instrument with closed and open-ended questions was used in this study and had been pre-tested prior to use. Survey questions pertained to a range of items related to urban life, smart cities, mobility, technologies, and so on. Survey respondents were also invited to participate in an in-depth interview and the interview protocol had been pre-tested prior to use in this study.

Data were also gathered systematically through group and individual discussions, conducted in parallel with this study, with diverse individuals in many sectors (e.g., students, faculty, and staff in higher education, business, community members and leaders, and city staff and leaders) in Canadian cities (e.g., Toronto, Vancouver, Victoria). Virtual conference polling was also employed, in the context of a paper presentation, as an exploratory way of learning dynamically in contemporary technology-enabled environments.

5.3 Data Analysis

Data were analyzed in this chapter guided by the research question posed which was then reformulated as a proposition for exploration, as follows.

P2: *The ambient emerges in urban life in the context of smart cities through interactions and awareness involving people and technologies.*

The use of descriptive statistics enabled the analysis of quantitative survey data pertaining to urban life and the ambient and elements such as culture, economies, and mobility, through exploration of items such as attuning to urban spaces in relation to heightening urban sensibilities. The correlation feature of the Real Statistics Resource Pack containing an-add in for Microsoft Excel (Zaiontz, 2022) enabled the conducting of a Spearman Correlation Coefficient for ordinal data in exploring the nature of relationships between items pertaining to urban life and the ambient. Content analysis was used for qualitative data gathered through open-ended survey questions, interviews, and group and individual discussions. As such, inductive analysis from data collection was combined with deductive analysis using terminology from the literature review.

Overall, an analysis was conducted for n=79 consisting of 42% females and 58% males for people ranging in age from their 20s to 70s.

6. RESULTS AND DISCUSSION

As shown in Table 7, assessments were conducted for *the ambient* in urban life using the proxies of *attuning to urban spaces* and *heightening urban sensibilities*.

For example, in response to the survey question about whether city-focused social media and other aware technologies give rise to elements such as *attuning to urban spaces* and *heightening urban sensibilities*, on a 7-point Likert type scale, where 1= Not at all and 7=Absolutely, for *attuning to urban*

Table 7. Assessments for the ambient using attuning to urban spaces and heightening urban sensibilities

Items	Assessments for the Ambient	Correlation
Attuning to urban spaces	17% (4); 33% (6); 50% (7)	.26
Heightening urban sensibilities	33% (5); 17% (6); 50% (7)	

Table 8. Assessments for ambient culture using sharing and urbanizing

Items	Assessments for Ambient Culture	Correlation
Sharing	33% (6); 67% (7)	.22
Urbanizing	50% (5); 17% (6); 33% (7)	

spaces, 17% of survey respondents selected the neutral position of 4, 33% responded toward the upper end of the scale at 6 (Sure), and 50% responded at position 7. For *heightening urban sensibilities*, 33% of survey respondents selected position 5 on the scale, 17% responded toward the upper end of the scale at 6 (Sure), and 50% responded at position 7.

Using the Real Statistics add-in for Microsoft Excel (Zaiontz, 2022), a Spearman correlation for ordinal data was conducted on the assessment results of the two items in Table 7, yielding a finding of .26. According to Creswell (2018), for correlations in the .20 to .35 range "there is only a slight relationship" and as such, findings "may be valuable to explore the interconnection of variables but of little value in prediction studies."

Exploring further, assessments for *ambient culture* are provided in Table 8 using the proxies of *sharing* and *urbanizing (e.g., adapting for urban uses)*. Again, in response to the survey question about whether city-focused social media and other aware technologies give rise to *sharing* and *urbanizing* (e.g., *adapting for urban uses*), on a 7-point Likert type scale, where 1= Not at all and 7=Absolutely, for *sharing*, 33% of survey respondents tended toward the upper end of the scale at 6 (Sure), and 67% responded at position 7. For *urbanizing (e.g., adapting for urban uses)*, 50% of survey respondents selected position 5 on the scale, 17% responded toward the upper end of the scale at 6 (Sure), and 33% responded at position 7.

A Spearman correlation was conducted on the assessment results of the two items in Table 8, yielding a finding of .22.

Exploring further still, assessments for *ambient economies* are provided in Table 9 using the proxies of *openness* and *technology-driven services.*

When asked to assess the extent to which factors such as openness are associated with smart cities, on a 7-point Likert type scale, where 1= Not at all and 7=Absolutely, 17% of survey respondents selected the neutral position of 4 on the scale, 33% tended toward the upper end of the scale at 6 (Sure), and 50%

Table 9. Assessments for ambient economies using openness and technology-driven services

Items	Assessments for Economies	Correlation
Openness	17% (4); 33% (6); 50% (7)	.21
Technology-driven services	33% (5); 17% (6); 50% (7)	

Table 10. Assessments for ambient mobility in relation to multiple modes of transport and walkability

Items	Assessments for Mobility	Correlation
Multiple modes of transport	17% (4); 17% (6); 66% (7)	.23
Walkability	33% (5); 50% (6); 17% (7)	

responded at position 7. For *technology-driven services,* 33% of survey respondents selected position 5, 17% responded toward the upper end of the scale at 6 (Sure), and 50% responded at position 7. When correlated, a Spearman correlation coefficient of .21 emerges between *openness* and *technology-driven services*.

And exploring further still, assessments for *ambient mobility* are provided in Table 10 using the proxies of *multiple modes of transport* and *walkability*. On a 7-point Likert type scale, where 1= Not at all and 7=Absolutely, when asked to assess the extent to which factor such as *multiple modes of transport* are associated with smart cities, 17% of survey respondents selected the neutral position of 4 on the scale, 17% tended toward the upper end of the scale at 6 (Sure), and 66% responded at position 7. For *walkability,* 33% of survey respondents selected position 5, 50% responded toward the upper end of the scale at 6 (Sure), and 17% responded at position 7.

A Spearman correlation was conducted on the assessment results for *multiple modes of transport* and *walkability* (Table 10), yielding a finding of .23.

Other findings of note include those emerging from two online polls conducted at the virtual Future Technologies Conference 2021 for a paper being presented on ambient theory (McKenna, 2022). When asked in the first poll, "what contributes to 'the ambient' in your city?", 19% of respondents selected the *people* option, 5% selected the *technologies* option, and 76% selected the option *people interacting with technologies*. Additionally, when asked in the second poll "where do you notice 'the ambient', as in, "heightened awareness" at play in your city?" 44% of respondents selected *cultural events*, 19% selected *economy (e.g., smart products, recommending of products, etc.),* 31% selected *urban mobility options* and 6% selected the *other* option. Poll results would seem to highlight the importance of people and of technologies and more crucially, people interacting with technologies. Further, the ambient figured relatively strongly at 31% when associated with urban mobility and even stronger still at 44% when associated with cultural events and significantly less so at 19% when considered in relation to the economy.

Qualitatively, it is worth noting the comments during a discussion with an individual in downtown Toronto who stated that "you're not just moving one way" describing how "a lot of times I'll ride my bike somewhere and I park my bike and I'll walk around" adding that "maybe I'll jump on a streetcar" and "take the subway" highlighting the multiple and "different modes of transportation" in the city, as in, multi-modality.

As depicted graphically in Fig. 2, the interconnectivity of ambient perspectives with approaches to culture, economy, and mobility gives way in urban life to the notion and potentials of ambient adaptability, ambient awareness, ambient communications, and ambient sharing.

Figure 2. Urban life and the ambient – perspectives in smart cities

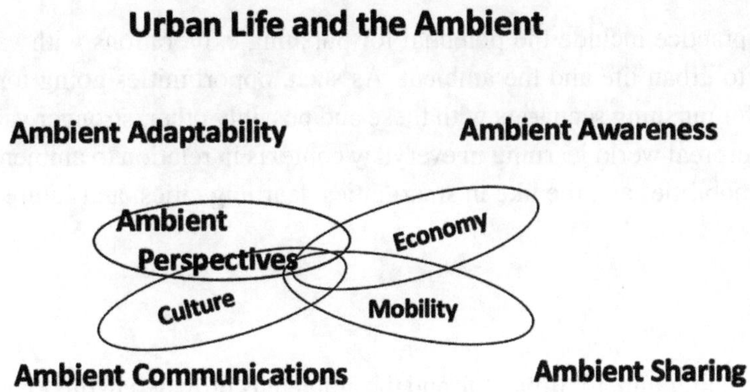

7. LIMITATIONS AND FUTURE DIRECTIONS

This section discusses possible limitations of this work followed by future research directions and opportunities for research and practice.

Limitations

Among the possible limitations of this chapter is that of the extent of the literature review which focuses on urban life and the ambient in smart cities in relation to the elements of culture, economies, and mobility, opening the way for other, more extensive literature reviews as the field evolves. Another limitation of this chapter is that of the small sample size underlying this work and as such, caution is encouraged in the interpretation of correlations from the assessment of proxies used for the ambient as well as for ambient culture, economies, and mobility. Going forward it may be possible to use other proxies or indicators, or other more direct variables, resulting in possibly stronger correlations.

Future Research Directions and Opportunities

Future directions for research include the potential for pursuing additional studies on the variables identified in this work relevant to urban life and the ambient. As such, opportunities going forward may emerge for the identification of other and possibly stronger variables relevant to the ambient that enable further learning in the context of ambient culture, ambient economies, ambient mobilities and the like in smart cities, learning cities, and future cities. Additional directions for further explorations emerge for ambient awareness, ambient adaptability, ambient communications, and ambient sharing.

Of note in Tables 7, 9, and 10 is the presence of neutral responses for *attuning to urban spaces*, *openness*, and *multiple modes of transport* identifying areas of interest for further exploration.

Findings from virtual conference polling may warrant further exploration in learning more about the ambient in relation to *people interacting with technologies* on the one hand and *cultural events* on the other hand as well as *urban mobility options*.

Future Practice Directions and Opportunities

Future directions for practice include the potential for pursuing explorations with variables identified in this work relevant to urban life and the ambient. As such, opportunities going forward for practice include the potential for pursuing scenarios with these and possibly other, stronger variables relevant to the ambient that support real world learning in everyday contexts in relation to ambient culture, ambient economies, ambient mobilities and the like in smart cities, learning cities, and future cities.

8. CONCLUSION

A review of the research literature for urban life and the ambient is provided in this chapter from multiple perspectives including a review of the research literature focusing on ambient culture, ambient economies, and ambient mobility. Issues, controversies, and problems emerging from the research literature provide guidance on the research question, reformulated as a proposition for exploration in this chapter. A key contribution of this work is the *perspectives on urban life and the ambient in smart cities conceptual framework* that is operationalized for use in this chapter using an exploratory case study approach together with an explanatory correlational design. The proposition under study in this chapter focuses on how the ambient emerges in urban life in the context of smart cities involving people and technologies and associated interactions and awareness. While correlations emerge for the ambient, ambient culture, ambient economies, and ambient mobility, employing the use of proxies, "only a slight relationship" emerges in each case in the form of a Spearman correlation for ordinal data, ranging from .21 to .26. As such, among the limitations of the study underlying this chapter is that of a small sample size so that caution is urged in the interpretation of correlations found and further research is encouraged. Nevertheless, a pathway to guide research and practice activities going forward is provided in terms of further validating the findings in this chapter while also encouraging activities to identify other, additional factors relevant to urban life and the ambient in relation to smart, learning, and future cities.

Key insights and ideas emerging from this chapter are provided, along with exercises encouraging further exploration, as follows:

Insights emerging from this chapter include the following:

1. The importance of city-focused social media and other aware technologies as factors giving rise to many possibilities such as *the ambient* in the form of *attuning to urban spaces* and *heightening urban sensibilities*
2. The importance of city-focused social media and other aware technologies as factors giving rise to *ambient culture* in the form of *sharing* and *urbanizing* (e.g., *adapting for urban uses*)
3. The potential for many more factors to emerge for understanding the ambient in smart cities, learning cities, and future cities such as *people interacting with technologies* and *cultural events*

Ideas emerging from this chapter include the following:

1. The opportunities potential for exploring the ambient from different perspectives in urban life and smart cities research

2. The opportunities potential for urban scenarios involving the ambient such as ambient culture, ambient economies, ambient mobilities and the like
3. The use of an interdisciplinary approach for the study and practice of urban life and the ambient in smart, learning, and future cities

Exercises for the reader to consider, encouraging further exploration include the following:

1. Identify one or more factors in your city that are giving rise to the ambient
2. Identify one or more ways in which the ambient would enhance the quality of urban life in your city
3. Describe one or more elements that contribute to making ambient culture, ambient economies, or ambient mobility possible in your city

The audience for his work will be potentially broad, of interest as it is to an interdisciplinary mix of people, including but not limited to community members, educators, the business community, researchers, urban planners, data scientists, and anyone concerned with urban life and the ambient, spanning culture, economies, and mobility along with associated practice and research opportunities and challenges pertaining to smart cities, learning cities, and future cities.

REFERENCES

Angela Xiao Wu, A. X. (2022). The ambient politics of affective computing. *Public Culture*, *34*(96), 21–45. doi:10.1215/08992363-9435427

Bauer, R. (2015). *How ambient intelligence will change our lives.* World Economic Forum. Retrieved 19 October 2021 from https://www.weforum.org/agenda/2015/10/how-ambient-intelligence-will-change-our-lives/

Bohn, J., Coroamă, V., Langheinrich, M., Mattern, F., & Rohs, M. (2005). Social, economic, and ethical implications of ambient intelligence and ubiquitous computing. In W. Weber, J. M. Rabaey, & E. Aarts (Eds.), *Ambient Intelligence*. Springer. doi:10.1007/3-540-27139-2_2

Creswell, J. W. (2018). *Educational research: Planning, conducting, and evaluating quantitative and qualitative research* (6th ed.). Pearson.

Cubitt, S., Lury, C., McQuire, S., Papastergiadis, N., Palmer, D., Pfefferkorn, J., & Sunde, E. (2021). Ambient images. *The Nordic Journal of Aesthetics*, *61*(2), 60–69.

Dainow, B. (2017). Smart city transcendent: Understanding the smart city by transcending ontology. *Orbit (Amsterdam, Netherlands)*, *1*, 1–15.

David, B., Zhou, Y., Xu, T., & Chalon, R. (2011). Mobile user interfaces and their utilization in a smart city. *Proceedings on the International Conference on Internet Computing (ICOMP).*

Doherty, G. J., du Bousquet, L., Campos, J. C., El Atifi, E. M., Falquet, G., Massink, M., & Santoro, C. (2006). Ambience and mobility. In S. W. Gilroy & M. D. Harrison (Eds.), Lecture Notes in Computer Science: Vol. 3941. Interactive Systems. Design, Specification, and Verification. Springer. , DSV-IS 2005. doi:10.1007/11752707_24

Eliot, D. (2022). *Towards a Theory of Ambient Capitalism: An Inquiry into How Data Creates Value* [Thesis]. Queen's University. Retrieved 18 August 2022 from https://qspace.library.queensu.ca/bitstream/handle/1974/30266/Eliot_David_A_202205_MA.pdf?sequence=5

Encarnação, J. L. (2008). Ambient mobility: Human environment interface and interaction challenges. In *Proceedings of the Working Conference on Advanced Visual Interfaces (AVI '08)*. Association for Computing Machinery. 10.1145/1385569.1385571

Fanea-Ivanovici, M., & Pană, M. (2020). From culture to smart culture. How digital transformations enhance citizens' well-being through better cultural accessibility and inclusion. *IEEE Access: Practical Innovations, Open Solutions*, *8*, 37988–38000. doi:10.1109/ACCESS.2020.2975542

Fraunhofer, I. A. O. (2021). *Ambient Mobility Lab*. Joint initiative of the Senseable City Lab, Massachusetts Institute of Technology (MIT) and the Mobility Innovation Lab, Fraunhofer Institute for Industrial Engineering, IAO. Retrieved September 15 from https://www.muse.iao.fraunhofer.de/en/projects/ambient-mobility-lab-copy.html

Freudendal-Pedersen, M., Kesselring, S., & Servou, E. (2019). What is smart for the future city? Mobilities and automation. *Sustainability*, *11*(1), 221. doi:10.3390u11010221

Gray, M. L. (2015). *Re-assembling the assembly line: Digital labor economies and demands for an ambient workforce*. Berkman Klein Luncheon Series. Berkman Center for Internet & Society, Harvard University. Retrieved 1 April 2018 from https://cyber.harvard.edu/events/luncheon/2015/11/Gray

Hopkins, T., Bae, S. S., Uhr, J., Banić, A., Zheng, C., & Do, E. Y. L. (2021). User interfaces in smart cities. In J. C. Augusto (Ed.), *Handbook of Smart Cities*. Springer. doi:10.1007/978-3-030-15145-4_94-1

Iaconesi, S., & Persico, O. (2017). Third X. In *Digital Urban Acupuncture*. Springer. doi:10.1007/978-3-319-43403-2_3

IFTF. (2016). *When everything is media: The future of ambient communications*. Institute for the Future SR-1883. Retrieved 7 October 2021 from https://www.iftf.org/fileadmin/user_upload/downloads/th/2016_IFTF_TH_WhenEverythingIsMedia_map_SR1883.pdf

Kaiying, C. L., Plewe, D. A., & Röcker, C. (2015). The ambience of ambient intelligence: An Asian approach to ambient systems? *Procedia Manufacturing*, *3*, 2155–2161. doi:10.1016/j.promfg.2015.07.355

Kjær, K. E. (2007). A survey of context-aware middleware. *Proceedings of the 25th Conference on IASTED International Multi-conference: Software Engineering*, 148–155.

Kordts, B., Gerlach, B., & Schrader, A. (2021). Towards self-explaining ambient applications. In *The 14th PErvasive Technologies Related to Assistive Environments Conference (PETRA 2021)*. ACM. 10.1145/3453892.3461325

Lynch, K. (1960). *The image of the city*. MIT Press.

Marzano, S. (2006). Ambient culture. In True Visions. Springer. doi:10.1007/978-3-540-28974-6_3

McCullough, M. (2013). *Ambient Commons: Attention in the Age of Embodied Information*. The MIT Press. doi:10.7551/mitpress/8947.001.0001

McKenna, H. P. (2015a). Ambient culture in 21st century urban educational environments: An exploration of awareness, learning, openness, and engagement. *Proceedings of the 9th International Technology, Education & Development (INTED2015) Conference*, 1502-1512.

McKenna, H. P. (2015b). Learning to Iterate in the City: Interactive Environments for Socially Responsive Urban Participation. In S. Carliner, C. Fulford & N. Ostashewski (Eds.), *Proceedings of EdMedia 2015-World Conference on Educational Media and Technology* (pp. 1694-1698). Association for the Advancement of Computing in Education (AACE). Retrieved August 17, 2022 from https://www.learntechlib.org/primary/p/151445/

McKenna, H. P. (2016). Emergent ambient culture in smart cities: Exploring the Internet of Cultural Things (IoCTs) and applications in 21st century urban spaces. *Proceedings of the 12th International Conference on Signal-Image Technology and Internet-Based Systems (SITIS 2016)*, 420-427.

McKenna, H. P. (2017a). Adaptive reuse of cultural heritage elements and fragments in public spaces: The Internet of Cultural Things and Applications as infrastructures for learning in smart cities. *13th International Conference on Signal-Image Technology & Internet-Based Systems (SITIS)*, 479-484. 10.1109/SITIS.2017.84

McKenna, H. P. (2017b). Re-conceptualizing jobs, work, and labour: Transforming learning for more creative economies in 21st century smart cities. *Proceedings of the 10th Annual International Conference of Education, Research and Innovation (iCERi2017)*, 8367-8376. 10.21125/iceri.2017.2251

McKenna, H. P. (2019). *Ambient Urbanities as the Intersection Between the IoT and the IoP in Smart Cities*. IGI Global. doi:10.4018/978-1-5225-7882-6

McKenna, H. P. (2021). *Seeing Smart Cities Through a Multi-Dimensional Lens: Perspectives, Relationships, and Patterns for Success*. Springer. doi:10.1007/978-3-030-70821-4

McKenna, H. P. (2022). Exploring the usefulness and usability of ambient theory for smart cities. In N. A. Streitz & S. Konomi (Eds.), Lecture Notes in Computer Science: Vol. 13325. *Distributed, Ambient and Pervasive Interactions. Smart Environments, Ecosystems, and Cities. HCII 2022*. Springer. doi:10.1007/978-3-031-05463-1_12

McKenna, H. P. (2022). Is ambient theory for smart cities even a theory: An affirmative assessment. In *Proceedings of the Future Technologies Conference (FTC) 2021, Volume 2. FTC 2021. Lecture Notes in Networks and Systems* (vol. 359). Springer. 10.1007/978-3-030-89880-9_41

Olgen, B., Mehrzad, F., Rahimi, N., & El Khatib, S. (2022). Sci-fi shows like 'Westworld' and 'Altered Carbon' offer a glimpse into the future of urban transportation. *The Conversation*. Retrieved 10 July 2022 from https://theconversation.com/sci-fi-shows-like-westworld-and-altered-carbon-offer-a-glimpse-into-the-future-of-urban-transportation-179916

Papastergiadis, N., Hannon, S., McQuire, S., Wyatt, D., & Carter, P. (2020). Ambient culture: Making sense of everyday participation in open public space. In A. de Dios & L. Kong (Eds.), *Handbook on the Geographies of Creativity*. Edward Elgar Publishing.

Patorniti, N., & Stevens, N. (2017). Unlocking the secrets of street ambience. *The Conversation*. Retrieved 17 March 2022 from https://theconversation.com/unlocking-the-secrets-of-street-ambience-72662

Qiao, S., Hiuang, G., & Yeh, A. G.-O. (2022). *Mobility as a service and urban infrastructure: From concept to practice*. Transactions in Urban Data, Science, and Technology. doi:10.1177/27541231221114171

Roux, T. (2019). Drifting between counting atoms and telling stories: Quantitative and qualitative ambient OOH media planning guidelines for emerging economies. *Journal of Global Marketing*, *32*(1), 49–63. doi:10.1080/08911762.2018.1431350

Sharma, P., Aittala, M., Schechner, Y. Y., Torralba, A., Wornell, G. W., Freeman, W. T., & Durand, F. (2021). What You Can Learn by Staring at a Blank Wall. *International Conference on Computer Vision*. 10.1109/ICCV48922.2021.00233

Stoop, E. (2003). Mobility: Freedom of body and mind. In The New Everyday: Views on Ambient Intelligence. 010 Publishers.

Yin, R. (2018). *Case study research and applications: Design and methods* (6th ed.). Sage Publications.

Zaiontz, C. (2022). *Real statistics using excel*. Retrieved from www.real-statistics.com

Zygiaris, S. (2013). Smart city reference model: Assisting planners to conceptualize the building of smart city innovation ecosystems. *J. Knowl. Econ.*, *4*(2), 217–231. doi:10.100713132-012-0089-4

ADDITIONAL READING

Arnade, C. (2022). Why I walk: Walking as learning. *Walking the World Blog*. Retrieved 1 May 2022 from https://walkingtheworld.substack.com/p/why-i-walk-part-1?s=r

Bushwick, S. (2021). A blank wall can show how many people are in a room and what they're doing: The technique is the latest method to glean a surprising amount of surveillance from a meager source. *Scientific American*, (October), 6.

Cheshire, J., & Uberti, O. (2021). *Atlas of the invisible: Maps and graphics that will change how you see the world*. Norton.

Eaton, K. (2012). Fujitzu gizmo turns heat, light into electricity, promises "power from everywhere". *FastCompany*. Retrieved 23 April 2022 from https://www.fastcompany.com/1708463/fujitsu-gizmo-turns-heat-light-electricity-promises-power- everywhere

Geotab. (2018). What is smart mobility? *Geotab Blog*. Retrieved 2 October 2021 from https://www.geotab.com/blog/what-is-smart-mobility/

Lacey, J. (2022). Cities are made from more than buildings and roads. They are also made by ambiences – how a city makes you feel. *The Conversation*. Retrieved 23 April 2022 from https://theconversation.com/cities-are-made-from-more-than-buildings-and-roads-they-are-also-made-by-ambiences-how-a-city-makes-you-feel-173429

Moorhead, P. (2021). Amazon's gives its take of an 'ambient' Alexa experience. *Forbes*. Retrieved September 24, 2021, from https://www.forbes.com/sites/patrickmoorhead/2021/07/22/amazons-gives-its-take-of-an-ambient-alexa-experience/?sh=58cfa0035a5a

Papastergiadis, N. (2013). *Ambient Perspectives*. Lyon HouseMuseum.

Stevens, G. (2020). Computers without boundaries: A deep dive into ambient computing. *StreetFight*. Retrieved 13 March 2022 from https://streetfightmag.com/2020/03/16/computers-without-boundaries-a-deep-dive-into-ambient-computing/

KEY TERMS AND DEFINITIONS

Adaptability: Adaptability in an urban context refers to the capacity to adjust to rapidly changing, complex situations and circumstances.

Ambient Adaptability: Ambient adaptability in the context of urban life in a smart city refers to the capacity to adjust to rapidly changing, complex situations and circumstances in the moment.

Ambient Awareness: Ambient awareness in an urban context refers to the capacity to perceive or have knowledge of the surroundings and the changing environment dynamically, in the moment.

Ambient Sharing: Ambient sharing in an urban context refers to the capacity to engage jointly with others, aided in some cases by technologies, in the moment.

Attention: Attention in an urban context refers to the capacity to notice or be observant of the surroundings and the environment more generally.

Attuning: Attuning in an urban context refers to the capacity to be more receptive to and aware of the surroundings and the environment more generally.

Awareness: Awareness in an urban context refers to the capacity to perceive, sense, or have knowledge of the surroundings and the environment more generally.

Creativity: Creativity in an urban context refers to spaces enabling or accommodating the use of imaginative and original ideas.

Sharing: Sharing in an urban context refers to the capacity to engage jointly with others in some way.

Walkability: Walkability in an urban context refers to the designing of spaces to accommodate walking.

Chapter 3
Urban Engagement and Participation in Smart, Learning, and Future Cities:
Approaches to Collaboration, Openness, and Privacy

ABSTRACT

The purpose of this chapter is to explore urban engagement and participation in the context of smart cities, learning cities, and future cities, focusing on the elements of collaboration, openness, and privacy. As such, a review of the research literature for urban engagement, urban participation, collaboration, and openness and privacy is provided in the context of smart cities. Issues, controversies, and problems are highlighted contributing to formulation of a conceptual framework for urban engagement and participation in smart cities. Using an exploratory case study approach combined with an explanatory correlational design, variables of interest are explored, such as collaborating and privacy, to learn more about the nature of their relationship in contributing to understandings of smart cities. Limitations of this chapter are identified, possibly opening opportunities, challenges, and directions for future research and practice. Insights and ideas emerging from this chapter are presented along with exercises encouraging further explorations.

1. INTRODUCTION

The importance of engagement as a mediating factor for people and their experiences across domains has been identified by researchers (Cogut, Webster, Marans, and Callewaert, 2019; O'Brien and Cairns, 2016). Zandbergen and Uitermark (2020) explore situated engagement in urban life through use of "data and sensing technologies" setting the stage for a rethinking of urban engagement possibilities. Przeybilovicz, Cunha, Geertman, Leleux, Michels, Tomor, Webster, and Meijer (2020) identify the underdeveloped

DOI: 10.4018/978-1-6684-4096-4.ch003

role of "citizen participation in smart city initiatives" while identifying types of interaction such as collaboration, said to be important for achieving quality of life (Mills, Izadgoshasb, and Pudney, 2021), among other urban benefits. Bloom, Lauriault, and Landry (2018) associate openness with collaboration, among a series of other elements in smart cities while van Zoonen (2016) explores privacy concerns in smart cities in terms of "personal data used for service purposes" versus that used for "surveillance purposes" as well as "impersonal data" used for these two purposes. As such, this work is significant in that it explores engagement and participation in relation to collaboration, openness, and privacy in the context of urban life and the ambient in smart, learning, and future cities, giving rise to the motivation for this chapter as described in the following objectives.

Objectives: The main objectives of this chapter are a) to explore engagement and participation in urban life in the context of smart cities; b) formulate a conceptual framework for urban life, engagement, and participation in smart cities; and c) explore the nature of the relationship between *collaborating* and *technology-driven services* as ways of learning more about engagement and participation in smart cities. These objectives give rise to the main research question under exploration in this chapter – *Why do engagement and participation matter in urban life in the context of smart cities?*

2. BACKGROUND

Solove (2008) claims that privacy is "a concept in disarray" and undertakes to develop a theory of privacy including a taxonomy. Gil-Garcia, Zhang, and Puron-Cid (2016) identify 14 dimensions of smartness in government, one of which is openness, and another is citizen engagement. Cardullo and Kitchin (2019) identify the need to "rethink 'smart citizens' and 'smart citizenship' and to remake smart cities if they are to truly become 'citizen centric'." Fedorowicz et al. (2020) note that for urban engagement, "methods exist on a spectrum" drawing on the work of Arnstein (1969) from "activities that 'inform' the public about a specific topic or event" to "activities that 'empower' community members by putting them in leadership roles." Przeybilovicz et al. (2020) explore "the dynamics of citizen participation in smart city initiatives" arguing that "their roles and influence" when considered "as key actors" in smart environments "is underdeveloped." Lister (2021) speaks of valuable engagement in the context of smart learning in the design of "urban digital citizen learning activities" that are especially amenable to creative participation. Doering, Schmidtner, and Timinger (2021) address collaboration between universities and smart cities to bring about innovation in response to "societal and infrastructural needs."

2.1 Definitions

Definitions are provided from the research literature for key terms used in this chapter.

- **Engagement (urban):** Engagement at the community level is describes by Fedorowicz, Arena, and Burrowes (2020) as "a process by which community members come together to reflect on and make decisions about the future of their community" and "can take place in person or online, and it can be an isolated, one-time event, or it can be an ongoing engagement."
- **Openness (urban):** Openness at the urban level is described by Bloom et al. (2018) as an Open Smart City "where residents, civil society, academics, private sector, and public officials collaboratively mobilize data and technologies when warranted in an ethical, accountable and transpar-

ent way in order to govern the city as a fair, viable and livable commons and balance economic development, social progress and environmental responsibility."

- **Privacy:** Privacy, according to Solove (2008) "is a plurality of different things" that "must be mapped like terrain, by painstakingly studying the landscape."

2.2 Overview

In the sections that follow in this chapter, a review of the research and practice literature for urban engagement, urban participation, collaboration, and openness and privacy is provided identifying issues, controversies, and problems. A conceptual framework for urban engagement and participation in smart cities is formulated based on the literature review and operationalized for use in this chapter. A reformulation of the research question under exploration as a proposition in this chapter is presented for study. The methodology for the underlying study in this chapter is described, followed by the presentation of results and a discussion of findings. Limitations are identified giving rise to future directions for research and practice. A conclusion with an overview of outcomes is presented including the identification of insights and ideas and the formulation of questions encouraging further exploration. The target audience for this chapter is also identified.

3. PERSPECTIVES ON URBAN ENGAGEMENT AND PARTICIPATION

This chapter provides a review of the research and practice literature for urban engagement, urban participation, collaboration, and openness and privacy in smart cities.

3.1 Urban Engagement in Smart Cities

O'Brien and Cairns (2016) provide an overview from a range of perspectives, across domains, of the importance of engagement "as a mediator of the user experience" in the form of a "variable between user characteristics, motivations, and preferences and individuals's ability to locate information, be entertained, learn, or connect with content or others" with the potential to "affect some kind of change in users for the better – be it affective, cognitive, or behavioural." Okner and Preston (2017) highlight the importance of "human connection" in smart cities claiming there to be a "symbiotic relationship between citizen engagement smart governance" in support of "human-centered planning and design." Alcaide Muñoz and Rodríguez Bolívar (2019) explore citizen engagement focusing on "actual practices of city governments in European" smart cities (SCs) and the use of information and communication technologies (ICTs) "to improve citizen involvement in public affairs." Findings (Alcaide Muñoz and Rodríguez Bolívar, 2019) reveal that "the context of SCs favours the use of social media to share citizens' opinions and ideas" for "improving public services, government effectiveness, and quality of life" even though results indicate that "few city governments show an active attitude in creating e-participation platforms for promoting citizen engagement" pertaining to the sharing of "ideas, comments, suggestions" and the like. Cogut et al. (2019) explore the moderating role of engagement in relation to "sustainability-related awareness and behavior" at the University of Michigan using data from the Sustainability Cultural Indicators Program. Cogut et al (2019) found that "a link between sustainability awareness and increases in sustainable behavior" although "this link is not present for all behaviors (i.e., use of sustainable travel/transportation)" while finding that "engagement does not amplify the awareness-behavior link."

Zandbergen and Uitermark (2020) explore "ways in which citizens negotiate urban life by means of data and sensing technologies" where, depending upon "specific situated interests and forms of engagement" it is found that "appealing visions of urban life" are possible. That is, based on ethnographic research in Amsterdam involving a project "which encourages citizens to collect and share air quality data", Zandbergen and Uitermark (2020) invoke "cybernetic experiences and practices" as in, "locating decision-making and control in the ad hoc, decentralised and real-time forms of engagement" and as such, this "may contain seeds of alternative ways of organizing urban life." Proposing an alternative to data visualization as an approach to civic engagement, Baibarac-Duignan and de Lange (2021) advance the notion of "controversing" of urban datification as a strategy for citizen engagement and participation focusing on recontextualization, meaning-making, and agency. According to Baibarac-Duignan and de Lange (2021), controversing would provide the space for "people to identify their own issues, come together temporarily as a public, imagine alternative possibilities and thus develop capacities for action" and additionally, enabling people to "develop the capacity to shape alternative urban futures." De Hoop, Moss, Smith, and Löffler (2021) use a case study approach to explore projects in European cities involving the development and use of digital devices focusing on three urban environments, namely, the ambient (e.g., noise, radiation), the built, and the natural in terms of the potential and limitations for knowledge generation and knowing in support of governance while being attentive to the importance of citizen engagement and non-digital knowledge generation and ways of knowing. Stelzle, Jannack, Holmer, Naumann, Wilde, and Noennig (2021) advance urban engagement protocols (UEP) in support of "digital-driven interaction between citizens and planners" as "a key component for future smart cities" which through testing, "proves the management possibilities of the U_CODE process in massive co-design urban development scenarios in a real environment." Chen, Gil-Garcia, and Gasco-Hernandez (2022) explore social sustainability in smart cities to complement existing research on environmental and economic sustainability, arguing for civic engagement as an effective method "to better understand and respond to all residents' social, economic, and environmental demands, enabling a more inclusive and equitable smart community."

3.2 Urban Participation in Smart Cities

Granier and Kudo (2016) explore the involvement of citizens in smart cities in Japanese "smart communities" highlighting the notion of co-production while describing forms of behavior change and identifying the importance of elements such as trust. Okner and Preston (2017) call for broader and more robust public participation, aided by online technologies while "eliminating traditional barriers." Cardullo and Kitchin (2019) formulate the "scaffold of smart city participation" drawing on the work of Arnstein (1969) as a way "to measure smart citizen inclusion, participation, and empowerment in smart city initiatives" in the city Dublin. Allen, Tamindael, Bickerton, and Cho (2020) test empirically whether "citizen e-participation in coproduction is able to increase the performance of service delivery" finding that e-participation seems to have "a stronger relative influence on complex problems that may involve multiple agencies" such as "damaged roads" rather "than with simple routine services." Through "an empirical study of citizen engagement in smart city governance in Brazil, the UK, and the Netherlands, Przeybilovicz et al. (2020) "demonstrate that the roles and functions undertaken by citizens are not static", such that "they participate in a dynamic mode that evolves and changes over time" with "three emerging patterns of contextually specific citizen interaction: contestation, acceptance and collaboration." Simonofski, Van Den Storme, and Meers (2020) use a design science approach to evaluate smart city participation through formulation of a holistic framework based on several frameworks from the research literature, demonstrating how "it is possible to evaluate to what extent smart city initiatives are citizen-oriented", applying the framework to Dublin, London, and Knokke-Heist in Belgium enabling recommendations for improvements of the smart city strategy in the Belgium case. Cardullo (2021) calls for "active and meaningful forms of participation" as well as "new data economics and digital rights" for cities. A critical examination of "'smart city' discourse" is provided by Cardullo (2021) focusing on "governance, initiatives, citizen participation and policies which place emphasis on the 'citizen' as an active recipient and co-producer of technological solutions to urban problems." It is worth noting that Pentland, Lipton and Hardjono (2021) call for a new deal for data and building a new economy around data, involving people. Among the four key themes surfaced by Chen et al. (2022) from the research literature is that of citizen participation in smart cities, claiming that in such contexts "initiatives influence and are affected by issues of equity, inclusion, well-being, and participation." Lister (2022) explores "experiences of participation" in the context of "smart learning journey activities" in relation to the expression by participants of their experience of technology in smart learning environments.

3.3 Collaboration in Smart Cities

Dommel (2005) describes the many challenges (e.g., from infrastructure and interoperability to security, trust, and privacy) associated with ambient collaboration that "embeds collaborative mechanisms in our surrounding environment and offers greater group cohesion through room or wall-based work areas" in support of "drastically widening interaction and multitasking scope with the computer in the background." Orel and Alonso Almeida (2019) explore "workplace ambiences in coworking spaces" with "a discussion on how the ambience that is tailored to a particular coworking space enhances the possibility of collaboration between two or more users of a selected workspace" finding that "factors of spatial design need to be adapted, and engagement strategies need to be constructed." Bartelt, Urbaczewski, Mueller, and Sarker (2020) employ a combination of social capital theory and knowledge transfer theory to qualitatively explore "how the living lab methodology used in the public sector within the smart cities context leads to better collaboration and innovation." Visvizi, Damiani, and Lytras (2020) encourage broader thinking beyond smart cities to include a rethinking of regional collaborations in the context of "a growing complexity of international, inter-regional and inter-continental collaborative arrangement" contributing to the notion of "smart clusters" arguing that "all stakeholders, including businesses, the society and the individuals are involved, and no one is immune to contemporary global transformations." Mills, Izadgoshasb, and Pudney (2021) provide a review of the research literature in exploring "the role of collaboration in a smart city" noting that although "collaboration is problematic in the public sector" it is nevertheless important "to address the intense complexity encountered by city governments" to "achieve city objectives of quality of life, efficiency, effectiveness, and economic and environmental sustainability." It is worth noting that Mills et al. (2021) speak in terms of "authentic collaboration" and "sustainable collaboration". Doering et al. (2021) propose an "adaptive reference model" in support of "collaboration and knowledge transfer between universities and smart cities" for testing particularly with "stakeholders involved in smart city projects" amenable possibly to the notion by McKenna of ambient collaboration (2019). Morrison-Smith, Chilton, and Ruiz (2021) developed AmbiTeam, an ambient display in support of the collaborative activities of researchers working on projects remotely. According to Morrison-Smith et al. (2021) "using AmbiTeam did have a quantifiable effect on researchers' perceptions of their collaborators' prioritization" and that "use of the system motivated researchers to work on their collaborative projects" an effect "known as 'the motivational presences of others,' one of the key challenges that make distance work difficult." Levinson (2022) reports on National Science Foundation (NSF) funding in the United States for smart city technologies involving collaborative activities between colleges, universities, industry partners, community organizations, municipalities, and K-12 schools. Lister (2022) employs collaboration as one of four smart learning journey system elements to explore participant experience variations of technology in examining "potential relationships between digital interactions."

3.4 Openness and Privacy in Smart Cities

Birch (2008) explores understandings of urban public and private space where "good public space" is said to be "malleable and allows its users to take or give it meaning and definition" and "experiments are ongoing" with "public/private partnerships" among other initiatives, highlighting in urban life "the importance of all public space elements" and "the public realm." McKenna, Arnone, Kaarst-Brown, McKnight, and Chauncey (2013) conceptualize ambient privacy focusing on "interactions and relationships" and "the nature and extent of the affordances and constraints in technology-pervasive environments, forming a balanced people, technology, and information dynamic" so as to "foster and enable autonomy, control, creativity, and innovativeness." Van Zoonan (2016) develops a framework "to hypothesize if and how smart city technologies and urban big data produce privacy concerns" among people. Gil-Garcia et al. (2016) identify openness as one of several dimensions of smartness in government, as in, "more transparent and more accountable" and being able to "use information in smarter ways." McKenna (2019) explores the notion of ambient privacy in advancing "a rethinking and innovating of the privacy construct." As such, ambient privacy is "organized around aware people and aware technologies, encompassing the physical and digital in terms of evolving forms of collaboration, openness, and sharing as ambient" (McKenna, 2019). Bushwick (2021) highlights privacy issues associated with emerging technologies that have the capability to "inconspicuously scan" a wall or "surface for shadows and reflections imperceptible to the human eye" and upon analysis, discern "details, including how many people are in the room – and what they are doing" and possibly access information "on activity around a corner" giving rise to concerns with issues of surveillance and privacy. Such methods for accessing information are referred to by Bennett Cyphers (Bushwick, 2021) as "side-channel attacks, or side-channel surveillance" where "routine use" is said to be "way off in the future, if ever." Streitz (2021) cites the work of Ceglowski (2019) in describing ambient privacy as "not a property of people, or of their data, but of the world around us." As such, Streitz (2021) argues that "cold spots" as "circumscribed areas with low-range connectivity, low emission, less connectivity" advanced by Van Kranenburg et al. (2020) "offer some autonomy and anonymity" in the form of "disposable identities" enabling "anonymous presence and self-determined transparent use of local resources, where no commercial data are collected" (Streitz, 2020). In this way, according to Streitz (2020), "cold zones provide ambient privacy in defined spaces of the city." According to Levinson (2022) reports on the funding by the National Science Foundation (NSF) in the United States of "an engineering center that will leverage a variety of technologies to improve the quality of urban environments and advance smart city technology" will be attentive to "critical questions regarding safety, equity, privacy and security" in support of "livable, safe, and inclusive communities." In an interview with the ACM Bulletin, Popa (2022) speaks of the notion of "confidential computing" as "the combination of hardware security via hardware enclaves and cryptographic techniques" that will "enable secure collaboration" between teams, predicting that "many data systems in the cloud (e.g., data analytics and machine learning) will migrate to use confidential computing platforms" going forward.

3.5 Literature Review Overview

An overview by author and date is provided in this section for the research and practice literature for engagement, participation, collaboration, and openness and privacy.

Table 1 provides an overview by author and date of the research and practice literature for urban life and the ambient pertaining to engagement, spanning the timeframe of 2016 to 2022 while drawing on the work of Arnstein (1969) pertaining to citizen participation and engagement.

The work of Arnstein (1969) informs engagement through a spectrum or ladder of methods for citizen engagement (Fedorowicz et al., 2020). O'Brien and Cairns (2016) explore engagement as a mediator of user experience across domains while Okner and Preston (2017) articulate the symbiotic relationship between citizen engagement and smart governance where technology is leveraged to augment community interaction. Alcaide Muñoz and Rodríguez Bolívar (2019) explore citizen engagement focusing on the actual practices of city governments in European smart cities. Cogut et al. (2019) explore the moderating role of engagement pertaining to sustainability-related awareness and behavior. Zandbergen and Uitermark (2020) explore ways that people negotiate urban life using sensing technologies, based on situated interests and forms of engagement. Baibarac-Duignan and de Lange (2021) advance the notion of "controversing" of urban datafication as a strategy for citizen engagement and participation focusing on recontextualization, meaning-making, and agency. De Hoop et al. (2021) explore knowledge generation and citizen engagement in digital environments and the potentials and limitations while being attentive to the importance of non-digital ways of knowing and knowledge generation. Stelzle et al. (2021) advance urban engagement protocols (UEP) for urban planning involving citizens and planners. Chen et al. (2022) argue for civic engagement as a method in support of social, economic, and environmental sustainability.

Table 2 provides an overview by author and date of the research and practice literature for urban life and the ambient pertaining to participation, spanning the timeframe of 2016 to 2022.

Granier and Kudo (2016) explore Japanese 'smart communities' for citizen participation where the potential for behavior change emerges as does the element of trust as an important factor. Okner and Preston (2017) advance the need for broader, robust participation in smart cities, removing traditional barriers, and aided by online and other technologies. Cardullo and Kitchin (2019) formulate the scaffold of smart citizen participation as a tool for measuring inclusion, participation, and empowerment in smart city initiatives. Allen et al. (2020) test e-participation as a form a citizen participation in terms of the relationship with service performance based on a smart city mobile platform finding a positive association. Simonofski et al. (2020) formulate a holistic framework for evaluation of citizen participation in smart cities. Przeybilovicz et al. (2020) argue that the role and influence of citizen participation in the initiatives of smart cities is underdeveloped where emergent patterns of interaction include acceptance, collaboration, and contestation. Cardullo (2021) articulates the need for active and meaningful citizen participation from governance to policies in developing technological solutions. Pentland et al. (2021) advance the need for a new deal and a new economy for data, involving people, while Chen et al. (2022) identify participation as a key theme in smart cities along with equity, inclusion, and quality of life. Lister (2022) focuses on the experiences of participation in smart learning environments involving technology.

Table 1. Overview of research for engagement in smart cities

Author(s)	Year	Engagement
Arnstein	1969	Citizen participation ladder
O'Brien & Cairns	2016	Mediator of user experience
Okner & Preston	2017	Citizen engagement & smart governance; tech augments interaction
Alcaide Muñoz & Rodríguez Bolívar	2019	Practices of city governments in European SCs
Cogut et al.	2019	Moderating role – sustainability-related awareness and behavior
Fedorowicz et al.	2020	Urban engagement spectrum
Zandbergen & Uitermark	2020	Use of data and sensing technologies to negotiate urban life
Baibarac-Duignan & de Lange	2021	Controversing – recontextualization, meaning-making, agency
De Hoop et al.	2021	Digital & non-digital knowledge generation in smart environments
Stelzle et al.	2021	Urban engagement protocols (UEP)
Chen et al.	2022	Civic engagement – social, economic, environmental sustainability

Table 3 provides an overview by author and date of the research and practice literature for urban life and the ambient pertaining to collaboration, spanning the timeframe of 2005 to 2022.

Table 2. Overview of research for participation in smart cities

Author(s)	Year	Participation
Grenier & Kudo	2016	Citizen participation, Japanese 'smart communities'; behavior change, trust
Okner & Preston	2017	Broaden & make more robust, removing traditional barriers, aided by tech
Cardullo & Kitchin	2019	Scaffold of smart citizen participation
Allen et al.	2020	Citizen coproduction as government-citizen interface & e-participation
Przeybilovicz et al.	2020	Role and influence of citizen participation in SCs is underdeveloped - acceptance, collaboration, contestation
Simonofski et al.	2020	Holistic framework for evaluation of citizen participation in smart cities
Cardullo	2021	Active, meaningful, new data economics, digital rights for tech solutions
Pentland et al.	2021	New deal for data, new economy for data involving
Chen et al.	2022	One of four key themes in smart cities
Lister	2022	Experiences of participation in smart learning environments; technology

Dommel (2005) provides an ambient collaboration environment (ACE) roadmap while highlighting the many challenges. McKenna (2019) advances the notion of ambient collaboration which may be amenable to the adaptive reference model proposed by Doering et al. (2021). Bartelt et al. (2018) provide a theory-based approach to understanding collaboration and innovation in the smart cities context of Denver, Colorado. Visvizi et al. (2020) argue for broader understandings of the need for collaboration in the context of complex arrangements and global transformations, thus the notion – from smart cities (SCs) to smart clusters. An adaptive reference model is proposed by Doering et al. (2021) for collaboration

Table 3. Overview of research for collaboration in smart cities

Author(s)	Year	Collaboration
Dommel	2005	Challenges of ambient collaboration (ambient + nomadic collaboration)
McKenna	2019	Ambient collaboration
Orel & Alonso Almeida	2019	Ambience of collaboration in coworking environments
Bartelt et al.	2020	Theory-based understanding of collaboration & innovation in SCs
Visvizi et al.	2020	SCs to smart clusters – complex arrangements & global transformations
Doering et al.	2021	Adaptive reference model for knowledge transfer
Mills et al.	2021	Complexities of quality of life, efficiency, effectiveness, sustainability
Morrison-Smith et al.	2021	Ambient displays, remote collaboration, motivational presence of others
Levinson	2022	NSF funding for SC technologies involving cross-sector collaborations
Lister	2022	Smart learning journey system element – participant experience variation

and knowledge transfer between smart cities and universities. Orel and Alonso Almeida (2019) explore the ambience of collaboration in coworking environments finding that spatial design and strategies for engagement matter. Mills et al. (2021) highlight the importance of collaboration in the public sector in addressing the complexities associated with quality of life, efficiency, effectiveness, and economic and environmental sustainability. Morrison-Smith et al. (2021) provide a discussion of ambient displays and their potential to "support remote collaborative work by recreating the motivational presence of others." Levinson (2022) reports on NSF funding for smart city technologies involving cross-sector collaborations. Lister (2022) employs collaboration as one of several smart learning journey system elements in the exploration of experience variations of technology.

Table 4 provides an overview by author and date of the research and practice literature for urban life and the ambient pertaining to openness and privacy, spanning the timeframe of 2008 to 2022.

Birch (2008) highlights the importance of public space elements in urban life. Claiming the privacy concept to be in disarray, Solove (2008) undertakes to formulate a theory of privacy. McKenna et al. (2013) undertake a conceptualizing of ambient privacy focusing on interactions and relationships and the affordances and constraints in technology-pervasive environments. Gil-Garcia et al. (2016) identify openness as one of several elements of smartness in government, contributing to accountability. Van Zoonan (2016) develops a framework for hypothesizing smart city technologies and urban big data privacy concerns for people. Bloom et al. (2018) describe the notion of an open smart city involving the collaborative use of data and technologies, ethically, transparently, and accountably that is responsible economically, environmentally, and socially. Ceglowski provides a definition of ambient privacy as a property of the world around us. McKenna (2019) explores the notion of ambient privacy for rethinking and innovating the privacy construct. Van Kranenburg et al. (2020) describe the notion of cold spots, offering less connectivity and some autonomy and anonymity. Bushwick (2021) explores privacy issues associated with technologies and surveillance capabilities while Streitz (2021) advances the use of cold zones for ambient privacy in defined spaces of the city. Levinson (2022) describes funding in the United States in support of technologies for smart cities (SCs) that take into consideration issues pertaining to safety, equity, privacy, and security. Popa (2022) predicts the use of confidential computing for secure collaboration between teams in the years to come.

Table 4. Overview of research for openness and privacy in smart cities

Author(s)	Year	Openness and Privacy
Birch	2008	Public space elements
Solove	2008	Theory of privacy
McKenna et al.	2013	Ambient privacy – conceptualizing; interactions & relationships
Gil-Garcia et al.	2016	Openness: dimension of smartness in government, accountability
Van Zoonan	2016	Framework for smart city tech and big data privacy concerns in SCs
Bloom et al.	2018	Open smart city, collaborative/responsible use of data and technologies
Ceglowski	2019	Ambient privacy definition as a property of the world around us
McKenna	2019	Ambient privacy – rethinking, innovating privacy construct
Van Kranenburg et al.	2020	Cold spots offering less connectivity; autonomy and anonymity
Bushwick	2021	Privacy issues and emerging technologies; surveillance
Streitz	2021	Use of cold zones for ambient privacy in defined spaces of the city
Levinson	2022	Tech for SCs attentive to safety, equity, privacy, and security
Popa	2022	Confidential computing for secure collaboration between teams

In terms of patterns and potential relationships, what emerges as presented in Table 5 from this review of the research and practice literature is a comparative overview for engagement, participation, collaboration, and openness and privacy.

Table 5. Patterns and potential relationships

Engagement	Participation	Collaboration	Openness and Privacy
Awareness	Acceptance	Adaptive reference model	Ambient privacy
Behavior	Active	Ambiences, co-workspaces	Autonomy & Anonymity
Data & sensing technologies	Behavior change	Ambient collaboration	Cold zones
Data: agency, make meaning	Collaboration	Ambient displays	Data privacy concerns
Human-centered design	Contestation	Complexities	Livable, safe communities
ICTs for involvement	Data, new economies	Efficiency	Open smart city -data/tech
Mediator of experience	Data, new deal	Effectiveness	- collaborative use
Moderating role	Digital rights	Knowledge transfer	- responsible use
Smart governance	Experience	Quality of life	Openness/Accountable/Gov
Social media platforms	Meaningful	Smart clusters	Privacy issues: surveillance
Sustainability	Technology solutions	Smart learning journey	Public space elements
Urban engagement spectrum	Trust	Sustainability	Theory of privacy

Common elements that emerge across columns of Table 5, include, but are not limited to, the ambient (co-working spaces, displays, collaboration, privacy), behavior, data, meaning-making and meaningfulness, sustainability, and technologies.

Table 6. Issues, controversies, and problems

	Issues	Controversies	Problems
Urban Engagement	• Behavior change for the better • e-platforms to engage public • Decision-making & control • Quality of Life (QOL)	• Data • Happiness • Social sustainability	• Limitations of awareness (for some sustainability behaviors) • Meaningful action
Urban Participation	• Behavior change • Citizen-centric • Co-production • Underdeveloped role of citizens	• Acceptance • Collaboration • Contestation • Data	• Active/Meaningful • Data Economics • Digital rights • Trust
Collaboration	• Infrastructure, interoperability • Privacy • Problematic in the public sector • Quality of life	• Authentic • Effectiveness • Efficiencies • Sustainability	• Complexities • Funding • Security • Trust
Openness & Privacy	• Autonomy • Control • Creativity • Innovativeness	• Anonymity • Future • Inclusive • Surveillance	• Equity • Liveable • Safe • Side-channel attacks

3.6 Issues, Controversies, Problems

From this review of the research and practice literature a range of issues, controversies, and problems emerge as identified in Table 6.

For urban engagement, issues, controversies, and problems pertain to behavior change for the better; decision-making and control in real-time forms of engagement; e-platforms by government as a way to engage the public; quality of life (QOL); data and sensing to mediate urban life; data; happiness; social sustainability; awareness (with limitations for some sustainability behaviors such as travel/transportation); and meaningful action.

For urban participation, issues, controversies, and problems pertain to behavior change; citizen-centric; co-production; the underdeveloped role and influence of citizens as key actors in smart city initiatives; dynamic modes of interaction as contestation, acceptance, and collaboration; data; active and meaningful forms of participation; as well as new data economics, digital rights, and trust.

For collaboration, issues, controversies, and problems pertain to infrastructure; interoperability; privacy; the public sector adoption of collaboration; quality of life; the notion of authentic collaboration; effectiveness; efficiencies; sustainable collaboration; intense complexities; funding; security; and trust.

For openness and privacy, issues, controversies, and problems pertain to autonomy; control; creativity; innovativeness; anonymity; inclusiveness; equity; livability; safe; and side-channel attacks and whether such surveillance potentials are a clear and present danger or "way off in the future."

Common elements that emerge across columns of Table 6, include, but are not limited to behavior change, data, meaningfulness, sustainability, and trust. Privacy issues emerge from this review of the literature along with problems of complexities and associated collaboration challenges such as authenticity and sustainability as well as involving people meaningfully in smart environments and spaces.

In summary, issues, controversies, and problems emerging from the literature review contribute to the surfacing of areas in need of further exploration, providing additional support for the research question under exploration in this chapter as stated in Section 1.

Figure 1. Conceptual framework for urban engagement and participation in smart cities

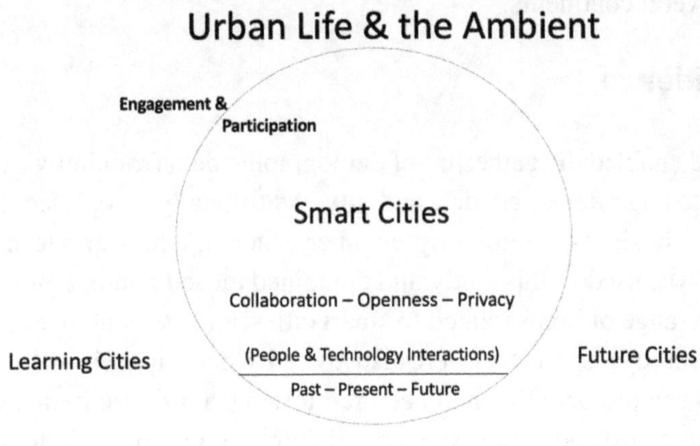

4. CONCEPTUALIZING URBAN LIFE, ENGAGEMENT, AND PARTICIPATION

Responding to the issues, controversies, and problems emerging from the literature review in the preceding section, a conceptual framework for urban engagement and participation in smart cities is formulated in Fig. 1. This framework builds upon and enhances the engagement and participation component of the urban life and the ambient conceptual framework for smart cities presented in Chapter 1. Incorporating learning cities and future cities, Fig. 1 focuses on collaboration, openness, and privacy, operating on an underlying continuum of the past-present-future, involving people and technology interactions.

The research question identified in Section 1 is reformulated as a proposition for exploration in this chapter, as follows:

P3: *Engagement and participation matter in urban life in the context of smart cities because they are influenced by the nature of collaboration, openness, and privacy*

The research design used in exploring the proposition under study in this chapter is described in the Methodology section.

5. METHODOLOGY

Sections 5.1 to 5.3 describe the research design for this chapter in terms of the process used, sources of evidence for data collection, and data analysis.

5.1 Process

In view of the contemporary nature of the inquiry in this chapter, an exploratory case study approach is used (Yin, 2018), combined with an explanatory correlational design, since the two are said to be complementary (Yin, 2018) in that the former may be "needed to examine the underlying processes that might explain a correlation." Using an online space to describe the study and invite people to participate, interest emerged from individuals in cities in multiple countries (Canada, United States, Greece, Ireland, Israel, etc.) reaching several continents.

5.2 Sources of Evidence

The online website used enabled the gathering of demographic data from individuals during sign up for the study, which included age range, gender, and city. Additionally, people could self-identify in one or more categories (e.g., business, community member, student, educator, etc.). A survey instrument, pre-tested prior to use, was used in this study and contained closed and open-ended questions. Survey questions pertained to a range of items related to smart cities, engagement, participation, collaboration, openness, and privacy. Survey respondents were also invited to participate in an in-depth interview which was guided by an interview protocol that had been pre-tested prior to use in this study.

Additionally, data were also gathered systematically through group and individual discussions, conducted in parallel with this study, with individuals in many sectors (e.g., students, faculty, and staff in higher education, business, community members and leaders, and city staff and leaders) in Canadian cities (e.g., Toronto, Vancouver, Victoria).

Virtual conference messaging and networking was also utilized (e.g., Whova platform), in the context of a conference paper presentation pertaining to the ambient (McKenna, 2021), as an exploratory way of learning about the nature of interest emerging in real time about smart cities and related issues and approaches, as well as collaboration potentials in contemporary virtual, global environments.

5.3 Data Analysis

Data analysis in this chapter is guided by the research question posed and the reformulated proposition for exploration, as follows:

P3: *Engagement and participation matter in urban life in the context of smart cities because they are influenced by the nature of collaboration, openness, and privacy.*

For the analysis of quantitative survey data, the use of descriptive statistics pertaining to urban participation and engagement along with other associated elements enabled exploration of items such as *collaborating* in relation to *technology-driven services.* The correlation feature of the Real Statistics Resource Pack containing an-add in for Microsoft Excel (Zaiontz, 2022) enabled the conducting of a Spearman Correlation Coefficient for ordinal data in exploring the nature of relationships between items pertaining to engagement and participation.

Content analysis was used for qualitative data gathered through open-ended survey questions, interviews, and group and individual discussions so that inductive analysis from data collection was combined with deductive analysis using terminology from the literature review.

Overall, an analysis was conducted for n=79 consisting of 42% females and 58% males for people ranging in age from their 20s to 70s.

Table 7. Assessments for urban engagement and participation: collaborating and tech-driven services

Items	Urban Engagement and Participation	Correlation
Collaborating	17% (5); 17% (6); 66% (7)	.49
Technology-driven services	33% (5); 17% (6); 50% (7)	

6. RESULTS AND DISCUSSION

Table 7 provides the results of assessments for *urban engagement* and *participation* using the proxies of *collaborating* and *technology-driven services.* In response to the survey question about whether city-focused social media and other aware technologies give rise to *collaborating*, on a 7-point Likert type scale, where 1= Not at all and 7=Absolutely, 17% of respondents selected positions 5 (Sort of) and 6 (Sure) on the scale while 66% responded at the upper end of the scale at position 7. When asked to what extent smart cities are associated with *technology-driven services*, 33% of survey respondents selected position 5 (Sort of), 17% responded toward the upper end of the scale at 6 (Sure), and 50% responded at the upper end position of 7.

Using the correlation feature of the Real Statistics add-in for Microsoft Excel (Zaiontz, 2022), a Spearman correlation coefficient for ordinal data was conducted on the assessment results of the two items, *collaborating* and *technology-driven services* (Table 7), showing a finding of .49. Correlations "above .35" are said to be "useful for limited prediction" (Creswell, 2018), while those in the .35 to .65 range are said to be "the typical values in prediction used to identify variable membership in the statistical procedure of factor analysis."

Exploring further, assessments for *urban engagement* and *participation* are provided in Table 8 using the proxies of *collaborating* and *privacy.* When asked to assess the extent to which factors such as privacy contribute to increased value for data in smart cities, 50% responded toward the upper end of the scale at 6 (Sure), and 50% responded at the upper end position 7.

Table 8. Assessments for urban engagement and participation: collaborating and privacy

Items	Assessments for the Urban Engagement and Participation	Correlation
Collaborating	17% (5); 17% (6); 66% (7)	.57
Privacy	50% (6); 50% (7)	

A Spearman correlation coefficient for ordinal data was conducted on the assessment results of the two items, *collaborating* and *privacy* (Table 8), showing a finding of .57.

Exploring further still, assessments for *urban engagement* and *participation* are provided in Table 9 using the proxies of *community participation* and *privacy.* When asked to assess the extent to which factors such as *community participation* contribute to the success of smart city project, 25% responded toward the upper end of the scale at 6 (Sure), and 75% responded at the upper end position 7.

A Spearman correlation coefficient for ordinal data was conducted on the assessment results of the two items, *community participation* and *privacy* (Table 9), showing a negative or inverse relationship of -.57.

Table 9. Assessments for urban engagement and participation: community participation and privacy

Items	Urban Engagement and Participation	Correlation
Community participation	25% (6); 75% (7)	-.57
Privacy	50% (6); 50% (7)	

The correlation for *collaborating* and *technology-driven services* at .49 is promising as is that for *collaborating* and *privacy* at .57. However, the inverse or negative relationship that emerges with a -.57 correlation for *community participation* and *privacy* highlights perhaps issues with privacy at the community level, as well as the nature of participation opportunities at the community level, echoing calls by Cardullo (2021) for "meaningful forms of participation."

Through use of a virtual conference messaging system (e.g., Whova), insights emerged into the nature of the "likes" and the number of attendees tuning in to the video presentation of a conference paper (McKenna, 2022) pertaining to ambient theory for smart cities. For example, based on a video presentation, one conference attendee sent a message expressing interest in the presentation and included an invitation to speak to a doctoral course on critiquing theories with the additional expression of interest in possible future collaborations. As such, the virtual conference environment, enriched with networking tools, accommodated an exploratory way of surfacing the nature of interest among researchers and educators, as well as possibly emergent forms of ambient networking and ambient collaboration.

As depicted graphically in Fig. 2, the interconnectivity of engagement and participation with approaches to collaboration, openness, and privacy give way in urban life to the notion and potentials of ambient engagement, ambient openness, ambient participation, and ambient privacy.

Figure 2. Urban life and the ambient –engagement and participation in smart cities

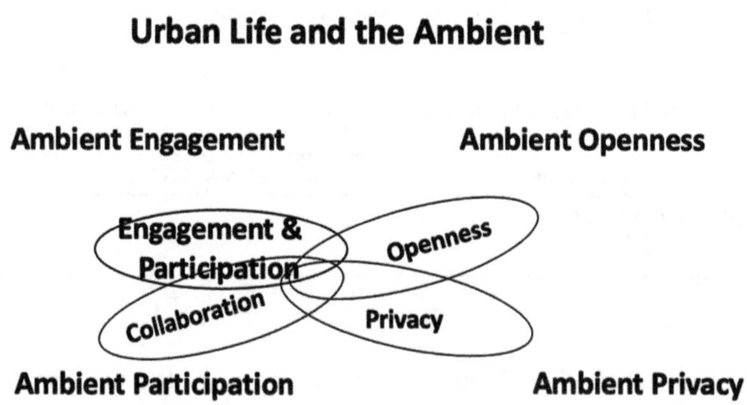

7. LIMITATIONS AND FUTURE DIRECTIONS

Possible limitations of this work are discussed in this section providing potential future directions, challenges, and opportunities for research and practice.

Limitations

Limitations of this chapter pertain possibly to the extent of the literature review, focusing on urban engagement and participation, collaboration, and openness and privacy in smart cities. Among other limitations of this chapter is that of the small sample size for the underlying study of this work, so that caution is encouraged in the interpretation of correlations calculated from the assessment of proxies used for engagement and participation. Going forward opportunities emerge possibly for the use of other proxies, or for more direct explorations of engagement and participation, resulting in perhaps stronger correlations with greater prediction potentials.

Future Research Directions and Opportunities

For research, future directions emerge in terms of the potential for conducting additional explorations on the variables identified in this chapter relevant to urban engagement and participation. In this way, opportunities for explorations going forward exist for the identification of other and possibly stronger variables relevant to urban engagement and participation that foster further learning in the context of collaboration, openness, privacy, and the like in smart cities, learning cities, and future cities. An avenue is also provided for further explorations of ambient engagement, ambient openness, ambient participation, and ambient privacy.

Emergent technologies in virtual conference environments may also provide opportunities among researchers for novel forms of real time feedback, insights, and networking influencing and enriching research directions going forward.

Future Practice Directions and Opportunities

Future directions for practice include the potential for pursuing explorations with variables identified in this work relevant to urban engagement and participation. In addition to collaboration, openness, and privacy, opportunities going forward for practice include the potential for exploring spaces with other, possibly stronger variables relevant to engagement and participation that foster real world interactions, action, and learning in everyday contexts in smart cities, learning cities, and future cities.

8. CONCLUSION

Challenges and opportunities associated with engagement and participation in urban life emerge from the review of the research literature provided in this chapter from multiple perspective involving openness and privacy and collaboration. A key contribution of this chapter is the formulation of a conceptual framework for urban engagement and participation in smart cities guided by the literature review where emerging issues, controversies, and problems provide further insight on the proposition explored while contributing to operationalization of the framework for use in this chapter. Through an exploratory case study approach, together with an explanatory correlational design, the proposition under study in this chapter is investigated. Based on survey data, correlations are conducted for urban engagement and participation employing the use of proxies, where, among the findings is that for *collaborating* and *privacy* in the form of a Spearman correlation coefficient of .57, said to be "useful for limited prediction." Limitations of the study underlying this chapter pertain to, among other things, the small sample size and as such, caution is urged in the interpretation of findings, opening opportunities for further research. Notwithstanding this and other possible limitations, future directions guiding research and practice activities going forward are provided in terms of further validating the findings in this chapter while also encouraging activities to identify other, additional factors relevant to urban engagement and participation and explorations of the notions of ambient collaboration, ambient privacy, ambient openness, and the like, in the context of smart, learning, and future cities.

Key insights and ideas emerging from this chapter are provided, along with exercises encouraging further exploration, as follows:

Insights emerging from this chapter include the following:

1. The importance of city-focused social media and other aware technologies as factors giving rise to many possibilities for *urban engagement* and *participation* in the form of *collaborating*
2. The importance of factors associated with *urban engagement* and *participation* in smart cities such as *technology-driven services*
3. The potential for many more factors to emerge for understanding *urban engagement* and *participation* in smart cities such as *privacy* and *community participation*

Ideas emerging from this chapter include the following:

1. Potential opportunities for exploring urban engagement and participation in relation to a variety of variables (e.g., collaboration, openness, and privacy) in the context of smart cities
2. The opportunities potential for the identification of other and additional variables for exploring urban engagement and participation in smart environments
3. The use of an interdisciplinary and multi-method approach for the study and practice of urban engagement and participation in smart, learning, and future cities

Exercises for the reader to consider, encouraging further exploration include the following:

1. Identify one or more factors in your city that are giving rise to engagement and participation
2. Identify one or more ways in which collaboration would enhance the quality of urban life in your city
3. Describe one or more elements that contribute to making engagement and participation possible in your city

The audience for this chapter will be interdisciplinary, including but not limited to community members and leaders, educators, businesses, researchers, urban planners, urban designers, and anyone concerned with engagement and participation in urban life and the ambient, while being attentive to elements such as collaboration, openness, and privacy along with associated practice and research opportunities and challenges in smart cities, learning cities, and future cities.

REFERENCES

Alcaide Muñoz, L., & Rodríguez Bolívar, M. P. (2019). Using tools for citizen engagement on large and medium-sized European smart cities. In *E-Participation in Smart Cities: Technologies and Models of Governance for Citizen Engagement. Public Administration and Information Technology* (Vol. 34). Springer. doi:10.1007/978-3-319-89474-4_2

Allen, B., Tamindael, L. E., Bickerton, S. H., & Cho, W. (2020). Does citizen coproduction lead to better urban services in smart cities projects? An empirical study on e-participation in a mobile big data platform. *Government Information Quarterly*, *37*(1), 101412. doi:10.1016/j.giq.2019.101412

Arnstein, S. R. (1969). A ladder of citizen participation. *Journal of the American Institute of Planners*, *35*(4), 216–224. doi:10.1080/01944366908977225

Baibarac-Duignan, C., & de Lange, M. (2021). Controversing the datafied smart city: Conceptualising a 'making-controversial' approach to civic engagement. *Big Data & Society*, *8*(2). Advance online publication. doi:10.1177/20539517211025557

Bartelt, V. L., Urbaczewski, A., Mueller, A. G., & Sarker, S. (2020). Enabling collaboration and innovation in Denver's smart city through a living lab: A social capital perspective. *European Journal of Information Systems*, *29*(4), 369–387. doi:10.1080/0960085X.2020.1762127

Birch, E. L. (2008). Public and Private Space in Urban Areas: House, Neighborhood, and City. In Handbook of Community Movements and Local Organizations. Springer. doi:10.1007/978-0-387-32933-8_8

Bloom, R., Lauriault, T. P., & Landry, J.-N. (2018). *Open smart cities in Canada: An assessment report*. OpenNorth.

Bushwick, S. (2021). A blank wall can show how many people are in a room and what they're doing: The technique is the latest method to glean a surprising amount of surveillance from a meager source. *Scientific American*, (October), 6.

Cardullo, P. (2021). *Citizens in the 'smart city': Participation, co-production, governance*. Routledge.

Cardullo, P., & Kitchin, R. (2019). Being a 'citizen' in the smart city: Up and down the scaffold of smart citizen participation in Dublin, Ireland. *GeoJournal*, *84*(1), 1–13. doi:10.100710708-018-9845-8

Cegłowski, M. (2019). The new wilderness. *Idle Words*. Accessed April 24, 2022. https://idlewords.com/2019/06/the_new_wilderness.htm

Chen, T., Gil-Garcia, J. R., & Gasco-Hernandez, M. (2022). Understanding social sustainability for smart cities: The importance of inclusion, equity, and citizen participation as both inputs and long-term outcomes. *Journal of Smart Cities and Society*, *1*(2), 135–148. doi:10.3233/SCS-210123

Cogut, G., Webster, N. J., Marans, R. W., & Callewaert, J. (2019). Links between sustainability-related awareness and behavior: The moderating role of engagement. *International Journal of Sustainability in Higher Education*, *20*(7), 1240–1257. doi:10.1108/IJSHE-09-2018-0161

Creswell, J. W. (2018). *Educational research: Planning, conducting, and evaluating quantitative and qualitative research* (6th ed.). Pearson.

de Hoop, E., Moss, T., Smith, A., & Löffler, E. (2021). Knowing and governing smart cities: Four cases of citizen engagement with digital urbanism. *Urban Governance*, *1*(2), 61–71. doi:10.1016/j.ugj.2021.12.008

Doering, C., Schmidtner, M., & Timinger, H. (2021). Collaboration for innovation between universities and smart cities. *2021 IEEE European Technology and Engineering Management Summit (E-TEMS)*, 82-86. doi:10.1109/E-TEMS51171.2021.9524896

Dommel, H. P. (2005). The challenges of ambient collaboration. *Richard Tapia Celebration of Diversity in Computing Conference*, 10-13. doi: 10.1145/1095242.1095248

Fedorowicz, M., Arena, O., & Burrowes, K. (2020). *Community engagement during the COVID-19 Pandemic and beyond: A guide for community-based organizations.* Washington, DC: Urban Institute. Retrieved 7 November 2021 from https://www.urban.org/sites/default/files/publication/102820/community-engagement-during-the-covid-19-pandemic-and-beyond.pdf

Gil-Garcia, J. R., Zhang, J., & Puron-Cid, G. (2016). Conceptualizing smartness in government: An integrative and multi-dimensional view. *Government Information Quarterly*, *33*(3), 524–534. doi:10.1016/j.giq.2016.03.002

Granier, B., & Kudo, H. (2016). How are citizens involved in smart cities? Analysing citizen participation in Japanese 'smart communities'. *Information Polity*, *21*(1), 61–76. doi:10.3233/IP-150367

Levinson, K. (2022). NSF invests $26M for smart city engineering research: Researchers will integrate real-time, hyperlocal technologies into streetscapes to build safe, equitable and secure communities. *GCN: The Technology Transforming State and Local Government.* Retrieved 4 September 2022 from https://gcn.com/cloud-infrastructure/2022/08/nsf-invests-26m-smart-city-engineering-research/375731/

Lister, P. (2022). Ways of experiencing technology in a smart learning environment. In N. Streitz & S. Konomi (Eds.), *Distributed, Ambient and Pervasive Interactions. HCII 2022*. Lecture Notes in Computer Science. Springer. doi:10.1007/978-3-031-05431-0_11

McKenna, H. P. (2019). Sharing-collaboration-openness: Innovating privacy for smarter urbanities. In *Ambient Urbanities as the Intersection Between the IoT and the IoP in Smart Cities*. IGI Global. doi:10.4018/978-1-5225-7882-6.ch008

McKenna, H. P. (2022). Is ambient theory for smart cities even a theory: An affirmative assessment. In *Proceedings of the Future Technologies Conference (FTC) 2021, Volume 2, Lecture Notes in Networks and Systems.* Springer. 10.1007/978-3-030-89880-9_41

McKenna, H. P., Arnone, M. P., Kaarst-Brown, M. L., McKnight, L. W., & Chauncey, S. (2013). Ambient privacy with wireless grids: Forging new concepts of relationships in 21st century information society. *International Journal for Information Security Research*, *3*(1-2), 408–417.

Mills, D. E., Izadgoshasb, I., & Pudney, S. G. (2021). Smart city collaboration: A review and an agenda for establishing sustainable collaboration. *Sustainability*, *13*(16), 9189. doi:10.3390u13169189

Morrison-Smith, S., Chilton, L. B., & Ruiz, J. (2021). AmbiTeam: Providing team awareness through ambient displays. *Proceedings of Graphics Interface, 2021*. Advance online publication. doi:10.20380/GI2021.04

O'Brien, H., & Cairns, P. (2016). Conclusions. In H. O'Brien & P. Cairns (Eds.), *Why Engagement Matters*. Springer. doi:10.1007/978-3-319-27446-1_10

Okner, T., & Preston, R. (2017). Smart cities and the symbiotic relationship between smart governance and citizen engagement. In H. Song, R. Srinivasan, T. Sookoor, & S. Jeschke (Eds.), *Smart Cities: Foundations, Principles, and Applications* (pp. 343–372). Wiley. doi:10.1002/9781119226444.ch12

Orel, M., & Alonso Almeida, M. M. (2019). The ambience of collaboration in coworking environments. *Journal of Corporate Real Estate, 21*(4), 273–289. doi:10.1108/JCRE-12-2018-0050

Pentland, A., Lipton, A., & Hardjono, T. (2021). *Building the New Economy: Data as Capital*. MIT Press. doi:10.7551/mitpress/13991.001.0001

Popa, R. A. (2022). Featured ACM member: Raluca Ada Popa. *ACM Bulletin*. Retrieved 20 September 2022 from https://www.acm.org/articles/people-of-acm/2022/raluca-ada-popa

Przeybilovicz, E., Cunha, M. A., Geertman, S., Leleux, C., Michels, A., Tomor, Z., Webster, C. W. R., & Meijer, A. (2020). Citizen participation in the smart city: Findings from an international comparative study. *Local Government Studies*. Advance online publication. doi:10.1080/03003930.2020.1851204

Simonofski, A., Van Den Storme, S., & Meers, H. (2020). Towards a holistic evaluation of citizen participation in smart cities. In *The 21st Annual International Conference on Digital Government Research (dg.o '20)*. Association for Computing Machinery. 10.1145/3396956.3396985

Solove, D. (2008). *Understanding privacy*. Harvard University Press.

Stelzle, B., Jannack, A., Holmer, T., Naumann, F., Wilde, A., & Noennig, J. R. (2021). Smart citizens for smart cities: A user engagement protocol for citizen participation. In M. E. Auer & T. Tsiatsos (Eds.), *Internet of Things, Infrastructures and Mobile Applications. IMCL 2019. Advances in Intelligent Systems and Computing* (Vol. 1192). Springer. doi:10.1007/978-3-030-49932-7_54

Streitz, N. A. (2021). From smart-only cities towards humane and cooperative hybrid cities. *Technology | Architecture + Design, 5*(2), 127-133. Doi:10.1080/24751448.2021.1967050

Van Kranenburg, R., Anania, L., Le Gars, G., Arniani, M., van Ditmar, D., Kaili, M., & Kavassalis, P. (2020). *Future urban smartness: Connectivity zones with disposable identities*. In J. C. Augusto (Ed.), *Handbook of Smart Cities* (pp. 1–29). Springer. doi:10.1007/978-3-030-15145-4_56-1

Van Zoonen, L. (2016). Privacy concerns in smart cities. *Government Information Quarterly, 33*(3), 472–480. doi:10.1016/j.giq.2016.06.004

Visvizi, A., Damiani, E., & Lytras, M. (Eds.). (2020). *From smart cities to smart clusters: Re-thinking regional collaboration and development*. Emerald Publishing.

Yin, R. (2018). *Case study research and applications: Design and methods* (6th ed.). Sage Publications.

Zaiontz, C. (2022). *Real statistics using excel*. Retrieved from www.real-statistics.com

Zandbergen, D., & Uitermark, J. (2020). In search of the smart citizen: Republican and cybernetic citizenship in the smart city. *Urban Studies (Edinburgh, Scotland), 57*(8), 1733–1748. doi:10.1177/0042098019847410

ADDITIONAL READING

Christensen, A. G. (2020). Composing urban green collaboration: Exploring coordination in long-term civic–municipal greenspace planning. *European Journal of Cultural and Political Sociology*, *7*(4), 477–501. doi:10.1080/23254823.2020.1772843

De Gregorio Hurtado, S. (2015). The implementation of the URBAN community initiative: A transformative driver towards collaborative urban regeneration? Answers from Spain. *European Journal of American Studies*. https://journals.openedition.org/ejas/11354

Manzini, E. (2019). Collaborative cities: What social innovation teaches us. *Six (Social Innovation Exchange)*. Retrieved 11 May 2019 from https://socialinnovationexchange.org/insights/collaborative-cities-what-social-innovation-teaches-us

McConnell Foundation. (2018). *Creating collaborative infrastructure for inclusive urban innovation.* Retrieved 11 May 2022 from https://mcconnellfoundation.ca/report/collaborative-infrastructure-inclusive-urban-innovation/

Sanul, G., & van Heur, B. (2018). Spaces of openness. *City, 22*(5-6), 801-819. doi:10.1080/13604813.2018.1549854

Sharaf, A. (2022). Concordia's Next Generation Cities Institute welcomes seven new partner organizations. *Next Generation Cities Institute News.* Retrieved 11 May 2022 from https://www.concordia.ca/news/stories/2022/04/04/concordias-next-generation-cities-institute-welcomes-seven-new-partner-organizations.html?c=/research/cities-institute/news

Tewdwr-Jones, M., & Wilson, A. (2022). Co-designing urban planning engagement and innovation: Using LEGO® to facilitate collaboration, participation and ideas. *Urban Planning*. Doi:10.17645/up.v7i2.4960

Tong, Z., Yang, H., Liu, C., Xu, T., & Xu, S. (2020). Quantification of the openness of urban external space through urban section. *Geo-Spatial Information Science*, *23*(4), 316–326. doi:10.1080/10095020.2020.1846464

Tosics, I. (2019). Towards metropolitan collaboration in urban areas. *URBACT: Driving Change for Better Cities Blog*. Retrieved 11 May 2022 from https://www.blog.urbact.eu/2019/10/metropolitan-collaboration/

Vakkayil, J. (2022). Order and openness in community-driven urban initiatives: Insights from a 'spot-fix.'. *Urban Studies (Edinburgh, Scotland)*, *59*(15), 3098–3113. Advance online publication. doi:10.1177/00420980221084246

KEY TERMS AND DEFINITIONS

Ambient Collaboration: Ambient collaboration in an urban context refers to working together in a dynamic way, possibly involving interactions with technology to create or improve one or more aspects of life in the city and surrounding regions.

Ambient Engagement: Ambient engagement in an urban context refers to dynamic and meaningful involvement with people, community, technologies, and services.

Ambient Networking: Ambient networking in an urban context refers to dynamic interactivities among people or technologies enabling information sharing, engagement, collaboration, and participation.

Ambient Openness: Ambient openness in an urban context refers to dynamic access to sharable data and other community resources and services.

Ambient Participation: Ambient participation in an urban context refers to active involvement in a dynamic way, when interacting with others and possibly also with and through technologies.

Ambient Privacy: Ambient privacy in an urban context refers to more dynamic, in the moment spaces for secure interaction, sharing, and collaboration.

Collaboration: Collaboration in an urban context refers to working together to create or improve some aspect of life in the city and surrounding regions.

Community Participation: Community participation in the context of smart cities refers to the involvement of people in their community in more aware, engaged, meaningful, and creative ways.

Participation: Participation in an urban context refers to active involvement with others and possibly also with and through technologies.

Technology-Driven Services: Technology-driven services in an urban context refer to services that are created, enhanced, augmented, reimagined, or reinvigorated with the aid of existing and emerging technologies.

Section 2

Evolving Infrastructures, Literacies, and Data Streams for Urban Life and the Ambient

This section introduces the reader to evolving infrastructures for urban life and the ambient in smart cities involving experiences, interactions, and augmentations to complement and extend physical infrastructures. New literacies for urban life and the ambient are required including those in areas such as the digital, algorithmic, play, and inclusion and the associated elements of data streams, information, knowledge, and computational intelligence (CI).

Chapter 4
Urban Life and Evolving Infrastructures in Smart Cities:
Experiences, Interactions, and Augmentations

ABSTRACT

The purpose of this chapter is to explore evolving understandings of urban life and infrastructure in smart cities in relation to experiences, interactions, and augmentations. As such, a review of the research literature is conducted focusing on infrastructures, experiences, interactions, and augmentations in smart, learning, and future cities. Issues, controversies, and problems emerging from the literature review are highlighted, contributing to formulation of a conceptual framework for urban life and the ambient featuring evolving infrastructures in smart cities. Using an exploratory case study approach combined with an explanatory correlational design, the conceptual framework is operationalized for use in this chapter. Through the lens of emerging infrastructures for experience, interactions, and augmentations, variables are identified, explored, and correlated to inform understandings of smart cities while identifying directions for future research and practice.

1. INTRODUCTION

Celani, Falcone, and d'Alessandro (2019) argue for "defining an upgrade to the meaning of infrastructures" enabling "the transformation from a classic Smart City to a model based on human dimension" involving an interweaving of the social with the physical and virtual. Streitz (2021) challenges the "smart-everything paradigm" of smart cities, arguing for "humane and cooperative cities" requiring a rethinking of "software infrastructures" in "resetting the priorities" in support of "a people-centered design approach" and one that would "provide appropriate, livable, sustainable, and resilient spaces." This chapter is significant in that, motivated by this earlier research literature, it explores evolving in-

DOI: 10.4018/978-1-6684-4096-4.ch004

frastructures in smart cities in relation to experiences, interactions, and augmentations giving rise to the following objectives.

Objectives: The main objectives of this chapter are to a) explore evolving infrastructures for experiences, interactions, and augmentations in the context of smart cities; b) formulate a conceptual framework for urban life and the ambient featuring evolving infrastructures in smart cities; and c) explore the nature of the relationship between variables such as *air quality* and *invisible infrastructures* as ways of learning more about and possibly rethinking infrastructures in smart cities in the context of rapidly changing needs and requirements. These objectives give rise to the main research question under exploration in this chapter – *How are urban infrastructures evolving in smart cities?*

2. BACKGROUND

Bauer (2015) points to the importance of identifying infrastructure inefficiencies, needs, and risks enabled by ambient intelligence which is described as "electronic environments that are sensitive and responsive to the presence of people." As such, ambient intelligence infrastructures are said to be "the amalgamation of neural networks, big data, IoT, wearables, and device user interfaces into services" giving way to spaces "that can automate processes and make recommendations" with the aim "to improve the users' quality of life" (Bauer, 2015). Qiao, Hiuang, and Yeh (2022) describe the notion of Mobility as a Service (MaaS) involving urban infrastructure consisting of transport-flow infrastructure, information-flow infrastructure, and computing-flow infrastructure with an ecosystem of traditional infrastructure along with "intelligent transportation concepts, such as high-speed communication networks and autonomous fleets."

2.1 Definitions

Definitions are provided from the research literature for key terms used in this chapter.

- **Augmentations:** Smith (2016) describes augmentations in the context of urban life and evolving infrastructures in relation to street signs and "a number of quasi-hidden elements of 'bundled infrastructure' in the form of signal boxes and conduit junctions for water, electricity, and traffic control."
- **Experience Infrastructures:** Dourish and Bell (2007) consider "space as infrastructure" and such spaces offer ways "through which we experience the world" in ubiquitous computing environments consisting of "the structures that lie below or beneath the surface of applications and interactions."
- **Infrastructures (urban):** Smith (2016) describes urban infrastructures as "a constant work in progress" that "shapes and facilitates daily life" and "can be analyzed as a materialization of ongoing communication."
- **Interactions (urban):** Dourish and Bell (2007) speak of the notion of "infrastructures of interaction" in urban environments and the complexities "between space, infrastructure, culture, and experience" in combination with the potential for technologies to introduce additional destabilizing and transformative forces.
- **Smart City:** Mohanty, Choppali, and Kougianos (2016), citing the work of Harrison et al. (2010), describe the smart city as one "connecting the physical infrastructure, the information-technology

infrastructure, the social infrastructure, and the business infrastructure to leverage the collective intelligence of the city."

2.2 Overview

This chapter consists of several sections containing a review of the research and practice literature for urban life and infrastructures in smart cities, experiences and smart cities, interactions and smart cities, and augmentations and smart cities and future cities. From this review, an overview of key issues, controversies, and problems is provided and a conceptual framework for urban life and the ambient featuring evolving infrastructures in smart cities is formulated and operationalized for use in this chapter. The main research question (Section 1) under exploration in this chapter is reformulated as a proposition for investigation. The methodology employed in the investigation of the proposition in this chapter is described followed by the presentation of results and a discussion of findings. Limitations of the study underlying this chapter are identified and future directions for research and practice are identified. The conclusion provides an overview of outcomes, the identification of insights and ideas, development of several exercise questions encouraging further exploration, and a description of the intended target audience for this chapter.

3. PERSPECTIVES ON EVOLVING INFRASTRUCTURES IN SMART CITIES

This section provides a review of the research and practice literature for urban life and infrastructures in smart cities, experiences and smart cities, interactions and smart cities, and augmentations in smart cities and future cities.

3.1 Urban Life and Infrastructures in Smart Cities

Dourish and Bell (2007) encourage consideration of "the human side of infrastructures" arguing that "we need to think architecturally about the mobile and wireless technologies that we develop and deploy." Fortin (2017) explores research methods for the study of digital infrastructural urban technology projects, making the case for "multidimensional approaches over unidimensional ones." Among the findings of the work by Fortin (2017) is that "a robust, and far-reaching longitudinal methodology" is preferred, enabling as it does "nuanced descriptions; a multi-site approach wherein the ethnographer is given the possibility of developing relationships with stakeholders over time" said to be important for "building bridges between them through publications and events, and using critical theory to look outside the box" to "support the reconciliation of multiple perspectives." Hatcher (2019) raises the question of whether hyperconvergence is "the cornerstone of future smart buildings?" where hyperconverged infrastructure is described as "a concept bringing together three major components of IT infrastructure (compute, storage, and networking), and using software to virtualize and manage them as a flexible resource." Alavi and Bahrami (2019) articulate the multi-dimensional nature of walking in the city where the data generated by people and their interactions forms an urban infrastructure. Citing instances of "severe weather events around the world this year", Wray (2021) makes the case for a back-up plan for smart cities where, for example, in New Orleans, LaGrue, the Chief Information Officer, "highlights how diversifying the way cities communicate is crucial" in support of redundancy when one or more systems break down. Hopkins, Bae, Uhr, Banić, Zheng, and Do (2021) address the importance of the interfaces of smart cities for people while defining the smart concept in a multi-purpose way, attentive to the presence of people, as in, "smart is welcoming", "smart is being cared for", "smart is having tools that enhance the lives of those living and working in the city" and so on.

3.2 Experiences and Smart Cities

Marzano (2006) explores questions such as "what experience should [Ambient Intelligence] deliver?" along with how, finding that what is needed is "a form and experience that users perceive as being relevant, meaningful, and understandable." Dourish and Bell (2007) speak in terms of the "experience of infrastructure" and "the infrastructure of experience" as in, "the embedding of a range of infrastructures into everyday space shapes our experience of that space" while it also "provides a framework through which our encounters with space take on meaning" involving the social and cultural as well as the technological. Additionally, Dourish and Bell (2011) point to the "messiness" of everyday life in instances when the Internet or other infrastructures fail. Albino, Berandi, and Dangelico (2015) address the definitions, dimensions, performance, and initiatives in smart cities, focusing on several measures, one of which is experiences. Dainow (2017) describes smart environments as "a lived experience" where the smart city is "experienced as a seamless experience as one moves from house to car to work, from one context to another" and people "will simply experience themselves as going about daily life" as in "an unbroken stream of changing contexts seamlessly merging from one to the other." Sari, Tedjasaputra, Ellen, Duh, and Lugmay (2017) describe the notion of semantic ambient media experiences (SAME) and refer to the "growing demand for new forms of experiences and services enabled through emerging technologies" such as ambient media where media is said to be "embedded in the human living environment." SAME (Sari et al., 2017) may take the form "in urban environments" of "real-time, context-aware, and interactive technologies; augmented and immersive realities; social media; and location-based services" enabled through "technological and sociocultural infrastructures." Using the example of smart university campuses, Vasileva, Rodrigues, Hughes, Greenhalgh, Goulden, and Tennison (2018) explore user experience with open data as a way of learning more about smart cities. Risks and challenges such as data privacy and protection along with "infrastructure hurdles" are acknowledged by Vasileva et al. (2018) as well as the potential for solving such issues so as to improve user experiences on campuses while providing an example for smart cities.

Edge, Boluk, Groulx, and Quick (2020) explore "diverse lived experiences in the smart city" using a "creative analytic practice" (CAP) approach. From the perspective of a smart city industrial experience, Fernandez, Sanchez, Velez, and Moreno (2021) undertake the design and application of a comprehensive framework designed to analyze Internet of Things (IoT) platforms, guided by a design science approach in an Australian context and it is said that the framework "can be viewed as a quality guideline for both researchers and practitioners who are developing and maintaining IoT platforms." Van der Graaf, Nguyen Long, and Veeckman (2021) provide an exploration of experience-based co-creation in smart cities "as a way to plan, execute, and evaluate public service design and delivery for contemporary and future cities." Extending the experience of smart cities to airport design, Fineout (2022) describes the importance of "environmentally sensitive geometries" focusing on a case study of the Mexico International Airport now under construction as an overlooked form of "sustainable practices". According to Fineout (2022), geometry "can have a profound effect upon the amount of energy consumption of a building" in both the construction process (e.g., "embodied energy") and in ongoing operations.

3.3 Interactions and Smart Cities

Lin, Jeng, and Chen (2013) use interaction design to explore user experiences in relation to sensing, the social, and the physical in reimagining more humane cities featuring "a framework for supporting a smart living environment." van Waart and Mulder (2014) explore meaningful interactions in a smart city focusing on interactive media through Moodly, an "art installation for city emotion insights"; Cosy, a "hybrid smart café"; and Nosey, "a mobile device for real time climate sensing" to detect air quality in support of insight, debate, and decision making. Smyth, Helgason, and Mitrović (2015) describe an installation comprised of "video scenarios that speculate about the experience of urban life" from the perspective of a summer school on urban interaction and design focusing on "public space and the relationship between people – with and through technology." Smyth et al. (2015) invite viewers to "consider how technology might shape the city of the future and subsequently, our relationship with the city, and with each other." Alavi and Bahrami (2019) focus on public spaces and interactions in the "pedestrian realm" that are said to be "mostly about spontaneous body management" and in smart environments where "interactivity should be able to respond to the complexity of humans' interactive experiences." Gómez-Carmona, Sádaba, and Casado-Mansilla (2019) explore the enhancing of interactions in smart cities at the street level focusing on smart modular furniture in the form of a "customizable kiosk" that is "specifically designed to fit into" the notion of "digital transformation and to respond to users' needs by offering contextualized information and services" involving digital and physical interaction with users, adaptable for different purposes.

A conceptual framework for human-smart environment interactions (HSEI) is advanced by McKenna (2020), where, guided by the framework, dimensions of smartness are explored in terms of their relationships. For example, McKenna (2020) found a very good Spearman correlation (.81) between citizen engagement (using interactive public spaces as a proxy) and openness, in support perhaps of "taking people and their voices into consideration in designing smart environments." Proctor and Proctor (2021) provide guidance on human sensation and perception in relation to human factors to consider in the design of human-computer interactions. Chew, Hespanhol, and Loke (2021) explore playful interactions "across the dimensions of smartness and disruption" focusing on "the overlap between smart and play in the design of street furniture", notably public benches, said to be "appropriated for playful, creative, and entertainment purposes, most recently through the addition of digital technology." Ourednik, Mlynář, Mutzner, and Alavi (2021) describe cities as "congregations of human *bodies*" and "sites of multimodal interaction in which sociality emerges on a daily basis" while going on to note that "cities and cyberspaces are material organizations." As such, Ourednik et al. (2021) claim that "the modern city concentrates a great diversity of people and amenities, allowing exponentially as many combinations of interactions" while "its transportation infrastructure further facilitates transition between *contexts* of interaction" arguing that "like cities, social media platforms provide opportunities for interaction, but they do so in a manner oriented toward the private profits of their owners" so that to "become better algorithmic cities" it will be important for social media platforms that "the common good becomes their primary focus." Hirsch, Economidou, Paraschivoiu, and Döring (2022) argue that "the collaboration of urban, material, and interaction designers is necessary for placemaking."

3.4 Augmentations: Smart Cities and Future Cities

Lin et al. (2013) advance the augmenting of human experience focusing on smart living aided by interactive technologies where the notion of "*humane* means that smart technology should augment human capabilities, such as mobility, memory, and health." Bello, Mydlarz, and Salamon (2018) point to the importance of sound analysis for smart cities "as a source of information about urban life" focusing on "the computational analysis of urban sounds" particularly "audio surveillance, and noise pollution monitoring" together with "how a combination of convolutional networks and data augmentation result in the current state of the art." Celani et al. (2019) explore the challenges of the augmented city arguing that infrastructure is a unifying element "able to reunite territories, giving back their identity." Leorke (2019) explores the "growing intersection of digital games and the 'smart city' model" in terms of the range of "ways that games can alternately support, challenge, or counter the push to instrumentalise, optimize, and 'program' the city through ubiquitous smart technologies and 'sentient' infrastructure." As such, Leorke (2019) is concerned with the ability to enable spaces to "*reappropriate, reconfigure,* and *augment*" involving citizens more meaningfully to "access the infrastructure, data, and platforms" to "support the smart city agenda" and "also move beyond its underpinning constraints and imperatives."

Fernandez, Sanchez, Velez, and Moreno (2020) describe the smart city in terms of an augmented space model involving "augmented reality (AR) and augmented space (AS) principles" said to include "augmented systems, channels, images, models, and maps" in formulation of a proposed augmented enterprise architecture framework. Chew et al. (2021) explore livability in smart cities based on "technology augmentation" involving "playful placemaking" with smart street furniture. Lee, Braud, Hosio, and Hui (2022) provide a review of human-city interaction involving augmented reality (AR) and other associated technologies giving rise to considerations of potentials and needs going forward. Lister (2022) speaks of the notion of augmented reality (AR) triggers in the context of smart learning journey activities and experiences in urban environments featuring "an 'AR interface' of choices" in technology mediated spaces. Gao, Wang, Lin, Luo, Ruan, and Yuan (2022) contribute a model to the detecting and learning about city intersection traffic contexts and weather conditions for autonomous vehicles (AVs).

3.5 Literature Review Overview

An overview by author and date is provided in this section for the research and practice literature for infrastructures, experiences, interactions, and augmentations.

Table 1 provides an overview by author and date of the research and practice literature for urban life and the ambient pertaining to infrastructures, spanning the timeframe of 2007 to 2022.

Dourish and Bell (2007) advance the importance of the human side of digital infrastructures while Bauer (2015) highlights the importance of identifying infrastructure inefficiencies, needs, and risks in support of the quality of life. Mohanty et al. (2016) describe smart cities as the connecting of infrastructures including the physical, information-technology, social, and business thus leveraging the collective intelligence of the city. Fortin (2017) advances multidimensional approaches to the study of digital infrastructural urban technology projects in support of multiple perspectives. Alavi and Bahrami (2017) describe the multidimensional nature of walking in the city giving rise to an urban infrastructure of people and their interactions. Celani et al. (2019) point to the need for upgrading the meaning of infrastructure in support of the human dimension to incorporate an interweaving of the social, virtual, and physical. Hatcher (2019) considers future smart buildings and the hyperconvergence of IT infrastructure involv-

Table 1. Overview of research for urban life and infrastructures in smart cities

Author(s)	Year	Infrastructures
Dourish & Bell	2007	Human side
Bauer	2015	Identify inefficiencies, need, risks in support of quality of life
Mohanty et al.	2016	Connecting of infrastructures – physical, info-tech, social, business
Fortin	2017	Multidimensional digital infrastructural urban technology projects
Alavi & Bahrami	2019	Walking multidimensionality, urban infrastructure of people interactions
Celani et al.	2019	Upgrade to the meaning of infrastructure – social, human dimensions
Hatcher	2019	Hyperconverged infrastructure – compute, storage, networking
Wray	2020	Redundancy of urban systems in support of communication
Hopkins et al.	2021	Interfaces attentive to people in smart cities
Streitz	2021	Humane, cooperative software – appropriate, livable, sustainable, resilient
Qiao et al.	2022	MaaS – transport, information, computing

ing compute, storage and networking, virtualized and managed by software. Wray (2020) makes the case for redundancy in urban systems to ensure communication when severe weather and other events occur. Hopkins et al. (2021) address the importance of developing interfaces attentive to people in smart cities while Streitz (2021) advocates for the rethinking of software infrastructures taking people into consideration in support of humane and cooperative cities. Qiao et al. (2022) describe integrated urban infrastructure systems involving transportation, information, and computing.

Table 2 provides an overview by author and date of the research and practice literature for urban life and the ambient pertaining to experiences, spanning the timeframe of 2006 to 2022.

Marzano (2006) explores smart environments in terms of experiences, finding those perceived by users as relevant, meaningful, and understandable to be important. Dourish and Bell (2007) refer to the experience of infrastructure and the messiness that occurs in everyday life (Dourish and Bell, 2011)

Table 2. Overview of research for urban life and experiences in smart cities

Author(s)	Year	Experiences
Marzano	2006	Experience as relevant, meaningful, and understandable
Dourish & Bell	2007	Experience of infrastructure
Dourish & Bell	2011	Messiness of everyday life when infrastructures fail
Albino et al.	2015	As a measure of SCs
Dainow	2017	Smart environments as lived experience
Sari et al.	2017	Semantic ambient media experiences (SAME) – tech, sociocultural
Vasileva et al.	2018	User experiences of open data on smart campuses as models for SCs
Edge et al.	2020	Lived experiences in SCs, CAP approach
Fernandez et al.	2021	Comprehensive framework to analyze IoT platforms in SCs
Van der Graff et al.	2021	Co-creation in SCs
Fineout	2022	Environmentally sensitive geometries for sustainable practices

Table 3. Overview of research for urban life and interactions in smart cities

Author(s)	Year	Interactions
Lin et al.	2013	Framework supporting a smart living environment
van Waart & Mulder	2014	Meaningful interactions for people aided by interactive technologies
Smyth et al.	2015	City, data, future & the experience of urban life going forward
Alavi & Bahrami	2019	Walking in future cities
Gómez-Carmona et al.	2019	Interactive and modular furniture at the street level
McKenna	2020	Human-smart environment interactions – public spaces and openness
Chew et al.	2021	Smartness and disruption of public benches designed for play
Ourednik et al.	2021	Multimodal interaction and sociality
Proctor & Proctor	2021	Human sensation, perceptions
Hirsch et al.	2022	Collaboration: urban, material, and interaction designers in placemaking

when system break down or failure occurs. Albino et al. (2015) focus on experiences as one of several measures of smart cities while Dainow (2017) describes smart environments as seamless lived experience. Sari et al. (2017) describe semantic ambient media experiences (SAME) afforded by real-time, context-aware, augmented and immersive emerging technologies and the technological and sociocultural infrastructures of urban environments. Vasileva et al. (2018) advance the improving of user experiences of open data in smart campuses as a way of guiding the improving of user experiences in smart cities. Edge et al. (2020) explore lived experiences in smart cities (SCs) using a creative analytic practice (CAP) approach focusing on the city of Kitchener in Ontario, Canada. Fernandez et al. (2021) describe a comprehensive framework to analyze the industrial experience of IoT platforms in smart cities. Van der Graaf et al. (2021) offer an exploration of experience-based co-creation in smart cities for public service design and delivery. Fineout (2022) describes more sustainable practices for architectural designs focusing on environmentally sensitive geometries.

Table 3 provides an overview by author and date of the research and practice literature for urban life and the ambient pertaining to interactions, spanning the timeframe of 2013 to 2022.

Lin et al. (2013) advance a framework supporting a smart living environment. van Waart and Mulder (2014) describe a range of designs for public spaces in support of meaningful interactions aided by interactive technologies involving art, a café, and climate data. Smyth et al. (2015) present a collection of videos depicting scenarios for the experience of urban life and the influence of people and technology interactions. Alavi and Bahrami (2019) highlight the importance of interactions from the perspective of walking in future cities and the associated complexities in smart environments. Gómez-Carmona et al. (2019) shed light on the deploying of new forms of interactive furniture at the street level identifying adoption barriers and opportunities afforded by such interactions. McKenna (2020) formulated a human-smart environment interactions framework in exploring dimensions of smartness, finding a good relationship between, for example, interactive public spaces and openness. Chew et al. (2021) explore the smartness and disruptiveness of public benches designed for playful interactions involving smart technologies. Ourednik et al. (2021) find cities to be sites of multimodal interaction and sociality. Proctor and Proctor (2021) focus on interactions in relation to sensation and perceptions in support of designing

Table 4. Overview of research for urban life and augmentation in smart cities and future cities

Author(s)	Year	Augmentations
Lin et al.	2013	Smart living, human capabilities – mobility, memory, and health
Bello et al.	2018	Sound analysis of data – audio surveillance & noise pollution monitoring
Celani et al.	2019	Challenges of the augmented city; infrastructure as unifying
Leorke	2019	Digital games, meaningfulness – reappropriate, reconfigure, augment
Fernandez et al.	2020	Augmented space model for AR and augmented space (AS) principles
Chew et al.	2021	AI technologies for playful placemaking with smart street furniture
Gao et al.	2022	Data for detecting & learning about traffic intersection contexts for AVs
Lee et al.	2022	Interaction design for AR in connected cities for human-city interaction
Lister	2022	Augmented reality triggers & smart learning journey activities & choices

for human-computer interactions. Hirsch et al. (2022) highlight collaboration among urban, material, and interaction designers in placemaking initiatives.

Table 4 provides an overview by author and date of the research and practice literature for urban life and the ambient pertaining to augmentations, spanning the timeframe of 2013 to 2022.

Lin et al. (2013) advance augmented smart living with interactive technologies that augment human capabilities pertaining to mobility, memory, and health. Bello et al. (2018) address the combining of convolutional networks and data augmentation in the analysis of urban sounds involving audio surveillance and noise pollution monitoring. Celani et al. (2019) explore the challenges of the augmented city where infrastructure is said to be a unifying element. Leorke (2019) explores digital games and the smart cities model to enable the reappropriation, reconfiguring, and augmenting of urban spaces for more meaningful involvement by citizens. Chew et al. (2021) focus on artificial intelligence (AI) technology augmentations as co-creators for playful placemaking with smart street furniture. Fernandez et al. (2020) describe an augmented space model for smart cities incorporating augmented reality (AR) and augmented space (AS) principles in support of an augmented enterprise architecture framework. Gao et al. (2022) address data augmentations pertaining to the detecting and learning about city intersection traffic contexts for autonomous vehicles (AVs). Lee et al. (2022) focus on interaction design for augmented reality (AR) in connected cities for human-city interaction going forward. Lister (2022) explores augmented reality (AR) triggers in the context of smart learning journey activities and experiences in urban environments featuring a technology mediated AR interface of choices.

In terms of patterns and potential relationships, what emerges from this review of the research and practice literature enables a comparative overview for augmentations, experiences, infrastructures, and interactions as shown in in Table 5.

Of note are the elements that emerge across columns of Table 5, that include, but are not limited to, context, human capabilities and concerns, livability, meaningfulness, placemaking, playfulness, social considerations, and technologies.

Table 5. Patterns and potential relationships

Augmentations	Experiences	Infrastructures	Interactions
Activities and choices	Ambient media	Appropriate	Collaboration
Augmented reality triggers	Augmented	Attentive to people, walking	Contexts
Augmented space model	Context-aware	Connecting, business	Human-smart EI
AS principles	Cultural	- info/tech, physical, social	Meaningful
City challenges	Everyday life, Lived	Human needs	Multimodal
Data, urban sounds	Immersive	Hyperconverged - CSN	Openness
Digital games & SC model	Infrastructure	Inefficiencies	Perceptions
Human capabilities	Meaningful	Liveable	Placemaking
- health	Measure of SCs	Multidimensional tech	Playful
- memory	Messiness	Quality of life	Public spaces + tech
- mobility	Real-time	Redundancy, communications	Sensation, Sensing
Livability	Relevant	Resilient	Smart furniture
Playful placemaking	Social	Risks	Sociality
Traffic intersections	Technological	Social	Street-level
- contexts	Understandable	Sustainable	Technologies

3.6 Issues, Controversies, Problems

The review of the research and practice literature enables the surfacing of a range of issues, controversies, and problems as identified in Table 6.

For augmentations, issues, controversies, and problems include, but are not limited to, data; playfulness; the social with the physical and virtual; smart street furniture; humane technologies for smart living where smart living is aided by such technologies; the upgrading of the meaning of infrastructures; urban audio surveillance; digital gaming to meaningfully engage people in public spaces; noise and other forms of pollution monitoring; technology sentient infrastructure; and traffic intersection contexts for autonomous vehicles (AVs).

Table 6. Infrastructure and associated issues, controversies, and problems in smart cities

	Issues	Controversies	Problems
Augmentations	• Data • Playfulness • Social, P/Virtual • Street furniture	• Humane tech for smart living • Smart living aided by tech • Infrastructure meaning upgrade • Urban audio surveillance	• Gaming, meaning, people • Noise/Pollution monitoring • Tech-sentient infrastructure • Traffic contexts for AVs
Experiences	• As a measure • Meaningful • Relevant • Understandable	• Ambient media • Lived experience • Seamless •Technological, social, cultural	• Data privacy & protection • Environmentally sensitive practices in architecture • Sustainable practices
Interactions	• Complexities • Human aspects • Interactive tech • Meaningfulness	• Adaptability • Digital transformations • Public spaces + openness • Playfulness – smart, disruptive	• Contextual info & services • Cross-sector collaborations • Multimodal • Social media +common good
Urban life & Infra-structures	• Human side • Interfaces • MD for MP • Walking & data	• Communication capabilities • Hyperconvergence for IT • Smartness as life enhancing • Smartness as multi-purpose	• Back-up plans – diverseness • Redundancy • Rethink software infra… • Upgrade meaning of infra…

Figure 1. Conceptual framework for urban life & the ambient with evolving infrastructures in smart cities

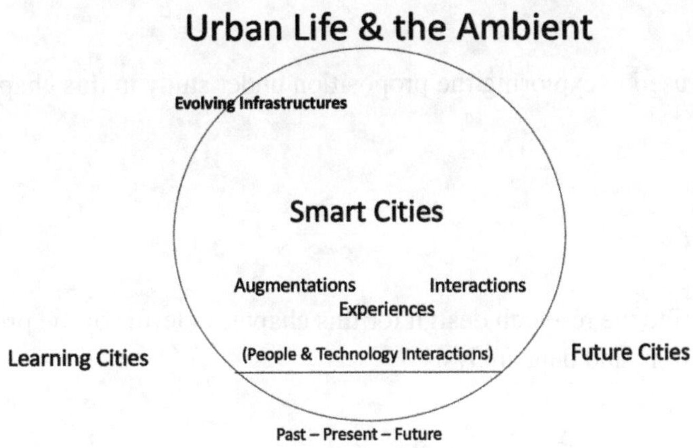

For experiences, issues, controversies, and problems include, as a measure; the importance of meaningfulness, relevance, and understandability; ambient media; lived experience; seamless; combining technological, social, and cultural infrastructures; data privacy and protection; environmentally sensitive practices in architecture; and sustainable practices.

For interactions, issues, controversies, and problems include, complexities; human aspects; interactive technologies for insights and sensing in support of meaningfulness; adaptability; digital transformations; public spaces and openness; playfulness in relation to smartness and disruptiveness; contextualized information and services; the need for cross-sector collaborations; multimodal nature of travel and mobility in urban spaces; and the importance of the common good for social media platforms.

For urban life and infrastructures, issues, controversies, and problems include, the human side; interfaces designed with people in mind; multidimensional (MD) approaches in support of multiple perspectives; urban data generation through walking in the city; the importance of robust communication capabilities; hyperconvergence for IT infrastructures (e.g., compute, storage, networks); smartness as life enhancing; smartness as multi-purpose; back-up plans featuring diverse approaches; redundancy of systems; rethinking of software infrastructures; and an upgrading of the meaning of infrastructure.

Common elements that emerge across columns of Table 6, include, but are not limited to data, meaningfulness, and sustainability.

4. URBAN LIFE AND EVOLVING INFRASTRUCTURES IN SMART CITIES

Responding to the issues, controversies, and problems emerging from the literature review in the preceding section, a conceptual framework for urban life and the ambient featuring evolving infrastructures in smart cities is formulated in Fig. 1.

The importance of evolving infrastructures for the augmenting of urban life experiences and interactions is featured to meet the complex and multidimensional needs of learning cities, smart cities, and future cities involving people and technology interactions over time.

The research question identified in Section 1 is reformulated as a proposition for exploration in this chapter, as follows:

P4: *Urban infrastructures are evolving in smart cities to include experiences, interactions, and augmentations*

The research design used in exploring the proposition under study in this chapter is described in the Methodology section.

5. METHODOLOGY

Sections 5.1 to 5.3 describe the research design for this chapter in terms of the process used, sources of evidence for data collection, and data analysis.

5.1 Process

The inquiry in this chapter uses an exploratory case study approach combined with an explanatory correlational design since the two are said to be complementary (Yin, 2018) in that the former may be "needed to examine the underlying processes that might explain a correlation." An online space is used to describe the study and invite people to participate. Interest emerged from individuals in a range of cities in multiple countries (e.g., Canada, United States, Greece, Ireland, Israel, etc.) reaching several continents.

5.2 Sources of Evidence

A webspace was used for the gathering of demographic data from individuals during sign up for the study, including information such as age range, gender, and city. Additionally, people could self-identify in one or more categories (e.g., business, community member, student, educator, etc.). A survey instrument, pre-tested prior to use, was used in this study and contained closed and open-ended questions. Survey questions pertained to a range of items related to smart cities, including augmentation, experiences, and interactions. Survey respondents were also invited to participate in an in-depth interview. Interviews were guided by an interview protocol, also pre-tested prior to use in this study.

Additionally, data were gathered systematically through group and individual discussions, conducted in parallel with this study, with individuals in many sectors (e.g., students, faculty, and staff in higher education, business, community members and leaders, and city staff and leaders) in Canadian cities (e.g., Toronto, Vancouver, Victoria).

5.3 Data Analysis

Data analysis in this chapter is guided by the research question posed which was then reformulated into a proposition for exploration, as follows:

P4: *Urban infrastructures are evolving in smart cities to include experiences, interactions, and augmentations.*

Table 7. Assessments for experiences and infrastructures: air quality and visible infrastructures

Items	Experiences and Urban Infrastructures	Correlation
Air quality	25% (4); 75% (7)	.57
Visible infrastructures	25% (5); 75% (7)	

For the analysis of quantitative survey data, the use of descriptive statistics pertaining to urban life and evolving infrastructures in smart cities in relation to augmentations, experiences, and interactions enabled exploration of items such as *air quality* in relation to *invisible infrastructures.* The correlation feature of the Real Statistics Resource Pack containing an-add in for Microsoft Excel (Zaiontz, 2022) enabled the conducting of a Spearman Correlation Coefficient for ordinal data in exploring the nature of relationships between items pertaining to evolving urban infrastructures.

Content analysis was used for qualitative data gathered through open-ended survey questions, interviews, and group and individual discussions where inductive analysis from data collection was combined with deductive analysis using terminology from the literature review. Overall, an analysis was conducted for n=79 consisting of 42% females and 58% males for people ranging in age from their 20s to 70s.

6. RESULTS AND DISCUSSION

Table 7 provides the results of assessments for evolving urban infrastructures and *experiences* using the proxies of *air quality* and *visible infrastructures.* In response to the survey question where participants were asked to what extent elements such as *air quality* contribute to the making of smart cities, on a 7-point Likert type scale, where 1= Not at all and 7=Absolutely, 25% of respondents selected position 4 on the scale (Neutral) and 75% responded at the upper end of the scale at position 7. When asked to what extent *visible infrastructures*, contribute to the making of smart cities, 25% of survey respondents selected position 5 on the scale (Sort of) and 75% responded at the upper end position of 7.

Using the correlation feature of the Real Statistics add-in for Microsoft Excel (Zaiontz, 2022), a Spearman correlation coefficient for ordinal data was conducted on the assessment results of the two items, *air quality* and *visible infrastructures* (Table 7), showing a finding of .57. As people experience their city in relation to *air quality* and *visible infrastructures* it could also be said that people are interacting with their city through the air they breathe and the mobility and other physical infrastructures they use.

Exploring further as shown in Table 8, focusing on evolving urban infrastructures, when asked to what extent various elements contribute to the making of smart cities, for i*nvisible infrastructures*, 50% of survey respondents selected position 5 on the scale (Sort of) and 50% responded at the upper end position of 7.

Using the correlation feature of the Real Statistics add-in for Microsoft Excel (Zaiontz, 2022), a Spearman correlation coefficient for ordinal data was conducted on the assessment results of the two items, *air quality* and *invisible infrastructures* (Table 8), showing a finding of .81. It is worth noting that the correlation seems to be significantly higher for *invisible infrastructures* than *visible infrastructures.* It may be that people appreciate and benefit from the *invisible infrastructures* that enable their smart phone for example to provide details about air quality in their city enabling them to make decisions and choices affecting their health and well-being. As people experience their city in relation to *air quality*

Table 8. Assessments for experiences and infrastructures: air quality and invisible infrastructures

Items	Experiences and Evolving Infrastructures	Correlation
Air quality	25% (4); 75% (7)	.81
Invisible infrastructures	50% (5); 50% (7)	

and *invisible infrastructures* providing air quality details and other types of data from mobile apps, it could also be said that people are interacting with their city through augmentations enabled by continually emerging and awareness-enabling technologies.

Correlations in the .35 to .65 range are said to be "the typical values in prediction used to identify variable membership in the statistical procedure of factor analysis" (Creswell, 2018) while those in the .66 – .85 range are said to be "very good" (Creswell, 2018). As such, the correlations that emerged for variables in this chapter provide opportunities going forward for further explorations for both research and practice with these and other possible variables.

Regarding interactions with technologies in urban spaces, it is worth noting that a higher education instructor commented that "on the digital level, keeping up to date on social media allows me to find out things that my eyes could not see while walking around." The experience of smartness in the city is described by the instructor as "wifi available for free" as well as "smart meters for footfall, air and water quality" and "urban displays" providing information about parking availability, among other examples cited. A community leader commented, as follows, "I truly believe that the 'smart city' umbrella is the chief catalyst for changes that are needed in cities" referring to "infrastructure updates" and noting that "government operations are due for a version upgrade as well" resulting in "a complete evolution of the current system" in support of "comprehensive change" with "the caveat" that "primary efforts should be focused around people and their needs, with everything else" such as technology, "playing a supporting role." As such, efforts to explore evolving infrastructures in smart cities in this chapter focus on variables pertaining to people and their experiences of smart cities, along with their assessment of such variables.

As depicted graphically in Fig. 2, the interconnectivity of infrastructures with experiences, interactions, and augmentations would seem to give way in the context of urban life and the ambient to the notion and potentials of ambient augmentations, ambient experiences, ambient interactions, and ambient media.

7. LIMITATIONS AND FUTURE DIRECTIONS

Limitations of this work are discussed in this section opening the way to future directions based on challenges and opportunities for research and practice.

Limitations

Limitations of this chapter pertain potentially to the literature review, focusing as it does on urban life and the ambient in relation to evolving infrastructures focusing on augmentations, experiences, and interactions in smart cities and this may be mitigated by extending the literature review going forward. Among other limitations of this chapter is that of the small sample size for the underlying study of this work, and as such, caution is encouraged in the interpretation of correlations calculated from the assess-

Figure 2. Urban life and the ambient – evolving infrastructures in smart cities

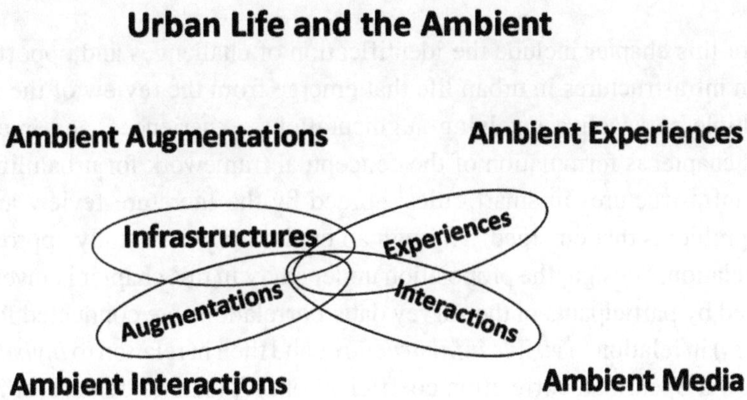

ment of visible and invisibles infrastructures and the proxy of air quality used for experiences. Going forward opportunities may emerge for the use of other possible proxies, or for additional variables to surface in the exploration of evolving urban infrastructures, resulting in possibly stronger correlations with greater prediction potentials.

Future Research Directions and Opportunities

For research, future directions emerge in terms of the potential for conducting additional explorations on the variables identified in this work relevant to evolving urban infrastructures as identified in Table 5. In this way, opportunities for explorations going forward exist for the further testing and validation of variables used in this work. Additionally, opportunities exist for the identification of other and possibly stronger variables relevant to evolving urban infrastructures that foster further learning in the exploration of dimensions such as ambient augmentations, ambient experiences, and ambient interactions, and the like in smart cities, learning cities, and future cities.

Of note in Tables 7 and 8 is the presence of neutral responses for *air quality* identifying potential areas of interest for further exploration.

Future Practice Directions and Opportunities

Future directions for practice include the potential for pursuing urban explorations guided by the variables identified in this work relevant to evolving urban infrastructures. Other examples of augmentations, experiences, and interactions provide opportunities for explorations going forward for practice, such as playful placemaking (Table 5). And the potential exists for exploring spaces with other, possibly stronger variables relevant to evolving urban infrastructures that foster real world action and learning in everyday contexts in smart cities, learning cities, and future cities, in addressing the issues, controversies, and problems presented in Table 6.

8. CONCLUSION

Key contributions of this chapter include the identification of challenges and opportunities associated with evolving urban infrastructures in urban life that emerge from the review of the research literature provided from multiple perspective involving augmentations, experiences, and interactions. Another contribution of this chapter is formulation of the conceptual framework for urban life and the ambient featuring evolving infrastructures in smart cities, guided by the literature review and the key issues, controversies, and problems that emerged. Through an exploratory case study approach, together with an explanatory correlational design, the proposition under study in this chapter is investigated. Based on assessments provided by participants in the survey data, correlations are conducted for *air quality* (as a proxy for *experiences*) in relation to *visible infrastructures* and then in relation to *invisible infrastructures* where, findings show a Spearman correlation coefficient of .57, (said to be "useful for limited prediction") and .81 (said to be "very good"), respectively (Creswell, 2018). A key limitation of this chapter is the small sample size and as such, caution is urged in the interpretation of findings, opening opportunities for further research. Notwithstanding possible limitations of this chapter, future directions guiding research and practice activities going forward are provided in terms of further testing and validation of the variables used in this chapter while also encouraging activities to identify other, additional factors relevant to evolving urban infrastructures and to augmentations, experiences, interactions, in the context of smart, learning, and future cities.

Key insights and ideas emerging from this chapter are provided, along with exercises encouraging further exploration, as follows:

Insights emerging from this chapter include the following:

1. The importance of factors contributing to smart cities giving rise to many possibilities for *invisible infrastructures* and *experiences* in relation to *air quality*
2. The importance of factors associated with *invisible infrastructures* in smart cities such as mobile apps for *air quality*
3. The potential for many more factors to emerge for understanding *evolving urban infrastructures* whether *visible* or *invisible* in smart cities in terms of augmentations, experiences, and interactions

Ideas emerging from this chapter include the following:

1. Potential opportunities and challenges for exploring evolving urban infrastructures in relation to a variety of variables (e.g., augmentations, experiences, and interactions) in the context of smart cities
2. The opportunities potential for exploring ambient dimensions (e.g., ambient experiences, ambient interactions, etc.) of evolving urban infrastructures in smart environments
3. The use of an interdisciplinary and multi-method approach for the study and practice of evolving urban infrastructures in smart, learning, and future cities

Exercises for the reader to consider, encouraging further exploration include the following:

1. Identify one or more factors in your city that are giving rise to invisible infrastructures

2. Identify one or more ways in which augmentations to urban infrastructures would enhance the quality of urban life in your city
3. Describe one or more elements that contribute to evolving urban infrastructures in your city

The audience for this chapter will be interdisciplinary in nature, including but not limited to community members and leaders, educators, businesses, researchers, urban planners, urban designers, infrastructure developers, and anyone concerned with evolving urban infrastructures in urban life, while being attentive to elements such as augmentations, experiences, and interactions along with associated practice and research opportunities and challenges in smart cities, learning cities, and future cities.

REFERENCES

Alavi, H., & Bahrami, F. (2019). Walking in smart cities. *Interactions (New York, N.Y.)*, *26*(2), 66–68. doi:10.1145/3301659

Albino, V., Berardi, U., & Dangelico, R. M. (2015). Smart cities: Definitions, dimensions, performance, and initiatives. *Journal of Urban Technology*, *22*(1), 3–21. doi:10.1080/10630732.2014.942092

Bauer, R. (2015). *How ambient intelligence will change our lives.* World Economic Forum. Retrieved 19 October 2021 from https://www.weforum.org/agenda/2015/10/how-ambient-intelligence-will-change-our-lives/

Bello, J. P., Mydlarz, C., & Salamon, J. (2018). Sound analysis in smart cities. In T. Virtanen, M. Plumbley, & D. Ellis (Eds.), *Computational Analysis of Sound Scenes and Events*. Springer. doi:10.1007/978-3-319-63450-0_13

Celani, P., Falcone, R., & d'Alessandro, E. (2019). The challenge of augmented city: New opportunities of common spaces. In *New Metropolitan Perspectives. ISHT 2018. Smart Innovation, Systems and Technologies* (vol. 100). Springer.

Chew, L., Hespanhol, L., & Loke, L. (2021). To play and to be played: Exploring the design of urban machines for playful placemaking. *Frontiers of Computer Science*, *3*, 635949. doi:10.3389/fcomp.2021.635949

Creswell, J. W. (2018). *Educational research: Planning, conducting, and evaluating quantitative and qualitative research* (6th ed.). Pearson.

Dainow, B. (2017). Smart city transcendent: Understanding the smart city by transcending ontology. *Orbit (Amsterdam, Netherlands)*, *1*, 1–15.

Dourish, P., & Bell, G. (2007). The infrastructure of experience and the experience of infrastructure: Meaning and structure in everyday encounters with space. *Environment and Planning. B, Planning & Design*, *34*(3), 414–443. doi:10.1068/b32035t

Dourish, P., & Bell, G. (2011). *Divining our digital future: Mess and mythology in ubiquitous computing*. MIT Press. doi:10.7551/mitpress/9780262015554.001.0001

Edge, S., Boluk, K., Groulx, M., & Quick, M. (2020). Exploring diverse lived experiences in the Smart City through Creative Analytic Practice. *Cities (London, England)*, *96*, 102478. doi:10.1016/j.cities.2019.102478

Fahmideh, M., Yan, J., Shen, J., Mougouei, D., Zhai, Y., & Ahmad, A. A. (2021). Comprehensive framework for analyzing IoT platforms: A smart city industrial experience. *Smart Cities*, *4*(2), 588–622. doi:10.3390martcities4020031

Fernandez, F., Sanchez, A., Velez, J. F., & Moreno, B. (2021). The augmented space of a smart city. *2020 International Conference on Systems, Signals and Image Processing (IWSSIP)*, 465-470. doi: 10.1109/IWSSIP48289.2020.9145247

Fineout, M. (2022). The Mexico City new international airport: A case study in environmentally Sensitive geometries. In H. Abdalla, H. Rodrigues, V. Gahlot, M. Salah Uddin, & T. Fukuda (Eds.), *Resilient and Responsible Smart Cities. Advances in Science, Technology & Innovation*. Springer. doi:10.1007/978-3-030-86499-6_4

Fortin, C. (2017). Delving deeper: Considerations on applying empirical research methods to infrastructural urban technology projects. *plaNext – Next Generation Planning, 4*, 76-93. Doi:10.23506/plnxt.2017.04.006

Gao, J., Wang, D., Lin, C.-P., Luo, C., Ruan, Y., & Yuan, M. (2022). Detecting and learning city intersection traffic contexts for autonomous vehicles. *Journal of Smart Cities and Society*, *1*(3), 213–239. doi:10.3233/SCS-220010

Gómez-Carmona, O., Sádaba, J., & Casado-Mansilla, D. (2019). Enhancing street-level interactions in smart cities through interactive and modular furniture. *Journal of Ambient Intelligence and Humanized Computing*. https://link.springer.com/article/10.1007/s12652-019-01577-8

Harrison, C., Eckman, B., Hamilton, R., Hartswick, P., Kalagnanam, J., Paraszczak, J., & Williams, P. (2010). Foundations for smarter cities. *IBM Journal of Research and Development*, *54*(4), 1–16. doi:10.1147/JRD.2010.2048257

Hatcher. (2019). Hyperconvergence, the cornerstone of future smart buildings? *Smart Buildings Magazine*. Retrieved 11 November 2021 from https://smartbuildingsmagazine.com/features/hyperconvergence-the-cornerstone-of-future-smart-buildings

Hirsch, L., Economidou, E., Paraschivoiu, I., & Döring, T. (2022). Material meets the city: A materials experience perspective on urban interaction design. *Interaction*, *29*(Jan-Feb), 58–63. doi:10.1145/3501358

Hopkins, T., Bae, S. S., Uhr, J., Banić, A., Zheng, C., & Do, E. Y. L. (2021). User interfaces in smart cities. In J. C. Augusto (Ed.), *Handbook of Smart Cities*. Springer. doi:10.1007/978-3-030-15145-4_94-1

Lee, L.-H., Braud, T., Hosio, S., & Hui, P. (2022). Towards augmented reality driven human-city interaction: Current research on mobile headsets and future challenges. *ACM Computing Surveys*, *54*(8).

Leorke, D. (2019). Reappropriating, reconfiguring and augmenting the smart city through play. In A. Nijholt (Ed.), *Making Smart Cities More Playable*. Gaming Media and Social Effects., doi:10.1007/978-981-13-9765-3_3

Lin, H.-C., Jeng, T., & Chen, C.-H. (2013). Reimaging humane cities: Interaction design, city sense, and smart living. *International Journal of Affective Engineering*, *12*(2), 155–159. https://www.jstage.jst.go.jp/article/ijae/12/2/12_155/_article/-char/en

Lister, P. (2022). Ways of experiencing technology in a smart learning environment. In *Distributed, Ambient and Pervasive Interactions. HCII 2022*. Lecture Notes in Computer Science. Springer.

Marzano, S. (2006). Ambient culture. In True Visions. Springer. doi:10.1007/978-3-540-28974-6_3

McKenna, H. P. (2020). Human-smart environment interactions in smart cities: Exploring dimensionalities of smartness. *Future Internet*, *12*(5), 79. doi:10.3390/fi12050079

Mohanty, S. P., Choppali, U., & Kougianos, E. (2016). Everything you wanted to know about smart cities: The Internet of things is the backbone. *IEEE Consumer Electronics Magazine*, *5*(3), 60-70. doi:10.1109/MCE.2016.2556879

Ourednik, A., Mlynář, J., Mutzner, N., & Alavi, H. (2021). Social media platforms are failed cities. *Interaction*, *28*(6), 62–66. doi:10.1145/3488234

Proctor, R. W., & Proctor, J. D. (2021). Sensation and perception. In G. Salvendy & W. Karwowski (Eds.), *Handbook of Human Factors and Ergonomics* (5th ed.). Wiley. doi:10.1002/9781119636113.ch3

Qiao, S., Hiuang, G., & Yeh, A. G.-O. (2022). *Mobility as a service and urban infrastructure: From concept to practice*. Transactions in Urban Data, Science, and Technology. doi:10.1177/27541231221114171

Sari, E., Tedjasaputra, A., Ellen, D. Y. L., Duh, H., & Lugmay, A. (2017). Smart Cities for Better Living with HCI and UX (SEACHI). *Proceedings of the 8th Workshop on Semantic Ambient Media Experiences (SAME 2016), International Series on Information Systems and Management in Creative eMedia (CreMedia)*.

Smith, M. L. (2016). Urban infrastructure as materialized consensus. *World Archaeology*, *48*(1), 164–178. Advance online publication. doi:10.1080/00438243.2015.1124804

Smyth, M., Helgason, I., & Mitrović, I. (2015). City | Data | Future: Envisioning interactions in hybrid urban spaces. In *C&C '15, Proceedings of the 2015 ACM SIGCHI Conference on Creativity and Cognition*. Association for Computing Machinery. 10.1145/2757226.2757380

Streitz, N. A. (2021). From smart-only cities towards humane and cooperative hybrid cities. *Technology|Architecture + Design*, *5*(2), 127-133. Doi:10.1080/24751448.2021.1967050

van der Graaf, S., Nguyen Long, L. A., & Veeckman, C. (2021). *Co-creation and Smart Cities: Looking Beyond Technology*. Emerald Publishing Limited. doi:10.1108/9781800436022

van Waart, P., & Mulder, I. (2014). Meaningful interactions in a smart city. In N. Streitz & P. Markopoulos (Eds.), Lecture Notes in Computer Science: Vol. 8530. *Distributed, Ambient, and Pervasive Interactions. DAPI 2014*. Springer. doi:10.1007/978-3-319-07788-8_57

Vasileva, R., Rodrigues, L., Hughes, N., Greenhalgh, C., Goulden, M., & Tennison, J. (2018). What smart campuses can teach us about smart cities: User experiences and open data. *Information*, *9*(10), 251. doi:10.3390/info9100251

Wray, S. (2021). Smart cities need a back-up plan: Cities today. *MyITU News*. Retrieved 14 November 2021 from https://www.itu.int/en/myitu/News/2021/11/08/11/26/Smart-cities-resilience-digital-services-Cities-Today

Yin, R. (2018). *Case study research and applications: Design and methods* (6th ed.). Sage Publications.

Zaiontz, C. (2022). *Real statistics using excel*. Retrieved from www.real-statistics.com

ADDITIONAL READING

Alvarez, R. (2017). The relevance of informational infrastructures in future cities. *Field Actions Science Reports. Special Issue on Artificial Intelligence and Robotics in the City*, *17*, 12–15.

Habibzadeh, M., Xiong, W., Zheleva, M., Stern, E. K., Nussbaum, B. H., & Soyata, T. (2017). Smart city sensing and communication sub-infrastructure. *2017 IEEE 60th International Midwest Symposium on Circuits and Systems (MWSCAS)*, 1159-1162. doi: 10.1109/MWSCAS.2017.8053134

IEC. (2015). *Orchestrating Infrastructure for Sustainable Smart Cities: White Paper*. IEC and Centre for European Policy Studies (CEPS). Retrieved 12 May 2022 from https://www.ceps.eu/wp-content/uploads/2015/01/iecWP-smartcities-LR-en.pdf

Kasznar, A. P. P., Hammad, A. W. A., Najjar, M., Linhares Qualharini, E., Figueiredo, K., Soares, C. A. P., & Haddad, A. N. (2021). Multiple dimensions of smart cities' jnfrastructure: A review. *Buildings*, *11*(2), 73. doi:10.3390/buildings11020073

NexusIntegra. (2022). *Smart infrastructures: Essential for smart cities*. Retrieved 12 May 2022 from https://nexusintegra.io/smart-infrastructures-essential-for-smart-cities/

Porumbescu, G. (2021). Infrastructure law's digital equity goals are key to smart cities that work for everyone. *The Conversation*. Retrieved 27 April 2022 from https://theconversation.com/infrastructure-laws-digital-equity-goals-are-key-to-smart-cities-that-work-for-everyone-170068

Schiener, D. (2020). The road to smart city infrastructure starts with research. *IOTA Blog*. Retrieved 12 May 2022 from https://blog.iota.org/the-road-to-smart-city-infrastructure-starts-with-research-a03498c33f86/

Tavmen, G. (2020). Data/infrastructure in the smart city: Understanding the infrastructural power of Citymapper app through technicity of data. *Big Data & Society*, *7*(2). Advance online publication. doi:10.1177/2053951720965618

UNCTAD. (2016). *Smart Cities and Infrastructure: Report of the Secretary General*. United Nations Economic and Social Council E/CN.16/2016/2. Retrieved 12 May 2022 from https://unctad.org/system/files/official-document/ecn162016d2_en.pdf

Wilson, M. (2022). *Google is betting big on ambient computing. But it's getting it wrong*. FastCompany. Retrieved 12 May 2022 from https://www.fastcompany.com/90751095/google-is-betting-big-on-ambient-computing-but-its-getting-it-wrong

KEY TERMS AND DEFINITIONS

Human-Smart Environment Interactions: Human-smart environment interactions refer to more aware people interacting in environments that are pervasively infused with awareness-enabling technologies.

Invisible Infrastructures: Invisible infrastructures in urban environments refer to digital, virtual, and other seemingly immaterial structures (e.g., social) and less visible, as in, embedded, networks.

Urban Interfaces: Urban interfaces refer to devices, surfaces, or other mechanisms enabling connection, interaction, and communication to occur.

Visible Infrastructures: Visible infrastructures in urban environments refer to streets, bridges, walkways, wiring, and other physical structures and networks.

Chapter 5
Literacies for Urban Life in Smart Cities:
Digital, Algorithmic, Play, and Inclusion

ABSTRACT

The primary purpose of this chapter is to explore literacies for urban life and the ambient in smart cities with a focus on the digital, the algorithmic, play, and inclusion. To this end, a review of the research literature is conducted in the context of smart cities, learning cities, and future cities. Issues, controversies, and problems emerging from the literature review are highlighted, contributing to formulation of a conceptual framework for literacies and urban life and the ambient in smart cities. Using an exploratory case study approach combined with an explanatory correlational design, variables are explored such as interactive public spaces and information and communication technologies (ICTs), along with the nature of their relationships. From the perspective of literacies for urban life, insights from correlations inform understandings of smart cities, learning cities, and future cities while highlighting challenges and opportunities going forward for research and practice.

1. INTRODUCTION

Regarding the algorithmic, Roche (2017) highlights the complexities and challenges "of integrating technical and human geo-sensing capabilities towards smart cities" and the need to address social issues. Play, according to Glas, Lammes, de Lange, Raessens, and de Vries (2019), in the context of the playful citizen, "can be considered a problem-solving force" while for de Lange (2019) in the context of the playful city, "play redefines roles and relations between professionals and citizens in processes of making cities." Morrison, Bjørnstad, Martinussen, Johansen, Kerspern, and Dudani (2020) focus on the digital dimension of urban life in addressing the notion of "design futures literacies" in the context of a world that is becoming "more complex and contested, economically and politically." Among literacies for inclusion in urban life, Qutab, Adil, Gardner, and Ullah (2022) refer to "social and digital inclusion" in relation to "citizen literacies." As such, this chapter is significant in that it addresses the range of

DOI: 10.4018/978-1-6684-4096-4.ch005

literacies required that are associated with the digital, the algorithmic, inclusion, and play in the context of smart environments and smart cities, forming the basis for exploration in this chapter, giving rise to the motivation for the main objectives that are described as follows.

Objectives: The main objectives of this chapter are to a) explore literacies for urban life and the ambient in smart cities; b) formulate a conceptual framework for literacies for urban life and the ambient in smart cities; and c) explore the nature of the relationship between variables associated with urban literacies as a way of learning more about smart cities, learning cities, and future cities. These objectives give rise to the main research question under exploration in this chapter – *Why are literacies for urban life in smart cities needed?*

2. BACKGROUND

Florida and McLean (2017) identify six principles for inclusive prosperity developed for "real estate and other businesses" to "adapt and use in their communities in the United States" and beyond. Indeed, it may be that these six principles for inclusive prosperity could contribute to literacies for inclusion in smart environments, more broadly. The notion of "algorithmic guardians" is discussed by Zambonelli, Salim, Loke, De Meuter, and Kanhere (2018) in the form of "software tools and algorithms" that are "capable of somehow protecting us from undesirable behavior on the part of third party algorithms" conceived as they are "as far more evolved instances of current personal assistants." Regarding play, De Lange (2019) describes a game (*Rezone the Game*) that involves a range of elements including "a physical board game with a number of 3D printed iconic buildings that represent the neighborhood" as well as "an augmented reality layer of real-time information about these buildings projected on a screen" along with a "computer algorithm programmed to let buildings descend into vacancy." Morrison et al. (2020) emphasize the importance of "design futures literacies" considered "in terms of rapid and long-lasting environmental change" where "resource use and reuse" must be reconsidered "in futures that are uncertain, contingent and emergent." Worthy of note are emerging initiatives such as that at the University of Waterloo (2022) involving "futures literacy" as "applied to cities" in the form of Future Cities involving "urban-focused research, knowledge mobilization, instruction, and civic engagement" providing "a forum for presentations, dialogue, and knowledge sharing."

2.1 Definitions

Definitions are provided from dictionaries, the research literature, and other sources for key terms used in this chapter.

- **Algorithm:** Merriam-Webster (2021) describes an algorithm as "a set of steps that are followed in order to solve a mathematical problem or to complete a computer process."
- **Inclusion:** Carnemolla, Robinson, and Lay (2021) describe inclusion as "a process of increasing participation and decreasing exclusion" involving "a responsibility of communities at a localised level."
- **Literacies:** UNESCO (2021) describes literacies as "a means of identification, understanding, interpretation, creation, and communication in an increasingly digital, text-mediated, information-rich and fast-changing world."

- **Play:** Vygotsky (1978) describes play as "an adaptive mechanism promoting cognitive growth" that "creates the zone of proximal development."

2.2 Overview

This chapter consists of several sections containing a review of the research and practice literature providing perspectives on literacies for urban life and the ambient in smart cities including digital literacies for urban life, algorithmic literacies for urban life, literacies for inclusion in urban life, and literacies for play in urban life. Key issues, controversies, and problems from the literature review are highlighted and a conceptual framework for literacies for urban life and the ambient in smart cities is formulated and operationalized for use in this chapter. The main research question under exploration in this chapter is reformulated as a proposition for investigation. The methodology for investigation of the proposition in this chapter is described followed by the presentation of results and a discussion of findings. Limitations of the study underlying this chapter are identified and future directions for research and practice are outlined. The conclusion provides an overview of outcomes, the identification of insights and ideas, development of several questions encouraging further exploration, and identification of the target audience for this chapter.

3. LITERACIES FOR URBAN LIFE AND THE AMBIENT IN SMART CITIES

This chapter provides a review of the research and practice literature for literacies including digital literacies for urban life and the ambient, algorithmic literacies for urban life and the ambient, literacies for inclusion in urban life and the ambient, and literacies for play in urban life and the ambient.

3.1 Digital Literacies for Urban Life and the Ambient

Ylipulli, Kangasvuo, Alatalo, and Ojala (2016) employ anthropological design fiction to explore the notion of future hybrid cities, envisioning "3D virtual city models and physical reality are intertwined seamlessly" focusing on the digital and physical, past and present, and near and far places; speculations on digital traces and bodies after death; digital city rights and ownership; and the sociocultural. From a human geography perspective, Roche (2017) provides an overview of the range of digital and spatial literacies associated with geographic information science, highlighting "spatial skills, spatial thinking and urban intelligence" in the movement toward smart cities. Borda and Bowen (2019) explore smart cities as "innovations across digital culture smart city platforms and participatory-centric approaches", identifying future opportunities and directions in relation to museums. McKenna (2019) explores digital literacies in the context of smart cities focusing on learning, play, and inclusion as ambient in smart cities in relation to aware people and aware technologies. Mouton, Ducey, Green, Hardcastle, Hoffman, Leslie, and Rock (2019) seek to extend literacies associated with the "digital divide" to health promotion and equity, calling to "integrate the 'healthy cities' tradition with the current interest on smart cities." Morrison et al. (2020) focus on how design "may engage in shaping futures through experimentation and exploration in the critical and productive engagement with techno-cultural life" involving "language and design (from lexis to discourse)", extending vocabulary by "situating lexis in cultural-historical contexts", and the development by students, in collaboration with government, of "four future digital urban living scenarios with trust as their central focus" having implications for practice, policy and design pedagogy. In the context of smart learning in the community, Lister (2020) explores ways of supporting the digital skills, competencies, and literacies of citizens. Elberzhager, Mennig, Polst, Scherr, and Stüpfert (2021) explore the notion of digital transformation in relation to smart cities in terms of a digital ecosystem, through formulation of a framework for the procedure, results, and lessons learned for sharing and further development. Qutab et al. (2022) undertake "to identify the range of literacies required in smart cities" together with "the roles of libraries, archives, and museums (LAM) in supporting citizen literacies for social and digital inclusion" claiming "the LAM sector" to be "one of the major stakeholders in the digital transformation sphere." Setlur (2022) discusses data literacies in the context of augmented analytics and the importance of supporting people "to augment AI skills with their own creativity and critical thinking." Tippens Reinitz, McCormack, Reeves, Robert, Arbino, Anderson, Hamman, Johnson, Kew-Fickus, Snyder, and Stevens (2022) identify the need for "data literacy training" noting that "data literacy is the next barrier to using data for teaching and learning" where data literacy is defined, citing the work of Gartner (2021) as "the ability to read, write, and communicate data in context, including an understanding of data sources and constructs, analytical methods and techniques applied, and the ability to describe the use case, application, and resulting value." Tippens Reinitz et al. (2022) also note that "data literacy and AI skills still lag behind the rapid adoption of big data analytics products."

3.2 Algorithmic Literacies for Urban Life and the Ambient

The AmbientTalk (Cutsem, Mostinckx, and Meuter, 2009) programming language is described as one capable of supporting "better programming" with 'clean' code writing capability. Boddington (2017) explores "how to produce realistic and workable ethical codes or regulations" in the rapidly evolving domain of artificial intelligence (AI) while presenting arguments, pitfalls, challenges, and opportunities. From a learning sciences perspective, Headrick Taylor (2017) explores emerging literacies associated with "learning on-the-move with technologies through urban spaces" focusing on "the digital traces" generated, contributing to "locative literacies as a way of understanding place-based, digital modes of reading and writing", using "the new analytic unit, learning along lines", providing perhaps a sort of groundwork for algorithmic literacies. As such, Headrick Taylor (2017) is able to present findings that "represent young people participating in an emerging terrain of digital literacies that are mobile, embodied, geospatial, and participatory" who "learned how to use mobile, location-aware technologies on the move through their neighborhood" resulting in "a uniquely youth-centric experience of a place" where people emerge "not merely as consumers or generators of texts but as being part of that text" and as such, "literacy agents of a text they populate." From a human geography perspective, Roche (2017) describes the notion of "algorithmic urban places" and the potential for "algorithmic-based planning using big data, personalized algorithmic wayfinding and navigation" as well as "algorithmic personalization of places" for improving urban life. Lido, Reid, and Osborne (2018) address literacies in relation to urban big data and "the potential for addressing city challenges, such as learning inclusion" using learning city frameworks while "promoting lifelong learning and lifewide literacies for more engaged citizenry." Zambonelli et al. (2018) address issues associated with algorithmic governance in smart cities, noting that "with the current literacy of the general public in computer science" there would seem to be "no analogue whatsoever for public understanding of algorithms." As such, Zambonelli et al. (2018) call for "better programming languages, inspectability of code, and users' computing literacy." From a social sciences perspective, Lee and Björklund Larsen (2019) provide "an increased analytical awareness of how algorithms are theorized in society and culture" in a wide range of settings (e.g., from farming to schools, to surveillance, etc.), involving an array of approaches, namely – "anxieties, pluralities, recalibrations, folds, aesthetics, accountability." Duin and Pedersen (2021) explore "algorithmic control" and "algorithmic culture" claiming that "the ambiguities bound up with algorithms become a lived phenomenon for everyday people" and as such encourage people generally and "students to think critically about the workings of algorithms." Herczeg, Ohlei, Schumacher, and Winkler (2021) describe examples of ambient learning spaces (ALS) as a "platform to foster aesthetic and creative learning" for "digitally-enabled teaching and learning environments" in the form of applications for location-based media (e.g., Peripheral Media – ActMotion; Tangible Media – InteractiveWall) and location-independent media (e.g., Mobile Media – MoLES and InfoGrid; Wearables – Smart Fashion) designed for "constructive teaching and self-directed aesthetic and algorithmic learning." Beer (2022) explores the notion of a "recursive society" involving "multiple feedback loops as a result of algorithmic sorting and data processes" giving rise to "difficulties of understanding the social" since "recursive algorithmic processes" are said to "have repeatedly shaped outcomes, practices, relations, and actions over time" resulting in "the social world itself being recursive."

3.3 Literacies for Inclusion in Urban Life and the Ambient

From a research and practice perspective, Florida and McLean (2017) speak of "inclusive prosperity" as "the idea that the opportunity and benefits of economic growth should be widely shared by all segments of society" and describe "what inclusive urban development can look like" based on "practical policies and initiatives." McKenna (2017) theorizes the notion of ambient inclusion and formulates an ambient inclusion framework for the exploration of innovative ICT-enabled spaces, services, and designs in smart cities (SCs) resulting in a glimpse of the framework in action, based on study findings, revealing benefits that include engagement, learning, and meaningful participation. Rebernik, Szajczyk, Bahillo, and Goličnik Marušić (2020) explore inclusion in relation to people with disabilities, proposing a Disability Inclusion Evaluation Tool (DIETool) and Disability Inclusion Performance Index (DIPI) "designed to guide cities through a maze of accessibility and disability inclusion related requirements set within the political, legislative and standardization frameworks." Tested in a European context, the tool is said to be "beneficial for providing diagnosis as to how disability friendly a city is" offering "an opportunity for designing informed corrective measures towards disability inclusive city design." Sessa (2020) identifies inclusion as one of several characteristics of the smart home "in order to improve the domestic routine, to optimize everyday life." Drawing on the work of Beretta (2015), Sessa (2020) notes that inclusion "means avoiding all forms of discrimination and valuing differences." UNESCO-ILL (2020) identifies learning cities among a list of key action points for lifelong learning. Carnemolla et al. (2021) provide a "scoping review of initiatives to support the inclusion of people with intellectual disabilities in civic and social activities" using the inclusive cities framework (Broadhead and Kierans, 2019) and five core areas of action. Correia, Feio, Teixeira, and Lourenço Marques (2021) explore inclusion in relation to smart city policymaking, advocating for the use of participatory methodologies to guide the practical application of involving citizens, based on knowledge from a development geographies approach. UNESCO-ILL (2021) describes learning cities as being "driven by the principle of inclusion" and as such "advance policies and practices that foster sustainable development, notably through lifelong learning programmes that promote equity, cohesion and peace." Chen, Gil-Garcia, and Gasco-Hernandez (2022) explore social sustainability in smart cities, complementing existing research on environmental and economic sustainability, noting that "inclusion and equity are crucial components that must be considered."

3.4 Literacies for Play in Urban Life and the Ambient

Misteli, Poulakos, Kapadia, and Sumner (2018) demonstrate a system enabling the experience of "emergent play within a mixed reality environment" consisting of "real and virtual objects" that "share a unified representation to allow joint interactions" in support of "game, story and agent-based play." Leorke (2019) explores the "growing intersection of digital games and the 'smart city' model" as "playful practices" enabling the potential to "*reappropriate*, *reconfigure*, and *augment*" in urban spaces. Glas et al. (2019) provide an introduction to the notion of the "playful citizen" and argue that "play is an important theoretical principle for comprehending new manifestations of civic engagement." De Lange (2019) argues that the "playful city", based on a review of the research literature, "complements the majority of smart city visions that equate 'smart' with automation and efficiency" and "vastly broadens our perspective of what 'smart citizenship' in relation to urban life could be." According to De Lange (2019), *Rezone the game* is adaptive and requires collaboration and is intended "to address complex questions like vacancy" where "no longer passive as users of the city, players temporarily become smart planners." Börjesson Rivera, Ringenson, and Pargman (2020) claim that "the playable city 'happens' when a city offers suitable (playful) affordances and citizens engage in and make use of them" irrespective of the presence of digital technologies while arguing for the "designing and building" of "sustainable playable cities." Fredericks (2020) presents "a smart engagement framework for playful city-making" comprised of "digital and physical mechanisms for greater collaboration." Papangelis, Saker, Lee, and Jones (2020) identify the gap in research pertaining to the "lived experience" of so called 'smart spaces' in terms of "the extent to which physical and digital environments are currently producing new forms of play and playfulness" giving rise to the need for a forum for researchers and practitioners in the form of a workshop "to engage with these issues that often remain hidden" in a space dominated by technology discussions. Bertran, Duval, Armengol, Chen, Dong, Dastoor, Bertran, and Isbister (2021) provide a portfolio of speculative playful urban technology ideas in urban spaces in human life as a way of exploring the potentials of play in smart cities, attentive to the social, emotional, cultural, and sustainable influencing design going forward. Papangelis, Saker, Leorke, Lee, and Jones (2022) introduce a special issue focusing on urban environments, calling for an updating of the smart cities notion so as to, encompass "a broader range of emerging and nascent technologies that are increasingly adopted into daily life through play and participation."

3.5 Literature Review Overview

An overview by author and date is provided in this section for the research and practice literature for urban life and the ambient pertaining to digital literacies, algorithmic literacies, literacies for inclusion, and play literacies.

Table 1 provides an overview by author and date of the research and practice literature for urban life and the ambient in smart cities pertaining to digital literacies, spanning the timeframe of 2016 to 2022.

Ylipulli et al. (2016) provide an envisioning of future hybrid cities involving a seamless interweaving of the physical and digital. From a human geography perspective, Roche (2017) provides an overview of digital and spatial literacies involving spatial skills and thinking and urban intelligence. Borda and

Table 1. Overview of research for digital literacies in smart cities

Author(s)	Year	Digital Literacies
Ylipulli et al.	2016	Envisioning future hybrid cities – digital & physical as seamless
Roche	2017	Human geography perspective, spatial skills, spatial thinking
Borda & Bowen	2019	Innovations across digital culture – SC platforms, participatory-centric
McKenna	2019	Emergent focusing on learning, inclusion & play as ambient in SCs
Mouton et al.	2019	Health promotion & equity in smart cities as healthy cities
Lister	2020	Learning cities – digital skills, literacies, and competencies of citizens
Morrison et al.	2020	Design, shaping futures, techno-cultural life, collaboration, trust
Elberzhager et al.	2021	Digital ecosystem for a SC district
Qutab et al.	2022	Range of SC literacies, role of LAMs in social and digital inclusion
Setlur	2022	Data literacies training; augment AI skills-creativity & critical thinking
Tippens Reinitz	2022	Data literacy training and data literacy and AI skills

Bowen (2019) provide an overview of innovations across digital culture focusing on smart city platforms and participatory-centric approaches. McKenna (2019) explores digital literacies as emergent in smart cities (SCs) focusing on learning, inclusion, and play as ambient. Mouton et al. (2019) extend the need for literacies to health promotion and equity in smart cities as healthy cities. Lister (2020) explores ways of supporting the digital skills, literacies, and competencies of citizens in learning cities. Morrison et al. (2020) focus on design in shaping futures through experimentation and exploration of techno-cultural life in collaborations between students and government where trust is said to be a key component. Elberzhager et al. (2021) explore digital transformation through formulation of a framework for developing a digital ecosystem for a smart city district. Qutab et al. (2022) identify the range of literacies required in smart cities and the role of libraries, archives, and museums in supporting literacies for citizens, pertaining to social and digital inclusion. Setlur (2022) argues for data literacies training and the importance of augmenting AI skills with the creativity and critical thinking of people. Tippens Reinitz et al. (2022) discuss the need for data literacy training for teaching and learning and the lag in data literacy and AI skills.

Table 2 provides an overview by author and date of the research and practice literature for urban life and the ambient in smart cities pertaining to algorithmic literacies, spanning the timeframe of 2009 to 2022.

The AmbientTalk programming language (Cutsem et al., 2009) is described as one with clean code writing capabilities in support of better programming as advanced by Zambonelli et al. (2018), who address concerns with algorithmic governance and computing literacies. Boddington (2017) provides a reference source for ethical codes and regulations to guide research and practice in the domain of artificial intelligence (AI). Headrick Taylor (2017) explores the analytic unit, learning along lines, through youth learning on-the-move with locative literacies involving physical and digital mobility in the city. Roche (2017) provides a human geography perspective on algorithmic urban places using big data for personalization and wayfinding for improving urban life. Lido et al. (2018) explore literacies in relation to urban big data for learning inclusion, promoting lifelong learning and lifewide literacies. Lee and Björklund Larsen (2019) introduce the theorizing of algorithms in society and culture across many sectors. Duin and Pedersen (2021) explore algorithmic control and culture encouraging critical thinking as an analytical skill. Herczeg et al. (2021) describe examples of ambient learning spaces (ALS) designed

Table 2. Overview of research for algorithmic literacies in smart cities

Author(s)	Year	Algorithmic Literacies
Cutsem et al.	2009	AmbientTalk - better programming, clean code writing capability
Boddington	2017	Ethical codes & regulations for AI
Roche	2017	Human geography, data, personalization, urban places, wayfinding
Headrick Taylor	2017	Locative literacies with learning along lines
Lido et al.	2018	Literacies and urban big data for learning inclusion, lifelong & lifewide
Zambonelli et al.	2018	Governance issues, computing literacies
Lee & Björklund Larsen	2019	Theorizing of algorithms in society & culture across many sectors
Duin & Pedersen	2021	Algorithmic control & algorithmic culture & critical thinking
Herczeg et al.	2021	Ambient learning spaces – constructive teaching & algorithmic learning
Beer	2022	Recursive algorithmic processes & the recursive society problem

for constructive teaching and algorithmic learning. Beer (2022) articulates the recursive society problem associated with recursive algorithmic processes, making the social difficult to understand.

Table 3 provides an overview by author and date of the research and practice literature for urban life and the ambient in smart cities pertaining to literacies for inclusion, spanning the timeframe of 2017 to 2022.

Florida and McLean (2017) describe the notion of inclusive prosperity in widely sharing the opportunity and benefits of economic growth across society, as in, inclusive urban development with practical policies and initiatives. McKenna (2017) advances an ambient inclusion framework for the exploration of innovative ICT-enabled spaces, services, and designs in smart cities (SCs). Rebernik et al. (2020) describe the proposed Disability Inclusion Evaluation Tool (DIETool) and Disability Inclusion Performance Index (DIPI) to diagnose "how disability friendly a city is" and provide "informed corrective measures towards disability inclusive city design." Sessa (2020) notes the characteristic of inclusion in the smart home for the improvement of the domestic routine in optimizing everyday life. UNESCO-ILL (2020) points to

Table 3. Overview of research for literacies for inclusion in smart cities

Author(s)	Year	Literacies for Inclusion
Florida & McLean	2017	Inclusive prosperity & urban development, for society, practical policies
McKenna	2017	Ambient inclusion framework, innovative ICT-enabled spaces in SCs
Broadhead & Kierans	2019	Inclusive cities framework
Rebernik et al.	2020	DIETool & DIPI for measuring disability inclusion in SCs
Sessa	2020	Smart home – improve domestic routines, optimize everyday life
UNESCO-ILL	2020	Lifelong learning; Learning cities
Carnemolla et al.	2021	Inclusion of people with intellectual disabilities in civic & social activities
Correia et al.	2021	Citizens & policymaking, participatory methodologies
UNESCO-ILL	2021	Learning cities & inclusion, sustainable development, lifelong learning
Chen et al.	2022	Inclusion & equity crucial components of social sustainability in SCs

the importance of lifelong learning and the notion of learning cities. Carnemolla et al. (2021) provide a scoping review of initiatives to support the inclusion of people with intellectual disabilities in civic and social activities, drawing on the inclusive cities framework (Broadhead and Kierans, 2019). Correia et al. (2021) advance the use of participatory methodologies to guide the inclusion of policymaking by citizens in smart cities, drawing on knowledge from a development geographies approach. UNESCO-ILL (2021) associates learning cities with the principle of inclusion while advancing policies and practices for sustainable development through lifelong learning, equity programs, and so on. Chen et al. (2022) highlight the crucial components of inclusion and equity for social sustainability in smart cities (SCs).

Table 4 provides an overview by author and date of the research and practice literature for urban life and the ambient in smart cities pertaining to play literacies, spanning the timeframe of 2018 to 2022, drawing on earlier work by Vygotsky (1978) who describes play as an adaptive mechanism, promoting cognitive growth.

Misteli et al. (2018) provide a system featuring emergent play in a mixed reality environment with real and virtual objects supporting joint interactions for game, story, and agent-based play. De Lange (2019) argues that the playful city notion complements visions of smart cities in relation to automation and efficiency while expanding understandings of what smart citizenship could be, as in, adaptive, collaborative. Glas et al. (2019) introduce the notion of the playful citizen as important for understanding new manifestations of civic engagement. Leorke (2019) explores digital games intersecting with the smart city model for playful practices in urban spaces. Börjesson Rivera et al. (2020) advance the importance of the sustainable playable city influencing design and building in smart cities going forward in support of the playful citizen. Fredericks (2020) advances a framework for playful city-making involving mechanisms, both physical and digital, for greater collaboration. Papangelis et al. (2020) use a conference workshop space as a platform for discussion of play and playfulness as lived experience in smart cities to complement discussions of technology. Bertran et al. (2021) explore the playful potential of urban spaces. Papangelis et al. (2022) call for an updating of the smart cities (SCs) model to include more emergent technologies being adopted into daily life through play and participation.

In terms of patterns and potential relationships, what emerges from this review of the research and practice literature is a comparative overview for algorithmic literacies, digital literacies, literacies for inclusion, and play literacies as presented in Table 5.

Elements that emerge across columns of Table 5, include, but are not limited to, the ambient, augmentations (e.g., analytics, etc.), collaboration, critical thinking, culture, data, digital skills (e.g., games, etc.), engagement, equity, human geography perspective, learning cities, lifelong learning, policy and policymaking, practice, sustainability (e.g., social to complement environmental, etc.), and the use of frameworks.

3.6 Issues, Controversies, Problems

The review of the research and practice literature for algorithmic literacies, digital literacies, literacies for inclusion, and play literacies enables the surfacing of a range of issues, controversies, and problems as identified in Table 6.

For the algorithmic, key issues, controversies, and problems pertain to complexities; challenges; computing literacy; the social; algorithmic-based planning; algorithmic personalization; better programming; guardians; ethical codes for AI; inspectability of code; locative literacies; and regulations for AI.

Table 4. Overview of research for play literacies in smart cities

Author(s)	Year	Play Literacies
Vygotsky	1978	Play as an adaptive mechanism promoting cognitive growth
Misteli et al.	2018	Emergent play in a mixed reality environment – game, story, agent-based
De Lange	2019	Playful city, automation, efficiency, smart citizenship, adaptive, collaborate
Glas et al.	2019	Playful citizen for new manifestations of civic engagement
Leorke	2019	Digital games & the SC model as playful practices in urban spaces
Börjesson Rivera et al.	2020	Sustainable playable city & the playful citizen
Fredericks	2020	Framework for playful city-making + mechanisms for greater collaboration
Papangelis et al.	2020	Play & playfulness as lived experience in SCs – a platform for discussion
Bertran et al.	2021	Portfolio of speculative playful urban technology ideas
Papangelis et al.	2022	Updating of SCs to include more emergent tech for play & participation

For the digital, key issues, controversies, and problems pertain to augmented reality; design futures literacies; digital city rights; trust; complexities; equity; participatory-centric approaches; rapid long-lasting environmental change; collaboration; innovation; ownership; and socio-cultural.

For inclusion, key issues, controversies, and problems pertain to the digital; prosperity; the social; urban development; action for inclusion; ambient inclusion and ICTs; optimizing the smart home; valuing differences; accessibility; DIPI performance index; people with disabilities; and policies and practices.

For play, key issues, controversies, and problems pertain to augmenting urban spaces; digital games and smart cities (SCs); playful practices; reconfiguring urban spaces; play as a problem-solving force;

Table 5. Patterns and potential relationships

Algorithmic Literacies	Digital Literacies	Literacies for Inclusion	Play Literacies
Ambient learning spaces	Ambient	Accessibility	Adaptive mechanism
Analytical awareness	Augmented analytics	Ambient inclusion	Augment
Better programming	Collaboration	Cities framework	Automation
- AmbientTalk	Creativity/Critical thinking	Citizen involvement	Civic engagement
- code writing	Data – Teaching/Learning	Cohesion	Cognitive growth
Computing literacies	Design	DIPI (disability index)	Collaboration
Control	Digital competencies/skills	Engagement	Complex questions
Critical thinking	Equity	Equity programs	Digital games
Culture	Future hybrid cities	Home & everyday life	Efficiency
Data	Health promotion	Learning cities	Emergent play, mixed reality
- personalization	Human geography	Lifelong learning	- agent-based
- urban places	LAMs, role	Participation, meaningful	- game/story
- wayfinding	- digital/social inclusion	Participatory methods	Playable city
Ethics in AI	Learning cities	Peace	Playful citizen
Governance issues	Policy	People, disabilities	Playful city
Human geography	Practice	Policymaking	Playful practices
Lifelong literacies	Shaping futures	Practical policies	Reappropriate
Lifewide literacies	Socio-cultural	Practical practices	Reconfigure
Location-based media	Spatial skills	Prosperity	Smart citizenship
Locative, place-based	Spatial thinking	Social sustainability	Smart city model
Quality of life	Techno-cultural	Sustainable development	Smart engagement framewk
Recursive society	Training	Urban development	Smart planners
Regulations	Trust	Valuing differences	Sustainable playable cities

Table 6. Issues, controversies, and problems

	Issues	Controversies	Problems
Algorithmic	• Complexities • Challenges • Computing literacy • Social	• Algorithmic-based planning • Algorithmic personalization • Better programming • Guardians	• Ethical codes for AI • Inspectability of code • Locative literacies • Regulations for AI
Digital	• Augmented reality • Design futures literacies • Digital city rights • Trust	• Complexities • Equity • Participatory-centric approaches • Rapid, long-lasting env change	• Collaboration • Innovation • Ownership • Socio-cultural
Inclusion	• Digital • Prosperity • Social • Urban development	• Action for inclusion • Ambient inclusion & ICTs • Optimizing the smart home • Valuing differences	• Accessibility • DIPI Index • People with disabilities • Policies & practices
Play	• Augment urban spaces • Digital games & SCs • Playful practices • Reconfigure urban spaces	• Play as problem solving force • Players as smart planners • Redefines roles & relations • Reappropriate urban spaces	• Collaboration • Complexity • Playful citizen • Playful culture

players as smart planners; redefines roles and relations; reappropriating urban spaces; collaboration; complexity; playful citizen; and playful culture.

It is worth noting the presence of overlap between the algorithmic, the digital, inclusion, and play dimensions and across issues, controversies, and problems where collaboration and complexities provide a few examples.

4. CONCEPTUALIZING LITERACIES FOR URBAN LIFE AND THE AMBIENT IN SMART CITIES

Responding to the issues, controversies, and problems emerging from the literature review in the preceding section, a conceptual framework for literacies for urban life and the ambient in smart cities is formulated, as presented in Fig. 1.

The importance of literacies for urban life and the ambient is featured in support of the algorithmic, the digital, inclusion, and play to meet the complex and multidimensional needs of learning cities, smart cities, and future cities involving people and technology interactions across time (past – present – future).

The research question identified in Section 1 is reformulated as a proposition for exploration in this chapter, as follows:

P5: *Literacies for urban life in smart cities are needed to support understandings of elements such as the digital, the algorithmic, play, and inclusion*

The research design used in exploring the proposition under study in this chapter is described in the Methodology section.

Figure 1. Literacies for urban life and the ambient in smart cities: conceptual framework

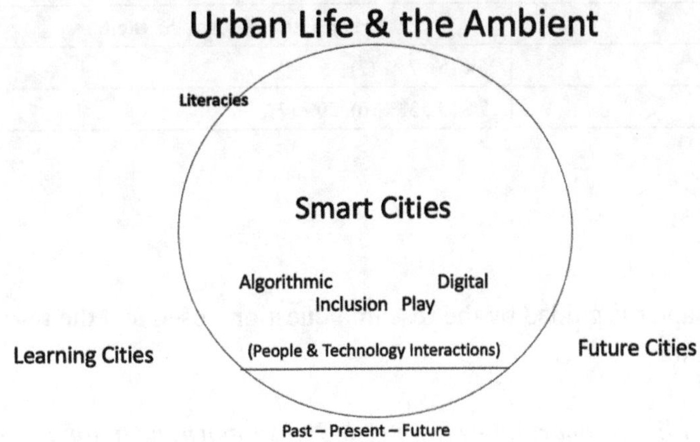

5. METHODOLOGY

Sections 5.1 to 5.3 describe the research design for this chapter in terms of the process used, sources of evidence for data collection, and data analysis.

5.1 Process

Considering the contemporary nature of the inquiry in this chapter, an exploratory case study approach is used (Yin, 2018) and combined with an explanatory correlational design since the two are said to be complementary (Yin, 2018) in that the former may be "needed to examine the underlying processes that might explain a correlation." Using an online space to describe the study and invite people to participate, interest emerged from individuals in a variety of cities in multiple countries (Canada, United States, Greece, Ireland, Israel, etc.) reaching several continents.

5.2 Sources of Evidence

An online website was used enabled the gathering of demographic data from individuals during sign up for the study, and this included age range, gender, and city. Additionally, people could self-identify in one or more categories (e.g., business, community member, student, educator, etc.). A survey instrument, pre-tested prior to use, was used in this study and contained closed and open-ended questions. Survey questions pertained to a range of items related to smart cities, that are relevant to the algorithmic, the digital, inclusion, and play. Survey respondents were also invited to participate in an in-depth interview which was guided by an interview protocol that had been pre-tested prior to use in this study.

Additionally, data were also gathered systematically through group and individual discussions, conducted in parallel with this study, across many sectors (e.g., students, faculty, and staff in higher education, business, community members and leaders, and city staff and leaders) in Canadian cities (e.g., Toronto, Vancouver, Victoria).

Table 7. Assessments for urban literacies and the digital in smart cities: education and placemaking

Items	Urban Literacies and the Digital	Correlation
Education	25% (6); 75% (7)	.57
Placemaking	17% (5); 33% (6); 50% (7)	

5.3 Data Analysis

Data analysis in this chapter is guided by the research question posed and the reformulated proposition for exploration, as follows:

P5: *Literacies for urban life in smart cities are needed to support understandings of elements such as the digital, the algorithmic, play, and inclusion.*

For the analysis of quantitative survey data, the use of descriptive statistics pertaining to literacies for urban life and the ambient in smart cities along with other associated elements enabled exploration of items such as *interactive public spaces* in relation to *ICTs (Information and Communication Technologies).* The correlation feature of the Real Statistics Resource Pack containing an add-in for Microsoft Excel (Zaiontz, 2022) enabled the conducting of a Spearman Correlation Coefficient for ordinal data in exploring the nature of relationships between items.

Content analysis was used for qualitative data gathered through open-ended survey questions, interviews, and group and individual discussions so that inductive analysis from data collection was combined with deductive analysis using terminology from the literature review. Overall, an analysis was conducted for n=79 consisting of 42% females and 58% males for people ranging in age from their 20s to 70s.

6. RESULTS AND DISCUSSION

Table 7 provides the results of assessments for *urban literacies* pertaining to *the digital* using the proxies of *education* and *placemaking.* In response to the survey question, to what extent elements such as *education* contribute to the success of a smart city project, on a 7-point Likert type scale, where 1= Not at all and 7=Absolutely, 25% of respondents selected position 6 (Sure) on the scale while 75% responded at the upper end of the scale at position 7. When asked to what extent smart cities are associated with factors such as *placemaking*, 17% of survey respondents selected position 5 (Sort of), 33% responded toward the upper end of the scale at 6 (Sure), and 50% responded at the upper end position 7.

Using the correlation feature of the Real Statistics add-in for Microsoft Excel (Zaiontz, 2022), a Spearman correlation coefficient for ordinal data was conducted on the assessment results of the two items, *education* and *placemaking* (Table 7), showing a finding of .57.

Exploring further, as shown in Table 8, when asked to what extent smart cities are associated with *interactive public spaces*, 17% of survey respondents selected position 5 (Sort of), 17% responded toward the upper end of the scale at 6 (Sure), and 66% responded at the upper end position 7. When asked to what extent smart cities are associated with *inclusiveness*, 34% of survey respondents selected position 3 (Maybe) and 66% responded at the upper end position 7.

Table 8. Assessments for urban literacies in smart cities: interactive public spaces and inclusiveness

Items	Urban Literacies and Inclusion	Correlation
Interactive Public Spaces	17% (5); 17% (6); 66% (7)	.36
Inclusiveness	34% (3); 66% (7)	

Table 9. Assessments for urban literacies in smart cities: interactive public spaces and ICTs

Items	Urban Literacies and Play	Correlation
Interactive Public Spaces	17% (5); 17% (6); 66% (7)	.57
ICTs	50% (5); 50% (7)	

Using the correlation feature of the Real Statistics add-in for Microsoft Excel (Zaiontz, 2022), a Spearman correlation coefficient for ordinal data was conducted on the assessment results of the two items, *interactive public spaces* and *inclusiveness* (Table 8), showing a finding of .36.

Continuing the exploration, as shown in Table 9, to consider *play* and possibly the *algorithmic,* when asked to what extent elements such as *ICTs* (Information and Communication Technologies) contribute to the making of smart cities, 50% responded at position 5 on the scale (Sort of) and 50% responded at the upper end position 7.

Using the correlation feature of the Real Statistics add-in for Microsoft Excel (Zaiontz, 2022), a Spearman correlation coefficient for ordinal data was conducted on the assessment results of the two items, *interactive public spaces* and *ICTs* (Table 9), showing a finding of .57.

However, when *education* is correlated with *interactive public spaces* a negative or inverse relationship emerges showing a Spearman correlation coefficient of -.33 (Table 10).

Similarly, when *education* is correlated with *ICTs* a negative or inverse relationship of -.57 emerges (Table 11). It may be that elements such as *education* in relation to *interactive public spaces* are still at an underdeveloped stage, possibly being assisted by efforts such as *placemaking*, which, when correlated with *education*, a stronger correlation emerges at .57 (Table 7).

Correlations in the .35 to .65 range are said to be "the typical values in prediction used to identify variable membership in the statistical procedure of factor analysis" (Creswell, 2018). As such, the correlations that emerged for variables in this chapter would seem to provide opportunities going forward for further explorations, testing, and validation for both research and practice with these and possibly other variables.

Qualitatively, it is worth noting that an educator commented that "learning in the most natural informal sense thrives on intrinsic motivation to find things out" adding that "ways of doing this to support

Table 10. Assessments for urban literacies in smart cities: education and interactive public spaces

Items	Urban Literacies and Smart Cities	Correlation
Education	25% (6); 75% (7)	-.33
Interactive Public Spaces	17% (5); 17% (6); 66% (7)	

Table 11. Assessments for urban literacies and the digital in smart cities: education and ICTs

Items	Urban Literacies and the Digital	Correlation
Education	25% (6); 75% (7)	-.57
ICTs	50% (5); 50% (7)	

initiatives are many but certainly lend themselves to the smart city as a playful and open territory of personal and community development." A community leader commented that "smart cities are those that are citizen-centric and inclusiveness-minded."

As depicted graphically in Fig. 2, the interconnectivity of literacies with the algorithmic, the digital, and inclusion and play give way in the context of urban life and the ambient to the notion and potentials of ambient learning, ambient inclusion, ambient literacies, and ambient play.

7. LIMITATIONS AND FUTURE DIRECTIONS

Possible limitations of this work are discussed in this section opening the way to challenges and opportunities for future research and practice directions.

Limitations

Possible limitations of this chapter pertain to the extent of the literature review, and this is mitigated by the potential to extend research literature reviews going forward to focus more broadly and deeply on urban literacies in smart cities, learning cities, and future cities in relation to the algorithmic, the digital, inclusion and play, and to other potential areas. The small sample size for the underlying study of this work is among other limitations of this chapter and as such, caution is encouraged in the interpretation of correlations calculated from the assessments of literacies and the variables used. Going forward, opportunities may emerge for the use of other variables or proxies, or for more direct exploration of

Figure 2. Urban life and the ambient – literacies in smart cities

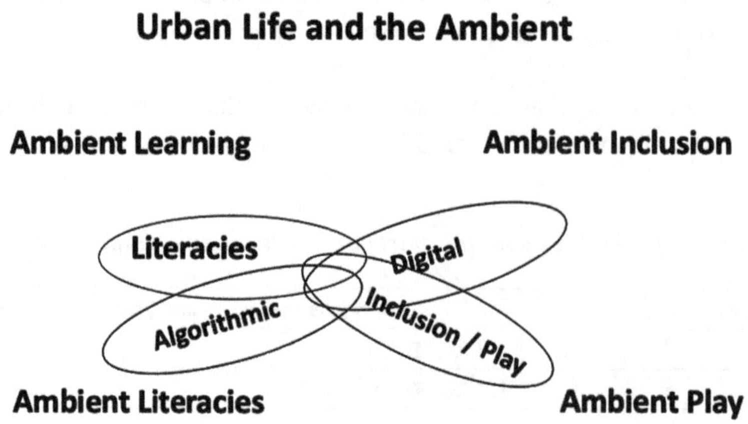

the algorithmic, the digital, inclusion and play, resulting in possibly stronger correlations with greater prediction potentials.

Future Research Directions and Opportunities

For research, future directions emerge in terms of the potential for conducting additional exploration, testing, and validation on the variables identified in this work relevant to urban literacies such as *education* and *placemaking*. Further, opportunities for explorations going forward exist for the identification of other and possibly stronger variables relevant to urban *interactive public spaces* and *ICTs* (information and communication technologies) that foster further learning, inclusiveness, and to other literacies identified in Table 5 for smart cities, learning cities, and future cities. A pathway is also provided for further explorations of ambient learning, ambient inclusion, ambient literacies, and ambient play.

Future Practice Directions and Opportunities

Future directions for practice include the potential for pursuing explorations with people pertaining to variables of interest to them identified in this work relevant to urban literacies and *placemaking*. Regarding the algorithmic, the digital, inclusion, and play, many possibilities are available for exploration such as computing literacies or playable city (Table 5). Opportunities going forward for practice include the potential for exploring spaces with other, possibly stronger variables relevant to *interactive public spaces* and *ICTs* that foster real world action and learning in everyday contexts in smart cities, learning cities, and future cities, in addressing the issues, controversies, and problems presented in Table 6.

8. CONCLUSION

A key contribution of this chapter is identification of the challenges and opportunities associated with literacies in urban life emerging from the review of the research literature provided from multiple perspective involving the algorithmic, the digital, inclusion, and play. Another contribution of this chapter is the conceptual framework for literacies for urban life and the ambient in smart cities that is formulated, based on the literature review where emerging issues, controversies, and problems provide guidance on the proposition explored enabling operationalization of the framework for use in this chapter. Through an exploratory case study approach, together with an explanatory correlational design, the proposition under study in this chapter is investigated. Based on survey data, correlations are conducted for urban literacies employing the use of proxies, where, among the findings is that for *placemaking* and *education* in the form of a Spearman correlation coefficient of .57, said to be "useful for limited prediction." Regarding limitations of the study underlying this chapter, the small sample size is to be noted and as such, caution is urged in the interpretation of findings, opening opportunities for further testing. Notwithstanding this and other possible limitations, future directions guiding research and practice activities going forward are provided in terms of further validating the variables explored in this chapter while also encouraging activities to identify other, additional factors relevant to urban literacies and to the notions of the algorithmic, the digital, inclusion, play, and so on, in the context of smart, learning, and future cities.

Key insights and ideas emerging from this chapter are provided here, along with exercises encouraging further exploration, as follows:

Insights emerging from this chapter include the following:

1. The importance of elements contributing to the success of a smart city project as factors giving rise to many possibilities for *urban literacies* in the form of *education*
2. The importance of factors associated with smart cities such as *placemaking* and *interactive public spaces* and *inclusiveness*
3. The potential for many more factors to emerge for understanding *urban literacies* in smart cities such as *play* and *the algorithmic*

Ideas emerging from this chapter include the following:

1. Potential opportunities for exploring urban literacies in relation to a variety of variables (e.g., the algorithmic, the digital, inclusion, and play) in the context of smart cities
2. The opportunities potential for the identification of other and additional variables for exploring urban literacies in smart environments
3. The use of an interdisciplinary and multi-method approach for the study and practice of urban literacies in smart, learning, and future cities

Exercises for the reader to consider, encouraging further exploration include the following:

1. Identify one or more factors in your city that are giving rise to the need for urban literacies
2. Identify one or more ways in which inclusiveness would enhance the quality of urban life in your city
3. Describe one or more elements that would support urban literacies in your city

The audience for this chapter is interdisciplinary in nature, including but not limited to community members and leaders, educators, businesses, researchers, urban planners, urban designers, and anyone concerned with improving literacies in urban life, while being attentive to elements such as the algorithmic, the digital, inclusion, and play along with associated practice and research opportunities and challenges in smart cities, learning cities, and future cities.

REFERENCES

Beer, D. (2022). The problem of researching a recursive society: Algorithms, data coils, and the looping of the social. *Big Data & Society*, *9*(2). Advance online publication. doi:10.1177/20539517221104997

Beretta, I. (Ed.). (2015). *L'umanesimo della smart city, Inclusione, innovazione,formazione*. Pensa MultiMedia Editore.

Bertran, F. A., Duval, J., Armengol, L. B., Chen, I., Dong, V., Dastoor, B., Bertran, A. A., & Isbister, K. (2021). A catalog of speculative playful urban technology ideas: Exploring the playful potential of smart cities. In Academic Mindtrek 2021. Association for Computing Machinery. doi:10.1145/3464327.3464374

Boddington, P. (2017). *Towards a Code of Ethics for Artificial Intelligence*. Springer. doi:10.1007/978-3-319-60648-4

Borda, A., & Bowen, J. P. (2019). Smart cities and digital culture: Models of innovation. In Museums and Digital Culture. Springer. doi:10.1007/978-3-319-97457-6_27

Börjesson Rivera, M., Ringenson, T., & Pargman, D. (2020). The sustainable playable city: Making way for the playful citizen. In A. Nijholt (Ed.), *Making Smart Cities More Playable. Gaming Media and Social Effects*. Springer. doi:10.1007/978-981-13-9765-3_5

Broadhead, J., & Kierans, D. (2019). *Inclusive cities: Framework - COMPAS*. Centre on Migration Policy and Society, University of Oxford. https://www.compas.ox.ac.uk/2019/inclusive-cities-framework/

Carnemolla, P., Robinson, S., & Lay, K. (2021). Towards inclusive cities and social sustainability: A scoping review of initiatives to support the inclusion of people with intellectual disability in civic and social activities. *City. Cultura e Scuola*, *25*(100398). Advance online publication. doi:10.1016/j.ccs.2021.100398

Chen, T., Gil-Garcia, J. R., & Gasco-Hernandez, M. (2022). Understanding social sustainability for smart cities: The importance of inclusion, equity, and citizen participation as both inputs and long-term outcomes. *Journal of Smart Cities and Society*, *1*(2), 135–148. doi:10.3233/SCS-210123

Correia, D., Feio, J., Teixeira, L., & Lourenço Marques, J. (2021). The inclusion of citizens in smart cities policymaking: The potential role of development studies' participatory methodologies. In N. Streitz & S. Konomi (Eds.), *Distributed, Ambient and Pervasive Interactions. HCII 2021. Lecture Notes in Computer Science* (Vol. 12782). Springer. doi:10.1007/978-3-030-77015-0_3

Creswell, J. W. (2018). *Educational research: Planning, conducting, and evaluating quantitative and qualitative research* (6th ed.). Pearson.

Cutsem, T. V., Mostinckx, S., & Meuter, W. D. (2009). Linguistic symbiosis between event loop actors and threads. *Computer Languages, Systems & Structures*, *35*(1), 80–98. doi:10.1016/j.cl.2008.06.005

de Lange, M. (2019). The playful city: Citizens making the smart city. In M. de Lange, R. Glas, S. Lammes, J. Raessens, & I. de Vries (Eds.), *The Playful Citizen: Civic Engagement in a Mediatized Culture* (Vol. 1, pp. 349–369). Amsterdam University Press. https://www.jstor.org/stable/j.ctvcmxpds.23

Duin, A. H., & Pedersen, I. (2021). Algorithmic writing futures. In *Writing Futures: Collaborative, Algorithmic, Autonomous. Studies in Computational Intelligence* (Vol. 969). Springer. https://link.springer.com/chapter/10.1007/978-3-030-70928-0_3

Elberzhager, F., Mennig, P., Polst, S., Scherr, S., & Phil Stüpfert, P. (2021). Towards a digital Ecosystem for a smart city district: Procedure, results, and lessons learned. *Smart Cities*, *4*(2), 686–716. doi:10.3390martcities4020035

Florida, R., & McLean, J. W. (2017). What inclusive urban development can look like. *Harvard Business Review*. Retrieved November 27, 2021, from https://hbr.org/2017/07/what-inclusive-urban-development-can-look-like

Fredericks, J. (2020). From smart city to smart engagement: Exploring digital and physical interactions for playful city-making. In A. Nijholt (Ed.), *Making Smart Cities More Playable. Gaming Media and Social Effects*. Springer. doi:10.1007/978-981-13-9765-3_6

Gartner. (2021). *A Data and Analytics Leaders's Guide to Data Literacy*. Retrieved 13 September 2022 from https://www.gartner.com/smarterwithgartner/a-data-and-analytics-leaders-guide-to-data-literacy

Glas, R., Lammes, S., de Lange, M., Raessens, J., & de Vries, I. (Eds.). (2019). *The Playful Citizen: Civic Engagement in a Mediatized Culture* (Vol. 1). Amsterdam University Press. https://www.jstor.org/stable/j.ctvcmxpds

Headrick Taylor, K. (2017). Learning along lines: Locative literacies for reading and writing the city. *Journal of the Learning Sciences*, *26*(4), 533–574. doi:10.1080/10508406.2017.1307198

Herczeg, M., Ohlei, A., Schumacher, T., & Winkler, T. (2021). Ambient learning spaces: Systemic learning in physical-digital interactive spaces. In Algorithmic and Aesthetic Literacy. Emerging Transdisciplinary Explorations for the Digital Age. Doi:10.25656/01:22299

Lee, F., & Björklund Larsen, L. (2019). *How should we theorize algorithms? Five ideal types in analyzing algorithmic normativities*. Big Data & Society, Special Issue on Algorithmic Normativities. doi:10.1177/2053951719867349

Lido, C., Reid, K., & Osborne, M. (2018). *Big data, lifelong learning and learning cities: Promoting city-discourse on social inequalities in learning. Briefing Paper 11*. Pascal International Observatory.

Lister, P. J. (2020). Smart learning in the community: Supporting citizen digital skills and literacies. In N. Streitz & S. Konomi (Eds.), Lecture Notes in Computer Science: Vol. 12203. *Distributed, Ambient and Pervasive Interactions. HCII 2020*. Springer. doi:10.1007/978-3-030-50344-4_38

McKenna, H. P. (2017). Re-conceptualizing social inclusion in the context of 21st century smart cities. In J. Choudrie, P. Tsatsou, & S. Kurnia (Eds.), *Social Inclusion and Usability of ICT-Enabled Services*. Routledge. doi:10.4324/9781315677316-4

McKenna, H. P. (2019). *Ambient Urbanities as the Intersection Between the IoT and the IoP in Smart Cities*. IGI Global. doi:10.4018/978-1-5225-7882-6

Merriam-Webster. (2021). *Algorithm*. Retrieved 28 November 2021 from https://www.merriam-webster.com/dictionary/algorithm

Misteli, P., Poulakos, S., Kapadia, M., & Sumner, R. W. (2018). Towards emergent play in mixed reality. *International SERIES on Information Systems and Management in Creative eMedia (CreMedia)*, 51-56. Retrieved 12 November 2021 from https://www.ambientmediaassociation.org/Journal/index.php/series/article/view/281

Morrison, A., Bjørnstad, N. Martinussen, E. S., Johansen, B., Kerspern, B., & Dudani, P. (2020). Lexicons, literacies and design futures. *Temes De Disseny*, *36*, 114-149. doi:10.46467/TdD36.2020.114-149

Mouton, M., Ducey, A., Green, J., Hardcastle, L., Hoffman, S., Leslie, M., & Rock, M. (2019). Towards 'smart cities' as 'healthy cities': Health equity in a digital age. *Canadian Journal of Public Health = Revue canadienne de sante publique, 110*(3), 331–334. doi:10.17269/s41997-019-00177-5

Papangelis, K., Saker, M., Lee, J. H., & Jones, C. (2020). Smart cities at play: Lived experiences, emerging forms of playfulness, and problems of participation. In *Extended Abstracts of the 2020 CHI Conference on Human Factors in Computing Systems (CHI EA '20)*. Association for Computing Machinery. 10.1145/3334480.3375146

Papangelis, K., Saker, M., Leorke, D., Lee, J. H., & Jones, C. (2022). Introduction to smart cities in transition: Challenges of Participation in urban environments. *International Journal of Human-Computer Interaction*, 1–3. Advance online publication. doi:10.1080/10447318.2021.2012385

Qutab, S., Adil, S. A., Gardner, L. A., & Ullah, F. S. (2022). The role of libraries, archives, and museums for metaliteracy in smart cities: Implications, challenges, and opportunities. In M. Taher (Ed.), *Handbook of Research on the Role of Libraries, Archives, and Museums in Achieving Civic Engagement and Social Justice in Smart Cities* (pp. 355–375). IGI Global. doi:10.4018/978-1-7998-8363-0.ch019

Rebernik, N., Szajczyk, M., Bahillo, A., & Goličnik Marušić, B. (2020). Measuring disability inclusion performance in cities using disability inclusion evaluation tool (DIETool). *Sustainability*, *12*(4), 1378. doi:10.3390u12041378

Roche, S. (2017). Geographic information science III: Spatial thinking, interfaces and algorithmic urban places – Toward smart cities. *Progress in Human Geography*, *41*(5), 657–666. doi:10.1177/0309132516650352

Sessa, M. (2020). Home smart home: A sociology of living in the age of reflective materialism. In R. Iannone & R. Gurashi (Eds.), *Smart Society: A Sociological Perspective on Smart Living*. Routledge.

Setlur, V. (2022). Navigating data literacy in the world of augmented analytics. *Forbes*. Retrieved September 19, 2022, from https://www.forbes.com/sites/tableau/2022/07/26/navigating-data-literacy-in-the-world-of-augmented-analytics/?sh=5f646455436b

Tippens Reinitz, B., McCormack, M., Reeves, J., Robert, J., Arbino, N., Anderson, J., Hamman, J., Johnson, C., Kew-Fickus, O., Snyder, R., & Stevens, M. (2022). *2022 EDUCAUSE Horizon Report, Data and Analytics Edition*. EDUCAUSE.

UNESCO. (2021). *Literacy*. United Nations Educational, Scientific, and Cultural Organization (UNESCO). Retrieved 28 November 2021 from https://en.unesco.org/themes/literacy

UNESCO-ILL. (2020). *Embracing a culture of lifelong learning: Contribution to the Futures of Education Initiative – Report: A transdisciplinary expert consultation*. UNESCO, Institute for Lifelong Learning. Retrieved 17 May 2022 from https://unesdoc.unesco.org/ark:/48223/pf0000374112

UNESCO-ILL. (2021). *Learning cities: Drivers of inclusion*. UNESCO Institute for Lifelong Learning. Retrieved 26 November 2021 from https://uil.unesco.org/lifelong-learning/learning-cities/learning-cities-drivers-inclusion-and-sustainability

University of Waterloo. (2022). Waterloo launches future cities initiative. In *News*. University of Waterloo. Retrieved 26 September 2022 from https://uwaterloo.ca/environment/news/waterloo-launches-future-cities-initiative

Vygotsky, L. (1978). *Mind in Society: The Development of Higher Psychological Processes*. Harvard University Press.

Yin, R. (2018). *Case study research and applications: Design and methods* (6th ed.). Sage Publications.

Ylipulli, J., Kangasvuo, J., Alatalo, T., & Ojala, T. (2016). Chasing digital shadows: Exploring future hybrid cities through anthropological design fiction. *NordiCHI'16 Proceedings of the 9th Nordic Conference on Human-Computer Interaction, 78*, 1-10. 10.1145/2971485.2993923

Zaiontz, C. (2022). *Real statistics using excel.* Retrieved from www.real-statistics.com

Zambonelli, F., Salim, F., Loke, S. W., De Meuter, W., & Kanhere, S. (2018). Algorithmic governance in smart cities: The conundrum and the potential of pervasive computing solutions. *IEEE Technology and Society Magazine, 37*(2), 80–87. doi:10.1109/MTS.2018.2826080

ADDITIONAL READING

Bousquet, C. (2018). Cities should improve their algorithms, not abandon technology. *Government Technology Blog.* Retrieved 13 May 2022 from https://www.govtech.com/computing/cities-should-improve-their-algorithms-not-abandon-technology.html

Dasgupta, S., & Hill, B. M. (2021). Designing for critical algorithmic literacies. In Algorithmic Rights and Protections for Children. doi:10.1162/ba67f642.646d0673

Horrigan, J. (2017). Fostering digital inclusion in smart cities. *items: Insights from the Social Sciences.* Retrieved 26 November 2021 from https://items.ssrc.org/parameters/fostering-digital-inclusion-in-smart-cities/

Kids Code Jeunesse. (2022). *The Algorithm and Digital Literacy Project.* Kids Code Jeunesse, CCUNESCO, and UNESCO. Retrieved 13 May 2022 from https://algorithmliteracy.org

Komninos, N., & Kakderi, C. (Eds.). (2019). *Smart Cities in the Post-Algorithmic Era: Integrating Technologies, Platforms and Governance*. Edward Elgar Publishing. doi:10.4337/9781789907056

Lipton, B. (2019). Five cities, five different approaches to algorithms in government: Jurisdictions of all sizes are dealing with the advent of unhuman decision-making. *Better Cities Project.* Retrieved 13 May 2022 from https://better-cities.org/criminal-justice-public-safety/five-cities-five-different-approaches-to-algorithms-in-government/

Mattern, S. (2021). Introduction cities, trees & algorithms. In *A City Is Not a Computer: Other Urban Intelligences* (pp. 1–17). Princeton University Press. doi:10.2307/j.ctv1h9dgtj.4

New London Group. (1996). A pedagogy of multiliteracies: Designing social futures. *Harvard Educational Review, 66*(1).

Oxford Urbanists. (2022). *New literacies for a new urbanism.* Retrieved 13 May 2022 from https://www.oxfordurbanists.com/new-urban-literacies

Pyles, D. G., Rish, R. M., & Warner, J. (Eds.). (2019). *Negotiating Place and Space through Digital Literacies*. Information Age Publishing.

KEY TERMS AND DEFINITIONS

Ambient Inclusion: Ambient inclusion in the context of smart urban environments refers to more adaptive, dynamic, and in-the-moment approaches to enabling access to resources, services, and support for involvement and participation.

Ambient Literacies: Ambient literacies refer to more dynamic ways of learning, knowing, communicating, and becoming aware in the context of smart urban environments.

Ambient Play: Ambient play refers to more adaptive, playful, and in-the-moment forms of interacting with each other and with technologies in urban spaces and regions.

Digital Literacies: Digital literacies in the context of urban life in smart cities refer to the broad range of skills required to interact and function in technology-pervasive environments.

Information and Communication Technologies (ICTs): ICTs in the context of smart cities refer to the broad range of technologies that enable and support the communication, transfer, storage, use, and management of information.

Interactive Public Spaces: Interactive public spaces in the context of smart cities refer to urban environments and regions that foster meaningful engagement.

Placemaking: Placemaking in urban environments refers to a cross-sector collaborative approach to the planning, design, use, reuse, and management of public spaces attentive to health and well-being and quality of life.

Chapter 6
Data, Information, Knowledge, Ways of Knowing, and Computational Intelligence in Smart Cities

ABSTRACT

The purpose of this chapter is to explore evolving understandings of data, information, and knowledge in smart cities including ways of knowing and computational intelligence. A review of the research literature is conducted in the context of smart cities. Issues, controversies, and problems emerging from the literature review are highlighted, contributing to formulation of a conceptual framework for urban life and the ambient with the evolving of data, information, and knowledge in smart cities. Using an exploratory case study approach combined with an explanatory correlational design, variables of interest are identified including information and communication technologies (ICTs), meaningfulness, and the use of visualizations while the nature of their relationships are assessed. Through the lens of emerging forms of data, information, and knowledge, efforts are made to inform understandings of smart, learning, and future cities while pointing to directions for research and practice.

1. INTRODUCTION

Schmitt (2015) develops the notion of "information cities" and employs "the metaphor of an urban metabolism" along with the notion of "the stocks and flows of a city related to its people", as well as to materials, water, energy, finances, health, density, and information. Iaconesi and Persico (2017) describe the idea of digital urban acupuncture (DUA) focusing on the flows of data and information in urban environments and "the resulting flows of knowledge, wisdom, emotion, and, in general, communication" in support of "insights about how to make interventions, creating a better, more inclusive, participatory, collaborative place." Bibri (2021) provides an extensive literature review of "sustainable cities and smart cities in terms of their integration as the leading global paradigm of urbanism" contributing to potentials

DOI: 10.4018/978-1-6684-4096-4.ch006

and understandings of "data-driven smart sustainable cities of the future." As such, this chapter is significant in that it explores data, information, knowledge, ways of knowing, and computational intelligence in the context of urban life and the ambient in smart cities, learning cities, and future cities, giving rise to the motivation for this work described in the following objectives.

Objectives: The main objectives of this chapter are to a) explore evolving understandings of data, information, and knowledge in the context of smart cities; b) formulate a conceptual framework for urban life and the ambient in relation to data, information, and knowledge in smart cities; and c) explore the nature of the relationship between variables associated with data, information, and knowledge as a way of learning more about evolving understandings of urban life and the ambient in smart cities. These objectives give rise to the main research question under exploration in this chapter – *How are data, information, and knowledge related in urban life in smart cities?*

2. BACKGROUND

Carrillo (2006) introduces the notion of knowledge cities, providing perspectives, examples, and approaches where chapter contributions pertain to the development of knowledge citizens, frameworks, and a taxonomy, to name a few. Batty (2013) advanced a new way of understanding cities, beyond just places and spaces, to systems and information flows across networks, and digital information. Schmitt (2015) proposed the notion of "information cities" emerging from the Future Cities Laboratory, ETH Zurich, where data, information, and knowledge are identified as the key components of information architecture for a city. Mattern (2019) claims that "our cities, past and present, mediate between various manifestations of intelligence" such as "legal codes and copper cables, inscriptions and imaginaries, algorithms and acoustics, public proclamations and system protocols." The United Nations Human Settlements Programme articulates the movement towards knowledge-based urban futures (UN-Habitat, 2022) in support of sustainable and equitable urbanization where responsible innovation, frontier technologies (e.g., drones, autonomous vehicles), and inclusion matter.

2.1 Definitions

Definitions from the research literature are provided for key terms used in this chapter.

- **Data:** Schmitt (2015) describes data "in the context of the city" as "the smallest entities of information" and "as values given to objects, expressions, functions, or properties."
- **Information:** According to Schmitt (2015), "information sets data in relation to each other" and "consists of data and connections." As such, information "is an important abstraction" and is referred to as "a virtual material, important for the information age and for the information society."
- **Knowledge:** Schmitt (2015) claims that knowledge emerges as "a result of connecting data and information."
- **Computational Intelligence:** While computational intelligence (CI) has been described (IEEE, 2021) as "the theory, design, application and development of biologically and linguistically motivated computational paradigms" drawing on the pillars of "neural networks, fuzzy systems and evolutionary computation" it is also said to be "an evolving field" which "encompasses comput-

ing paradigms like ambient intelligence, artificial life, cultural learning, social reasoning" and the like, extending to deep learning as "a core method for artificial intelligence."

2.2 Overview

This chapter consists of several sections containing a review of the research and practice literature for urban life and data, information, knowledge, and knowing and computational intelligence (CI) in smart cities. From this review, a highlighting of key issues, controversies, and problems is provided and a conceptual framework for evolving understandings of data, information, and knowledge in smart cities in relation to urban life and the ambient is formulated and operationalized for use in this chapter. The main research question under exploration in this chapter is reformulated as a proposition for investigation. The methodology for investigation of the proposition in this chapter is described, followed by the presentation of results and a discussion of findings. Limitations of the study underlying this chapter are identified and future directions for research and practice are outlined. The conclusion provides an overview of outcomes, the identification of insights and ideas, development of several exercise questions encouraging further exploration, and identification of the target audience for this chapter.

3. PERSPECTIVES ON DATA, INFORMATION, AND KNOWLEDGE FOR URBAN LIFE AND THE AMBIENT IN SMART CITIES

This chapter provides a review of the of the research and practice literature for evolving understandings of data, information, knowledge, and knowing and computational intelligence (CI) in smart cities in relation to urban life and the ambient.

3.1 Urban Life and Data

McKenna (2017) addresses "issues associated with the sharing and using of data for learning and solution making in smart cities" through the constructs of awareness, learning, openness, and meaningful engagement, in exploring everyday understandings of data and civic technologies through people and their voices, contributing to theorizing of the ambient data concept in everyday urban public spaces. Glaeser, Kominers, Luca, and Naik (2018) explore the "promises and limitations of improved measures of urban life" for "urban research and policy making" finding that "big data can meaningfully help big cities and improve research on cities" with the caveat that "it is used with thoughtful care." Osborne, Houston, and Lido (2018) reveal that learning cities over time involved elements such as place, culture, health, well-being, social cohesion, and inclusion while contemporary rendering of big data gives way to "complex sources of information created and stored by organisations, including traditional databases, and more technologically diverse sources of data, such as social media, geographic mapping data and weather." Bibri (2019) argues that "smart sustainable cities are becoming knowable, controllable, and tractable in new dynamic ways" and are said to be "responsive to the data generated about their systems and domains" through "reacting to the analytical outcome" in support of "optimizing and enhancing operational functioning, management, planning, design, development, and governance in line with the goals of sustainable development." Webb, Holford, Hodge, Milana, and Waller (2020) note that "researchers are recognizing that learning cities need to develop their own capacities, datasets and measurements that build on local engagement" while "drawing on adult education concepts, such as public pedagogy and intertwining data collection about education and learning." Cheshire and Uberti (2021) employ the visual use of data to provide views of the past, present, and future in the form of maps and graphics as a way of seeing the world. Bibri (2021) explores "data-driven smart sustainable cities" providing a review of the research literature, along with "social and environmental sustainability" while highlighting concerns with the future of "policy and governance" and whether it "will become too technocentric and technocratic." Karimi, Haghi Kashani, Akbari, and Mahdipour (2021) provide a systematic review of approaches to the handling of big data in smart cities over an eight year period (2013-2021), highlighting in findings factors such as scalability, availability, and accuracy while making suggestions for researchers going forward pertaining to challenges, open issues, and future trends. Wallace (2022) outlines the challenges and opportunities associated with Connected Vehicle Data (CVD) for smart cities, claiming it to be important for planning and decision making in real time while addressing issues of trust, cybersecurity, and the like.

3.2 Information and Urban Life in Smart Cities

Gil-Garcia and Aldama-Nalda (2013) describe the notion of achieving smarter city potentials as being associated with factors such as connectedness, responsiveness, efficiency, and sustainability. As such, Gil-Garcia and Aldama-Nalda (2013) argue that through "information integration initiatives" it is possible that "cities can achieve some of the goals and objectives extensively identified with smartness" irrespective of challenges "if there is enough political support" from mayors and other government executives, contributing to infrastructure, funding, and administrative support. de Jong, Joss, Schraven, Zhan, and Weijnen (2015) identify information cities as one of several key categories from the research literature with as yet, "limited resonance as [a] stand-alone concept" based on usage in the research literature to date" although it is said to have connections with the much stronger sustainable city category and as such, jointly with other categories such as intelligent city and ubiquitous city, a digital sub-network emerges. Schmitt (2015) claims that "information city" is a description of "the extension of information architecture to the urban scale" and as such, "might become a metaphor for the structuring and ordering of vast amounts of data, created increasingly by the city's inhabitants and its infrastructure." However, Schmitt (2015) acknowledges that the Future Cities Laboratory (FCL), working with the problem of "more data and more computing power will not solve the old questions" that cities are struggling with, since "each level of more analytical accuracy is increasing the computation costs exponentially" and as such, employs approaches such as "pre-specific modelling."

Thompson, Greenhalgh, Muldoon-Smith, Charlton, and Dolnik (2016) use city information modelling (CIM) case studies in the exploration of planning and governance in smart cities focusing on four themes pertaining to data – accessibility and availability, accuracy and consistency, manageability, and integration "designed to provoke discussion in relation to the exploitation and improvement of data modelling and visualization in the urban planning discipline." Noting that "data alone does not guarantee delivery of a sustainable urban future", Thompson et al. (2016) point to the need for research going forward to focus on the "interpretation and use of data." Monteiro, Costa, Pina, Santos, and Ferrão (2018) describe an information system for smart cities in support of sustainable urban planning in the form of an urban building database (UBD) for energy modelling. McKenna (2019) refers to ambient information architectures as "more dynamic, adaptive, and open" while Ismagilova, Hughes, Dwivedi, and Raman (2019) provide an information systems perspective on smart cities, highlighting advances in research while noting issues ranging from security to privacy to migrants "and how their citizenship rights could improve the sustainability of cities." From a management science perspective Attour, Dominguez-Péry, and Bendavid (2021) identify current and future trends for smart cities research using a bibliometric analysis focusing on the role of information technology (IT) while identifying three streams of current and future research – design, technological challenges in fostering the development of smart cities; digital innovations and knowledge; and the ethics of data management. Lytras, Visvizi, Chopdar, Sarirete, and Alhalabi (2021) propose an approach "to capture users' views and perceptions of smart city services and applications" as a way to "enrich the decision- and policy-making processes" using scale development focusing on the five dimensions of technology anxiety, work-life interface, engagement orientation, support orientation, and quality of life. The study is said by Lytras et al. (2021) to have "theoretical and managerial implications" pertaining to guidance on the management of information by "stakeholders involved in the making of a smart city" as well as for "technology experts and policymakers." Noting that "collecting, processing, and distributing information has always been a core function of government", Landsbergen, Girth, and Westover-Muñoz (2022) explore urban governance in relation to the managing of smart city information through formulation of "a new framework to identify how cities can select the appropriate governance rules to facilitate the political, financial, and operational sustainability of their IDEs" as in integrated data exchanges, as data platforms, in support of smart city efforts in "curating, managing, and disseminating information."

3.3 Urban Life and Knowledge in Smart Cities

Based on a bibliometric analysis of the research literature, de Jong et al. (2015) explore the conceptual differences and interrelationships between key city categories, one of which is that of knowledge city, which they identify as a distinct concept having "an interconnectedness" with "sustainable city, eco-city, and low carbon city" while having "largely economic innovation-oriented connotations" emphasizing the importance of understanding such terms in relation to "their implications for urban development and regeneration policy and practice." Negre, Rosenthal-Sabroux, and Gascó (2015) propose a "knowledge-based conceptual vision of the smart city" focusing on information and knowledge created by people in support of improving decision-making while enhancing business processes. Dattakumar and Sharma (2016) claim that in the rush to create smart cities "we may have omitted a synergistic but distinct notion of creating knowledge societies." As such, Dattakumar and Sharma (2016) explore the relationships between smart cities and knowledge societies presenting a maturity model involving human values, participation, usage, and access on one axis and governance, infrastructure, education and human capital, and innovation on the other. Dattakumar and Sharma (2016) conclude by proposing that going forward, research be focused upon "intelligent digital infrastructure and its role in reducing knowledge disparities and encouraging sustainable development." McFarlane and Söderström (2017) identify steps for the making of alternative smart cities in the form of knowledge-intensive smart urbanism with the aim of producing "usable knowledge."

Introducing a special issue of *Big Data & Society* focusing on knowledge production, Bonde Thylstrup, Flyverbom, and Helles (2019) point to the move away from claims made in favor of "datafication" or Big Data to explorations focused on more "contextual and conceptual enquiries into the particular conditions of knowledge production to which digital transformations and processes of datafication give rise" with considerations of questions such as "what counts as knowledge?" and "what are the resources that go into the production of knowledge?" and the methods and procedures and so on. From an information systems (IS) perspective, Ismagilova et al. (2019) note that the smart city is also referred to as a "Knowledge-based City" where emergent themes of a literature review contribute to "advances in bodies of knowledge" while pointing to areas for further research. Kurgan and Brawley (2019) provide insight into the "techniques and technologies of knowledge about the tools and practices through which cities are understood" such as mapping, both "digital and predigital" as mechanisms to "actively redraw, rebuild, and remake cities." Borba. Palacios, Luciano, and Chaves (2020) explore associations between knowledge management (KM) and smart cities using a semi-systematic literature review while proposing a research agenda on knowledge management for smart cities. Looking at the research literature for environmental open data in urban platforms, Gessa and Sancha (2020) explore the data life cycle of "acquisition, access, analysis, and application" finding that while "interest exists in providing environmental information" it would seem that "the capability to transform the data [and] information into knowledge relevant for cities is rather low." Allam et al. (2022) employ a bibliometric analysis and mapping to explore smart urban governance in terms of knowledge structure and trends finding "the need for citizen participation in urban policies" as well as "sustained innovation for e-participation, e-governance, and policy frameworks" with implications for researchers and practitioners in efforts pertaining to policymaking, governance, and inclusion.

3.4 Urban Life, Knowing, and Computational Intelligence in Smart Cities

Campbell (2012) explores "the way cities learn" and argues for the "learning side of urban development" and the importance of making "learning part of the mission of governance" and "acquiring knowledge" while using the power of connections and connecting for learning and innovation. Caprotti (2017) addresses the challenges for 'knowing' the smart city in view of the invisible nature of underground cables and the like. Kurgan and Brawley (2019) argue that "technology mediates how we know and experience cities" adding that "the nature of this mediation has always been deeply political" with "the deployment of data" said to be "at the forefront of projects to grasp and reshape urban life." Through what is said to be "a deeply interdisciplinary conversation", Kurgan and Brawley (2019) provide a work that "considers the role of technology in generating, materializing, and contesting urban epistemologies." For example, Kurgan and Brawley (2019) describe algorithms as "another way of knowing cities" such that "learning how to read data and its effects–and developing strategies and tactics for challenging its authority and its biases" is said to be "one of the most essential political tasks of our moment." It is worth noting that Kurgan and Brawley (2019) claim that "cities have never been just collections of people and buildings" and are also "dynamic networks of relationships that generate data, operate with data, and are transformed by data–for better and for worse." As such, Kurgan and Brawley (2019) argue that "data needs to be understood, harnessed, confronted, and critically examined." And so, from an "archaeology of urban intelligence" perspective, Mattern (2019) notes that "yesterday's cities–even our earliest human settlements–were just as smart, although theirs was an intelligence less computational and more material and environmental." According to Mattern (2019), "we inhabit a data space defined by various level of intersecting protocols that direct our connections, facilitate or close off access, and thus subtly shape the geographies–both informational and physical–we are then able to explore." Karvonen (2020) identifies "knowing" as one of several ways "that techno-politics will be undertaken in the coming decades" along with "governing and imagining cities." For Karvonen (2020), techno-politics refers to "the rise of smart cities and the digitalization of urban life" in combination with "the myriad debates about climate and environment, social cohesion and democracy, and economic prosperity and

Table 1. Overview of research for data in smart cities

Author(s)	Year	Data
Iaconesi & Persico	2017	Digital urban acupuncture, flows of data & information, collaborative
McKenna	2017	Ambient data – awareness, learning, openness, meaningful engagement
Glaeser et al.	2018	Big data for big cities improves research if used with thoughtful care
Osborne et al.	2018	Big data & technologically diverse sources in learning cities
Bibri	2019	Smart sustainable cities, dynamic, responsive, data analytics, planning
Webb et al.	2020	Datasets for learning cities, public pedagogy, data collection & learning
Karimi et al.	2021	Leveraging of big data in SCs – accuracy, availability, scalability, time
Bibri	2021	Data-driven smart sustainable cities, policy & governance, technocentric
Cheshire & Uberti	2021	Visual use of data to show past, present, future in maps of the world
Wallace	2022	Practice perspective on Connected Vehicle Data (CVD) for SCs

well-being." Karvonen (2020) adds that "contentious debates" extend as well to "decarbonization and resilience, citizen empowerment and protest, surveillance, and security, health and well-being, identity and culture, entrepreneurship and economic development" and the list goes on. According to Karvonen (2020), the "new urban science" involves the use of "digital tools and methods to know cities more intimately" citing Batty (2013), Townsend (2015), and Derudder and van Meeteren (2019) to name a few, and the potential "to increase urban knowledge through the application of big data, ubiquitous sensing, geospatial and social network analyses, algorithms, machine learning, and artificial intelligence." However, Puron-Cid and Gil-Garcia (2022) question the costs in the long term of smart cities in relation to ICT (information and communication technologies) infrastructure.

3.5 Literature Review Overview

An overview by author and date is provided in this section for the research and practice literature for data, information, knowledge, and knowing and computational intelligence (CI).

Table 1 provides an overview by author and date of the research and practice literature for urban life and the ambient pertaining to data, spanning the timeframe of 2017 to 2022.

Iaconesi and Persico (2017) describe the notion of digital urban acupuncture (DUA) focusing on urban flows of data and information, in support of insights for collaborative, participatory, inclusive, and other initiatives. McKenna (2017) explores the notion of ambient data with a focus on awareness, learning, openness, and meaningful engagement based on everyday understandings. Glaeser et al. (2018) argue that thoughtful care is required in the use of big data to meaningfully help big cities and improve research on cities. Osborne et al. (2018) describe big data as providing more technologically diverse data sources in learning cities. Bibri (2019) claims that smart sustainable cities are becoming dynamic and responsive, among other things through data analytics for planning, governance, and the like. Webb et al. (2020) address data in relation to learning cities and the need for datasets "that build on local engagement" as well as "public pedagogy and intertwining data collection about education and learning." Bibri (2021) provides an exploration of data-driven smart sustainable cities with concerns about the future of policy and governance and the possibility of techno-centrism. Cheshire and Uberti (2021) employ the visual

Table 2. Overview of research for information in smart cities

Author(s)	Year	Information
Gil-Garcia & Aldama-Nalda	2013	Information integration initiatives for smarter cities & political support
de Jong et al.	2015	Information cities as a limited category & sub-network of sustainable
Schmitt	2015	Information city, data generation by inhabitants and infrastructure
Thompson et al.	2016	City information modelling for urban planning & governance
Monteiro et al.	2018	Urban building database as an info system for energy modelling in SCs
McKenna	2019	Ambient information architectures – dynamic, adaptive, open
Ismagilova et al.	2019	IS perspective, security & privacy issues, migrants & sustainability
Attour et al.	2021	Research trends – design, innovations & ethics
Lytras et al.	2021	Information management in SCs – anxiety, engagement, QoL
Landsbergen et al.	2022	Managing smart city information for urban governance

use of data to provide a view of the past, present, and future of the world in maps and graphics. Karimi et al. (2021) provide a review of approaches to the handling of big data in smart cities, highlighting the prevalence of factors such as scalability, availability, time, and accuracy while making suggestions for future research addressing challenges, open issues, and future trends. Wallace (2022) advances a practice perspective on the importance and value of Connected Vehicle Data (CVD) for smart cities involving technology, people, and urban planners.

Table 2 provides an overview by author and date of the research and practice literature for urban life and the ambient pertaining to information, spanning the timeframe of 2013 to 2022.

Gil-Garcia and Aldama-Nalda (2013) advance information integration initiatives in achieving the notion of smarter cities (e.g., connectedness, responsiveness, efficiency, and sustainability) with political support from the city Mayor enabling the needed infrastructure, funding, and administrative support. de Jong et al. (2015) point to information cities as one of several key but limited categories from the research literature where jointly with other categories such as smart city, digital city, and so on, a sub-network of sustainable city emerges. Schmitt (2015) speaks in terms of the notion of information city where data is generated by inhabitants and infrastructure in vast quantities. Thompson et al. (2016) advance the use of city information modelling (CIM) case studies in the exploration of urban planning and governance in smart cities, attentive to the data themes of – accessibility and availability, accuracy and consistency, manageability, and integration – while pointing to the need for future research to focus on the use and interpretation of data. Monteiro et al. (2018) demonstrate use of an information system for smart cities in support of sustainable urban planning in the form of an urban building database (UBD) for energy modelling. McKenna (2019) characterizes ambient information architectures as more dynamic, adaptive, and open while Ismagilova et al. (2019) provide an information systems (IS) perspective on smart cities, highlighting security and privacy issues and the potential for migrants to improve the sustainability of cities. Attour et al. (2021) focus on current and future trends for smart cities research pertaining to the role of information technology (IT) identifying – design, technological challenges; digital innovations and knowledge; and the ethics of data management. Lytras et al. (2021) devise a scale development approach to information management in smart cities capturing the perspectives of people in the city involving aspects such as technology anxiety, engagement, quality of life and so on. Landsbergen et al.

Table 3. Overview of research for knowledge in smart cities

Author(s)	Year	Knowledge
de Jong et al.	2015	Knowledge city, smart city, sustainable city
Negre et al.	2015	Knowledge-based conceptual vision of SCs
Dattakumar & Sharma	2016	Relationships between knowledge societies & SCs
McFarlane & Söderström	2017	Knowledge-intensive smart urbanism
Bonde Thylstrup et al.	2019	Novel forms of knowledge production in a world of the digital & datafied
Ismagilova et al.	2019	Knowledge-based city
Kurgan & Brawley	2019	Techniques & technologies, tools & practices, mapping, redraw/build/make
Gessa & Sancha	2020	Data life cycle for environmental info, transform to knowledge is low
Borba et al.	2020	Knowledge management contribution to SCs & research agenda
Allam et al.	2022	Knowledge structure & trends – smart urban governance, policy, inclusion

(2022) provide guidance on the managing of smart city information for urban governance, attentive to the political, financial, and operational sustainability of integrated data exchanges (IDEs) as data platforms for "curating, managing, and disseminating information."

Table 3 provides an overview by author and date of the research and practice literature for urban life and the ambient pertaining to knowledge, spanning the timeframe of 2015 to 2022.

de Jong et al. (2015) explore the knowledge city category finding it to be a distinct concept while identifying the importance of relationships with smart city and sustainable city. Negre et al. (2015) advance a vision of smart cities that is knowledge-based, focusing on information and knowledge created by people in support of decision-making and business processes. Dattakumar and Sharma (2016) explore relationships between knowledge societies and smart cities. McFarlane and Söderström (2017) advocate for a move from technology-intensive to knowledge-intensive smart urbanism as alternative smart cities. Bonde Thylstrup et al. (2019) provide an introduction to explorations that "sketch the contours of novel forms of knowledge production in a digital and datafied world." Ismagilva et al. (2019) speak in terms of a knowledge-based city while Kurgan and Brawley (2019) describe the techniques and technologies and the tools and practices of knowledge in understanding cities using mapping, both digital and predigital, in the redrawing, rebuilding, and remaking of cities. Borba et al. (2020) provide a look at the contributions of knowledge management to smart cities and offer a research agenda going forward. Gessa and Sancha (2020) find that the data life cycle for environmental information seems to be low in terms of the capabilities of transforming such information into relevant knowledge for cities. Allam et al. (2022) explore knowledge structures and trends for smart urban governance through bibliometric analysis and mapping with implications for efforts pertaining to policymaking, governance, and inclusion.

Table 4 provides an overview by author and date of the research and practice literature for urban life and the ambient pertaining to knowing and computational intelligence (CI), spanning the timeframe of 2012 to 2022.

Campbell (2012) focuses on "the way cities learn" and the importance of learning as a "part of the mission of governance" as well as "acquiring knowledge" and leveraging connections and connecting for learning and innovation. Batty (2013), Townsend (2015), and Derudder and van Meeteren (2019) are cited by Karvonen (2020) on digital tools and methods for knowing cities. Caprotti (2017) highlights

Table 4. Overview of research for knowing and computational intelligence in smart cities

Author(s)	Year	Knowing and Computational Intelligence
Campbell	2012	Knowledge acquisition & the ways cities learn & relationships
Batty	2013	Digital tools and methods for knowing cities
Townsend	2015	Digital tools and methods for knowing cities
Caprotti	2017	Challenges for knowing the smart city and the invisible components
Derudder & van Meeteren	2019	Digital tools and methods for knowing cities
Kurgan & Brawley	2019	Know/experience cities tech-mediated, reshaping urban life epistemologies
Mattern	2019	Archeology of urban intelligence – material, environmental knowing
Karvonen	2020	Knowing as techno-politics, governance, imagining, contested debates
IEEE	2021	CI, evolving, encompassing computing paradigms: AI, cultural, social
Puron-Cid & Gil-Garcia	2022	Long term costs of ICT infrastructure

the challenges for knowing the smart city when components may be invisible while Kurgan and Brawley (2019) argue that how we know and experience cities is mediated by technology, is political, and the data is reshaping urban life and epistemologies. Mattern (2019) provides an archeology of urban intelligence (UI) perspective, noting that smartness has always existed in human settlements, although less computational and more material and environmental. Karvonen (2020) identifies knowing as techno-politics for governing and imagining cities in support of many debates such as those on climate and the environment, and in support of contested debates on decarbonization and resilience, citizen empowerment, surveillance, security, health and well-being, to name a few. Computational intelligence (CI) is described (IEEE, 2021) as an evolving field encompassing computing paradigms such as ambient intelligence, artificial intelligence (AI), cultural learning, and social reasoning to name a few. Puron-Cid and Gil-Garcia (2022) question the financial feasibility of smart city ICT infrastructure in the long term.

In terms of patterns and potential relationships, what emerges from this review of the research and practice literature is a comparative overview for data, information, knowledge, and knowing and computational intelligence (CI) as presented in Table 5.

Elements that emerge across columns of Table 5, include but are not limited to accessibility, ambient in the form of data and intelligence, big data, the use of digital and other tools, governance, management, security, sustainability, technologies, and techno-centricities of one form or another.

3.6 Issues, Controversies, Problems

This review of the research and practice literature enables the surfacing of a range of issues, controversies, and problems as identified in Table 6.

For data, issues, controversies, and problems pertain to access; while big data is said to require "thoughtful care" if it is to "improve research on cities"; sharing; the potential for policy and governance to become too technocentric; cybersecurity; governance and policy; technocratic; trust; accuracy; availability; the need for developing datasets for learning cities; and scalability.

Table 5. Patterns and potential relationships

Data	Information	Knowledge	Knowing and CI
Access	Accessibility	Citizen participation	Archeology of UI
Ambient data	Accuracy	Data life cycle	CI - ambient intelligence
Big data for big cities	Administrative support	- low transform of env	- artificial intelligence
Collaborative	Ambient information	Decision making	- cultural learning
Cybersecurity	Availability	Digital infrastructure	Digital tools & methods
Data flows	Connectedness	Eco-city	- AI, algorithms, ML
Digital urban acupuncture	Consistency	Knowledge-based cities	- application of big data
Dynamic	Efficiency	Low carbon city	- geospatial
Ethics	Funding	Management	- social network analysis
Learning cities	Information city	Production	- ubiquitous sensing
Responsive	Infrastructure	Regeneration policies	- urban knowledge/epist…
Sharing	Integration initiatives	Structure	Financial feasibility of ICTs
Smart sustainable cities	IS perspective	Sustainability	Technology-mediated life
- data-driven	- migrants, sustainability	Techniques, tools, maps	- citizen empowerment
- governance	- privacy	- redraw	- climate; decarbonization
- policy	- security	- rebuild	- economic dev/prosperity
- technocentric	Manageability	- remake	- environment, health
Trust	Modelling	Technologies	- governing, imagining
Visual use of data	Responsiveness	Trends	- resilience, social cohesion
- past, present, future	Sustainability	Urban relevance	- security
- world mappings	Urban governance	Usability	- surveillance

For information, issues, controversies, and problems pertain to computational costs; connectedness; integration initiatives; sustainability; accessibility and availability; dissemination; manageability; privacy; accuracy and consistency; efficiency; security; and urban governance.

For knowledge, issues, controversies, and problems pertain to areas for further research; production; management; sustainability; intelligent digital infrastructure; governance; inclusion; mapping as reconstruction; capabilities for the rendering of citizen participation; structure; urban relevance; and making knowledge usable.

Table 6. Issues, controversies, and problems

	Issues	Controversies	Problems
Data	• Access • Big data & care • Sharing • Technocentric	• Cybersecurity • Governance & policy • Technocratic • Trust	• Accuracy • Availability • Datasets for learning cities • Scalability
Information	• Computational costs • Connectedness • Integration • Sustainability	• Accessibility/Availability • Dissemination • Manageability • Privacy	• Accuracy & Consistency • Efficiency • Security • Urban governance
Knowledge	• Further research • Production • Management • Sustainability	• Digital infrastructure • Governance • Inclusion • Mapping as reconstructing	• Citizen participation • Structure • Urban relevance • Usability
Knowing & CI	• AI • Algorithms • Big data • Machine learning	• Reading data & its effects • Technology as mediator • Tech/urban epistemologies • Use of digital tools	• Geospatial analysis • New urban science • Social network analysis • Ubiquitous sensing

Figure 1. Conceptual framework for data, information, and knowledge in relation to urban life and the ambient in smart cities

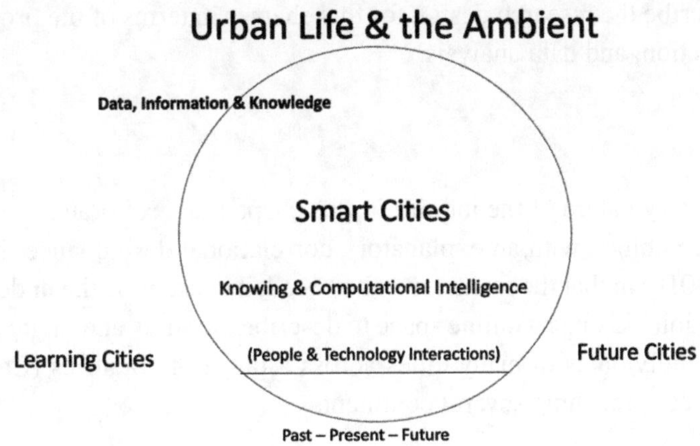

For knowing and computational intelligence, issues, controversies, and problems pertain to artificial intelligence (AI); algorithms as a way of knowing cities; big data; machine learning, "learning how to read data and its effects"; technology as mediator for knowing; role of technologies in contesting urban epistemologies; use of digital tools; geospatial analysis; use of the "new urban science" for understanding cities; social network analysis; and ubiquitous sensing.

Of note is the presence of overlap between data, the digital, information, knowledge, knowing and computational intelligence (CI) and across issues, controversies, and problems where accuracy and governance provide a few examples.

4. CONCEPTUALIZING DATA, INFORMATION, AND KNOWLEDGE FOR URBAN LIFE AND THE AMBIENT IN SMART CITIES

Responding to the issues, controversies, and problems emerging from the literature review in the preceding section, a conceptual framework for data, information, and knowledge in relation to urban life and the ambient in smart cities is formulated in Fig. 1, involving knowing and computational intelligence building upon elements in preceding chapters pertaining to people and technology interactions over time, as in, the past, present, and future while being attentive to learning cities and future cities.

The research question identified in Section 1 is reformulated as a proposition for exploration in this chapter, as follows:

P6: *Data, information, and knowledge are related in urban life in smart cities as shown through ways of knowing and computational intelligence*

The research design used in exploring the proposition under study in this chapter is described in the Methodology section.

5. METHODOLOGY

Sections 5.1 to 5.3 describe the research design for this chapter in terms of the process used, sources of evidence for data collection, and data analysis.

5.1 Process

Based on the contemporary nature of the inquiry in this chapter, an exploratory case study approach is used (Yin, 2018) and combined with an explanatory correlational design since the two are said to be complementary (Yin, 2018) in that the former may be "needed to examine the underlying processes that might explain a correlation." Using an online space to describe the study and invite people to participate, interest emerged from individuals in many cities across a range of countries (Canada, United States, Greece, Ireland, Israel, etc.) touching several continents.

5.2 Sources of Evidence

An online website was used for the study and during sign up, the gathering of demographic data from individuals occurred which included age range, gender, and city. Additionally, people could self-identify in one or more categories (e.g., business, community member, student, educator, etc.). A survey instrument, that had been pre-tested prior to use, was employed in this study and contained closed and open-ended questions. Survey questions pertained to a range of items related to smart cities such as data, information, knowledge, knowing, and computational intelligence. Survey respondents were also invited to participate in an in-depth interview which was guided by an interview protocol that had been pre-tested prior to use in this study.

Additionally, data were also gathered systematically through group and individual discussions, conducted in parallel with this study, with individuals in many sectors (e.g., students, faculty, and staff in higher education, business, community members and leaders, and city staff and leaders) in several Canadian cities (e.g., Toronto, Vancouver, Victoria).

5.3 Data Analysis

Data analysis in this chapter is guided by the research question posed which is reformulated into a proposition for exploration, as follows:

P6: *Data, information, and knowledge are related in urban life as shown through ways of knowing and computational intelligence.*

For the analysis of quantitative survey data, the use of descriptive statistics pertaining to urban data, information, and knowledge along with other associated elements enabled exploration of items such as *access to public data* in relation to *information.* The correlation feature of the Real Statistics Resource Pack containing an-add in for Microsoft Excel (Zaiontz, 2022) enabled the conducting of a Spearman Correlation Coefficient for ordinal data in exploring the nature of relationships between items pertaining to data, information, and knowledge.

Table 7. Assessments for urban data and information

Items	Urban Data and Information	Correlation
Access to Public Data	25% (4); 25% (5); 25% (6); 25% (7)	.63
Information	25% (5); 25% (6); 50% (7)	

Content analysis was used for qualitative data gathered through open-ended survey questions, interviews, and group and individual discussions so that inductive analysis from data collection was combined with deductive analysis using terminology from the literature review. Overall, an analysis was conducted for n=79 consisting of 42% females and 58% males for people ranging in age from their 20s to 70s.

6. RESULTS AND DISCUSSION

Table 7 provides the results of assessments for *urban data* and *information* using the proxy of *access to public data* for *urban data.* In response to the survey question about the extent to which factors such as *access to public data* contribute to the livability of a smart city, on a 7-point Likert type scale, where 1= Not at all and 7=Absolutely, 25% of respondents selected positions 4 (Neutral), 5 (Sort of), 6 (Sure), and 7 on the scale. When asked to what extent elements such as *information* contribute to the making of a smart city, 25% of survey respondents selected position 5 on the scale (Sort of), 25% responded toward the upper end of the scale at 6 (Sure), and 50% responded at the upper end position 7.

Using the correlation feature of the Real Statistics add-in for Microsoft Excel (Zaiontz, 2022), a Spearman correlation coefficient for ordinal data was conducted on the assessment results of the two items, *access to public data* and *information* (Table 7), showing a finding of .63.

Exploring further, as shown in Table 8, when asked to what extent elements such as *ICTs* (Information and Communication Technologies) contribute to the making of a smart city, 50% of survey respondents selected position 5 (Sort of) and 50% responded at the upper end of the scale at position 7.

Using the correlation feature of the Real Statistics add-in for Microsoft Excel (Zaiontz, 2022), a Spearman correlation coefficient for ordinal data was conducted on the assessment results of the two items, *information* and *ICTs* (Table 8), showing a finding of .94.

Exploring further still, as shown in Table 9, when asked to what extent factors such as *meaningfulness* contribute to increased value for data in smart cities, 75% of survey respondents selected position 6 (Sure) and 25% responded at the upper end of the scale at position 7.

Using the correlation feature of the Real Statistics add-in for Microsoft Excel (Zaiontz, 2022), a Spearman correlation coefficient for ordinal data was conducted on the assessment results of the two items, *ICTs* and *meaningfulness* (Table 9), showing a finding of .57.

Table 8. Assessments for urban information and ICTs

Items	Urban Information and ICTs	Correlation
Information	5% (5); 25% (6); 50% (7)	.94
ICTs	50% (5); 50% (7)	

Table 9. Assessments for urban information and knowledge: ICTs and meaningfulness

Items	Urban Information and Knowledge	Correlation
ICTs	50% (5); 50% (7)	.57
Meaningfulness	75% (6); 25% (7)	

Continuing the exploration, as shown in Table 10, when asked to what extent elements such as *visual ways to show success in real time* contribute to the success of a smart city project, 25% of survey respondents selected position 6 on the scale (Sure) and 75% responded at the upper end of the scale at position 7.

Using the correlation feature of the Real Statistics add-in for Microsoft Excel (Zaiontz, 2022), a Spearman correlation coefficient for ordinal data was conducted on the assessment results of the two items, *meaningfulness* and *visual ways to show success in real time* (Table 10), showing a finding of .33.

In considering the correlations that emerged for data, information, and knowledge and associated proxies, Creswell (2018) notes that for correlations in the .20 to .35 range "there is only a slight relationship" such that, this "coefficient may be valuable to explore the interconnection of variables but of little value in prediction studies." Correlations in the .35 to .65 range are said to be "the typical values in prediction used to identify variable membership in the statistical procedure of factor analysis" (Creswell, 2018). For correlations of .86 and above, as in the case of Table 8 for *ICTs* and *Information*, Creswell (2018) cautions that "when two or more variables are related, correlations this high are seldom achieved, and if they result, then two variables actually measure the same underlying trait and should probably be combined in data analysis."

The correlations that emerged for variables in this chapter would seem to provide opportunities going forward for further explorations for both research and practice while opening the way for the use of possibly other variables or proxies. Regarding the challenges associated with knowing as identified by Caprotti (2017) in relation to the invisible aspects of smart cities, opportunities may emerge in relation to elements such as *visual ways to show success in real time* or visual ways to work with smart city data more generally.

Qualitatively, regarding the question of what contributes to the making of a smart city, an educator noted the "usefulness of what might be provided through technology and access to information" adding that "ways of doing this to support initiatives are many but certainly lend themselves to the smart city as a playful and open territory of personal and community development" while "people who have heard the term" smart city, "think of big data and machine learning, often with relevance to sustainability." A community leader commented on the need for "finding fun and meaningful ways to engage the public" as well as to "visualize this data, and 'simplify' it for the larger public" and figure out "how to make

Table 10. Assessments for urban data, information, and knowledge: meaningfulness and visual ways to show success in real time

Items	Data, Information and Knowledge	Correlation
Meaningfulness	75% (6); 25% (7)	.33
Visual Ways to Show Success in Real Time	25% (6); 75% (7)	

Figure 2. Urban life and the ambient – data, information, and knowledge in smart cities

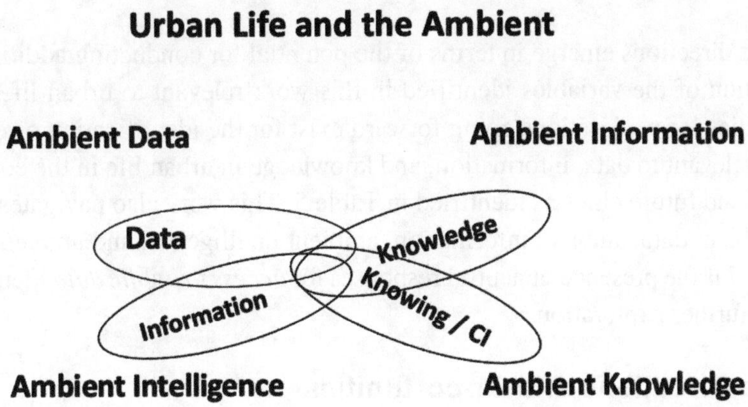

[smart cities] about more than data for the general public to be interested and understand more." Another community leader spoke of a "personal connection as a former city leader" enabling one to "still get to be 'in the know' of happenings in the city." Another educator described ways of knowing in the city in terms of "the physical level" and "the number of people who are out and about on a sunny weekend, or in the evenings" while "on the digital level" the capability for "keeping up-to-date with social media and online newspapers allows me to find out things that my eyes could not see while walking around."

As depicted graphically in Fig. 2, the interconnectivity of data with information, knowledge, and knowing would seem to be aided in urban life by computational intelligence (CI) giving way to the notion and potentials of ambient data, ambient information, ambient intelligence, and ambient knowledge.

7. LIMITATIONS AND FUTURE DIRECTIONS

Possible limitations of this work are discussed in this section, opening the path for future directions and opportunities for research and practice.

Limitations

Limitations of this chapter pertain possibly to the extent of the literature review, focusing as it does on urban life and the ambient in relation to data, information, knowledge, and knowing and computational intelligence (CI) in smart cities. Among other limitations of this chapter is that of the small sample size for the underlying study of this work, so that caution is encouraged in the interpretation of correlations calculated from the assessment of variables used for data, information, and knowledge. Going forward opportunities may emerge possibly for the use of other variables, or for more direct exploration instead of the proxies that have been used in this chapter, resulting in possibly stronger correlations with greater prediction potentials.

Future Research Directions and Opportunities

For research, future directions emerge in terms of the potential for conducting additional explorations, testing, and validation of the variables identified in this work relevant to urban life and the ambient. As such, opportunities for explorations going forward exist for the identification of other and possibly stronger variables relevant to data, information, and knowledge in urban life in the context of smart cities, learning cities, and future cities as identified in Table 5. This work also navigates a path for further explorations of ambient data, ambient information, ambient intelligence, and ambient knowledge.

Of note in Table 7 is the presence of neutral responses for *access to public data* identifying a potential area of interest for further exploration.

Future Practice Directions and Opportunities

Future directions for practice include the potential for pursuing explorations with variables identified in this work relevant to urban life and the ambient, focusing on data, information, and knowledge, as presented in Table 5. Additionally, opportunities going forward for practice include the potential for exploring spaces with other, possibly stronger variables relevant to urban life that foster real world action and knowing in everyday contexts in smart cities, learning cities, and future cities, in addressing the issues, controversies, and problems presented in Table 6.

8. CONCLUSION

The primary contributions of this chapter pertain to identification of the challenges and opportunities associated with urban life and the ambient that emerge from the review of the research literature provided from multiple perspective involving data, information, and knowledge. Another contribution of this chapter is the formulation of a conceptual framework for data, information, and knowledge in smart cities in relation to urban life and the ambient, based on the literature review where emerging issues, controversies, and problems provide guidance on the proposition explored while contributing to operationalization of the framework for use in this chapter. Through an exploratory case study approach, together with an explanatory correlational design, the proposition under study in this chapter is investigated. Based on survey data, several correlations are conducted, such as that for *access to public data* and *information* where, among the findings is that of a Spearman correlation coefficient of .63, said to be "useful for limited prediction." Limitations of the study underlying this chapter pertain to, among other things, the small sample size and caution is urged in the interpretation of findings, opening opportunities for further research. Notwithstanding this and other possible limitations, future directions guiding research and practice activities going forward are provided in terms of further validating the findings in this chapter while also encouraging activities to identify other, additional factors relevant to urban life and the ambient and to the notions of ambient data, ambient information, ambient knowledge, and ambient intelligence, in the context of smart, learning, and future cities.

Key insights and ideas emerging from this chapter are provided, along with exercises encouraging further exploration, as follows:

Insights emerging from this chapter include the following:

1. The importance of the relationship between urban data and information in the form of *access to public data* and *information* as factors contributing to livability and the making of a smart city
2. The importance of factors such as *ICTs* and *meaningfulness* associated with *data, information,* and *knowledge* in smart cities
3. The potential for many more factors to emerge for understanding *urban data, information,* and *knowledge* in smart cities such as *visual ways to show success in real time*

Ideas emerging from this chapter include the following:

1. Potential opportunities for exploring urban life and the ambient in relation to a variety of variables (e.g., data, information, and knowledge) in the context of smart cities
2. The opportunities potential for the identification of other and additional variables for exploring urban life and the ambient in relation to ways of knowing in smart environments
3. The use of an interdisciplinary and multi-method approach for the study and practice of urban life focusing on data, information, and knowing in smart, learning, and future cities

Exercises for the reader to consider, encouraging further exploration include the following:

1. Identify one or more ways in your city that data and information matter
2. Identify one or more ways in which data and information would enhance the quality of urban life in your city
3. Describe one or more elements that contribute to ways of knowing in your city

The audience for this chapter is intended to be interdisciplinary in nature, including but not limited to community members and leaders, educators, businesses, researchers, urban planners, urban designers, and anyone concerned with data, information, knowledge, ways of knowing and computational intelligence (CI) in urban life, along with associated practice and research opportunities and challenges in smart cities, learning cities, and future cities.

REFERENCES

Allam, Z., Sharifi, A., Bibri, S. E., & Chabaud, D. (2022). Emerging trends and knowledge structures of smart urban governance. *Sustainability*, *14*(9), 5275. doi:10.3390u14095275

Attour, A., Dominguez-Péry, C., & Bendavid, Y. (2021). Information technologies, knowledge and innovation in smart cities: Current and future trends for management research. *Systèmes d'information & management*, *26*, 3-18. . doi:10.3917/sim.214.0003

Batty, M. (2013). *The New Science of Cities*. The MIT Press. doi:10.7551/mitpress/9399.001.0001

Bibri, S. E. (2019). The anatomy of the data-driven smart sustainable city: Instrumentation, datafication, computerization and related applications. *Journal of Big Data*, *5*(1), 59. doi:10.118640537-019-0221-4

Bibri, S. E. (2021). Data-driven smart sustainable cities of the future: An evidence synthesis approach to a comprehensive state-of-the-art literature review. *Sustainable Futures*, *3*, 100047. doi:10.1016/j.sftr.2021.100047

Bonde Thylstrup, N., Flyverbom, M., & Helles, R. (2019). Datafied knowledge production: Introduction to the special theme. *Big Data & Society*, *6*(2). Advance online publication. doi:10.1177/2053951719875985

Borba, D., Palacios, R. A., Luciano, E. M., & Chaves, M. S. (2020). The contribution of knowledge management to smart cities for innovation: Proposal for a prescriptive framework and a research agenda. *International Journal of Innovation*, *8*(3), 516–540. doi:10.5585/iji.v8i3.18543

Campbell, T. (2012). *Beyond Smart Cities: How Cities, Learn, Network, and Innovate*. Routledge.

Carrillo, F. J. (Ed.). (2006). *Knowledge Cities: Approaches, Experiences, and Perspectives*. Routledge. doi:10.4324/9780080460628

Cheshire, J., & Uberti, O. (2021). *Atlas of the invisible: Maps and graphics that will change how you see the world*. Norton.

Creswell, J. W. (2018). *Educational research: Planning, conducting, and evaluating quantitative and qualitative research* (6th ed.). Pearson.

Dattakumar, A., & Sharma, R. S. (2016). Smart cities and knowledge societies: Correlation, causation or distinct? *2016 IEEE International Conference on Management of Innovation and Technology (ICMIT)*, 193-197. 10.1109/ICMIT.2016.7605032

de Jong, M., Joss, S., Schraven, D., Zhan, C., & Weijnen, M. (2015). Sustainable–Smart– Resilient–Low Carbon–Eco–Knowledge Cities; Making sense of a multitude of concepts promoting sustainable urbanization. *Journal of Cleaner Production*, *109*, 25–38. Advance online publication. doi:10.1016/j.jclepro.2015.02.004

Derudder, B., & van Meeteren, M. (2019). Engaging with 'urban science'. *Urban Geography*, *40*(4), 555–564. doi:10.1080/02723638.2019.1585138

Gessa, A., & Sancha, P. (2020). Environmental open data in urban platforms: An approach to the big data life cycle. *Journal of Urban Technology*, *27*(1), 27–45. doi:10.1080/10630732.2019.1656934

Gil-Garcia, J. R., & Aldama-Nalda, A. (2013). Making a city smarter through information integration: Angel Network and the role of political leadership. *2013 46th Hawaii International Conference on System Sciences*, 1724-1733. doi: 10.1109/HICSS.2013.387

Glaeser, E. L., Kominers, S. D., Luca, M., & Naik, N. (2018). Big data and big cities: The promises and limitations of improved measures of urban life. *Economic Inquiry*, *56*(1), 114–137. doi:10.1111/ecin.12364

Iaconesi, S., & Persico, O. (2017). Digital urban acupuncture. In *Digital Urban Acupuncture*. Springer. doi:10.1007/978-3-319-43403-2_9

IEEE. (2021). *What is computational intelligence?* IEEE Computational Intelligence Society (CIS). Retrieved 7 December 2021 from https://cis.ieee.org/about/what-is-ci

Ismagilova, E., Hughes, L., Dwivedi, Y. K., & Raman, K. R. (2019). Smart cities: Advances in research—An information systems perspective. *International Journal of Information Management*, *47*, 88–100. doi:10.1016/j.ijinfomgt.2019.01.004

Karimi, Y., Haghi Kashani, M., Akbari, M., & Mahdipour, E. (2021). Leveraging big data in smart cities: A systematic review. *Concurrency and Computation*, *33*(21), e6379. doi:10.1002/cpe.6379

Karvonen, A. (2020). Urban techno-politics: Knowing, governing, and imagining the city. *Science as Culture*, *29*(3), 417–424. doi:10.1080/09505431.2020.1766011

Kurgan, L., & Brawley, D. (Eds.). (2019). Ways of Knowing Cities. Columbia University Press.

Landsbergen, D., Girth, A., & Westover-Muñoz, A. (2022). *Governance rules for managing smart city information*. Urban Governance. doi:10.1016/j.ugj.2022.05.003

Lytras, M. D., Visvizi, A., Chopdar, P. K., Sarirete, A., & Alhalabi, W. (2021). Information management in smart cities: Turning end users' views into multi-item scale development, validation, and policy-making recommendations. *International Journal of Information Management, Elsevier*, *56*(C), 102146. doi:10.1016/j.ijinfomgt.2020.102146

Mattern, S. (2019). Ether and ore: An archaeology of urban intelligence. In Ways of Knowing Cities. Columbia University Press.

McFarlane, C., & Söderström, O. (2017). On alternative smart cities. *City*, *21*(3-4), 312–328. doi:10.1080/13604813.2017.1327166

McKenna, H. P. (2017). Civic tech and ambient data in the public realm. In N. Streitz & P. Markopoulos (Eds.), Lecture Notes in Computer Science: Vol. 10291. *Distributed, Ambient and Pervasive Interactions. DAPI 2017*. Springer. doi:10.1007/978-3-319-58697-7_23

McKenna, H. P. (2019). *Ambient Urbanities as the Intersection Between the IoT and the IoP in Smart Cities*. IGI Global. doi:10.4018/978-1-5225-7882-6

Monteiro, C. S., Costa, C., Pina, A., Santos, M. Y., & Ferrão, P. (2018). An urban building database (UBD) supporting a smart city information system. *Energy and Building*, *158*, 244–260. doi:10.1016/j.enbuild.2017.10.009

Negre, E., Rosenthal-Sabroux, C., & Gascó, M. (2015). A knowledge-based conceptual vision of the smart city. *2015 48th Hawaii International Conference on System Sciences*, 2317-2325. doi: 10.1109/HICSS.2015.279

Osborne, M., Houston, M., & Lido, C. (2018). The role of big data in elucidating learning cities ancient, present and future. In J. R. Stenger (Ed.), *Learning Cities in Late Antiquity: The Local Dimension of Education*. Routledge. doi:10.4324/9781315097848-2

Puron-Cid, G., & Gil-Garcia, J. R. (2022). Are smart cities too expensive in the long term? Analyzing the effects of ICT infrastructure on municipal financial sustainability. *Sustainability*, *14*(10), 6055. doi:10.3390u14106055

Schmitt, G. (2015). *Information Cities*. Apple Books.

Thompson, E. M., Greenhalgh, P., Muldoon-Smith, K., Charlton, J., & Dolnik, M. (2016). Planners in the future city: Using city information modelling to support planners as market actors. *Urban Planning, 1*(1), 79-94. doi:10.17645/up.v1i1.556

Townsend, A. (2015). Cities of data: Examining the new urban science. *Public Culture, 27*(2), 201–212. doi:10.1215/08992363-2841808

UN-Habitat. (2022). *Envisaging the Future of Cities: World Cities Report 2022*. United Nations Human Settlements Programme. Retrieved 29 September 2022 from https://unhabitat.org/sites/default/files/2022/06/wcr_2022.pdf

Wallace, E. (2022). The importance of connected vehicle data for smart cities. *RTInsights*. Retrieved 26 September 2022 from https://www.rtinsights.com/the-importance-of-connected-vehicle-data-for-smart-cities/

Webb, S., Holford, J., Hodge, S., Milana, M., & Waller, R. (2020). Learning cities and implications for adult education research. *International Journal of Lifelong Education, 39*(5-6), 423–427. doi:10.1080/02601370.2020.1853937

Yin, R. (2018). *Case study research and applications: Design and methods* (6th ed.). Sage Publications.

Zaiontz, C. (2022). *Real statistics using excel*. Retrieved from www.real-statistics.com

ADDITIONAL READING

Allam, Z., & Dhunny, Z. A. (2019). On big data, artificial intelligence and smart cities. *Cities (London, England), 89*, 80–91. doi:10.1016/j.cities.2019.01.032

Coelho, V. N., Coelho, I. M., Oliveira, T. A., & Ochi, L. S. (2019). *Smart and Digital Cities: From Computational Intelligence to Applied Social Sciences*. Springer. doi:10.1007/978-3-030-12255-3

de Hoop, E., Moss, T., Smith, A., & Emanuel Löffler, E. (2022). *Knowing and governing smart cities: Four cases of citizen engagement with digital urbanism*. Urban Governance. doi:10.1016/j.ugj.2021.12.008

Hayes, K., Ghosh, S., Gnenz, W., Annett, J., & Bryne, M. B. (2021). Smart city Edmonton. In J. C. Augusto (Ed.), *Handbook of Smart Cities*. Springer. doi:10.1007/978-3-030-69698-6_17

Highroad Team. (2020). 5 things you don't know about smart cities, but should. *HighRoad Blog*. Retrieved 14 May 2022 from https://blog.highroad.center/5-things-you-dont-know-about-smart-cities-but-should

Ismagilova, E., Hughes, L., Dwivedi, Y. K., & Raman, K. R. (2019). Smart cities: Advances in research—An information systems perspective. *International Journal of Information Management, 47*, 88–100. doi:10.1016/j.ijinfomgt.2019.01.004

Mihaljevic, H., Larsen, C. J., Meier, S., Nekoto, W., & Zirfas, F. M. (2021). Privacy-centred data-driven innovation in the smart city. Exemplary use case of traffic counting. *Urban, Planning and Transport Research, 9*(1), 425–448. doi:10.1080/21650020.2021.1950044

Russo, M., & Feng, G. (2020). The risks and rewards of data sharing for smart cities. *BCG Henderson Institute Newsletter*. Retrieved 14 May 2020 from https://www.bcg.com/publications/2020/smart-cities-need-to-understand-the-risks-and-rewards-of-data-sharing-part-3

Stone, M., Knapper, J., Evans, G., & Aravopoulou, E. (2019). Information management in the smart city. *The Bottom Line (New York, N.Y.)*, *31*(3/4), 234–249. doi:10.1108/BL-07-2018-0033

Tusikov, N. (2019). *Governing knowledge and data in smart cities.* The City Institute at York University. Retrieved 14 May 2022 from https://www.yorku.ca/cityinstitute/projects/current-projects/governing-knowledge-and-data-in-smart-cities/

KEY TERMS AND DEFINITIONS

Knowing: Knowing in an urban context refers to awareness capabilities associated with people and with technology as well as such capabilities associated with an interacting of people and technologies.

Meaningfulness: Meaningfulness in an urban context refers to having purpose and functionality in relation, for example, to data and information.

Real Time: Real time in an urban context refers to anything occurring in-the-moment or, as it is happening.

Success: Success in an urban context in relation to smart city projects and initiatives refers to meaningfully involving people, with clear purpose, and achieving one or more functional outcomes.

Visualizations: Visualizations in an urban context refer to representations that are visual in nature in the form of videos, graphs, artistic representations, and the like, depicting events, datasets, and so on in meaningful ways.

Section 3

From AI to Smart Homes, E-Vehicles, Robotics, and Extended Realities as Ambient Urban Living

This section introduces the reader to a range of areas influencing urban life and the ambient in smart cities including artificial intelligence (AI), machine learning (ML), and deep learning (DL). Areas such as smart homes; electric vehicles (EVs) and solar potentials; extended realities involving augmented, mixed, and virtual applications; and robotics are explored.

Chapter 7
Urban Life and Artificial Intelligence, Machine Learning, Deep Learning, and Ambient Learning in Smart Cities

ABSTRACT

The purpose of this chapter is to explore evolving understandings of artificial intelligence (AI) in smart cities as well as machine learning, deep learning, and ambient learning. A review of the research literature is conducted in the context of smart cities. Issues, controversies, and problems emerging from the literature review are highlighted, contributing to formulation of a conceptual framework for artificial intelligence in urban life and the ambient in smart cities. Using an exploratory case study approach combined with an explanatory correlational design, variables of interest are identified, such as collaborating and creative opportunities, along with assessments of the nature of the relationship. Through the lens of emerging manifestations of artificial intelligence, machine learning, and deep learning, efforts are made to inform understandings of the ambient in smart cities, and more particularly ambient learning, while pointing to directions for future research.

1. INTRODUCTION

Marzano (2006) describes concepts such as artificial intelligence (AI) as "terms that trigger the imagination." Cugurullo (2020) provides an exploration of "how the development of AI intersects with the development of the city" both in theory and in practice. For Cugurullo (2020), urban artificial intelligence takes the form of "artifacts operating in cities, which are capable of acquiring and making sense of information on the surrounding urban environment, eventually using the acquired knowledge to act rationally according to pre-defined goals, in complex urban situations when some information might be missing or incomplete." As such, Cugurullo (2020) argues that an inquiry of a philosophical and multidisciplinary nature is needed that is "proactive" to "inform the development of AI, as it now intersects

DOI: 10.4018/978-1-6684-4096-4.ch007

with the development of the city." Zhang and He (2020) highlight the importance of a sustainable future incorporating both humanity and nature using "life-oriented approaches." Wu, Ji, Yu, Zeng, Wu, and Shidujaman (2021) focus on the creativity potential of people and AI interactivities with the notion of AI creativity while advancing the Human-AI co-creation model with an emphasis on collaboration. In response, the work in this chapter is significant in that it considers that human capabilities are not static in that creativity, innovation, and discoveries are emergent and ongoing and as such, builds on the range of perspectives in the research literature on AI to explore urban life and the ambient in the context of smart cities giving rise to the motivation for this with the following objectives.

Objectives: The main objectives of this chapter are to a) explore urban life and the ambient in relation to artificial intelligence (AI), machine learning (ML), deep learning (DL), and ambient learning in the context of smart cities; b) formulate a conceptual framework for artificial intelligence (AI) for urban life and the ambient in smart cities; and c) explore the nature of the relationship between AI and smart cities as a way of learning more about ambient learning. These objectives give rise to the main research question under exploration in this chapter – *How and why does artificial intelligence, including machine learning and deep learning, manifest in urban life and the ambient in smart cities?*

2. BACKGROUND

Miailhe and Hodes (2017) provide an overview of the emergence and history of artificial intelligence (AI), describing artificial narrow intelligence (ANI) as 'weak' where machines "operate strictly within the confine of the scenarios for which they are programmed" and artificial general intelligence (AGI) as "strong" as in "an autonomous machine's ability to perform any intellectual tasks that a human can perform." Accordingly, current AI capabilities are said to be of the ANI type involving machine learning algorithms, excluding AGI capability and Artificial Super Intelligence (ASI) that are both said to be in the near to distant future (Miailhe and Hodes, 2017). Amabile (2020), an influential creativity researcher in the management domain argues for the study of "AI and computer-assisted human intelligence" by organizational researchers focused on creativity and innovation to determine "the ways in which they may yield creative breakthroughs" as well as "how those innovations might impact–and be impacted by–workers, consumers, organizations, and society." Shneiderman (2022) envisions an integration of AI (artificial intelligence) with HCI (Human-computer interaction) approaches so as "to amplify, augment, and enhance human abilities" in support of creativity and much more.

2.1 Definitions

Definitions from the research literature are provided for key terms used in this chapter.

- **Ambient Learning:** Bick, Kummer, Pawlowski, and Veith (2007) articulate the notion of ambient learning in terms of mobile learning in digital environments in support of "contextualized, personalized knowledge for learners" dynamically, anywhere, anytime.
- **Artificial Intelligence (AI):** According to the National Artificial Intelligence Initiative Act of 2020 (Division E, Sec. 5001) of the United States (NAII, 2021), artificial intelligence refers to "a machine-based system that can, for a given set of human-defined objectives, make prediction, recommendations or decisions influencing real or virtual environments."

- **Artificial Intelligence (AI) Creativity:** Amabile (2020) defines AI creativity as "the production of highly novel, yet appropriate, ideas, problem solutions, or other outputs by autonomous machines."
- **Machine Learning (ML):** ML is described by Jordan (2019) as "an algorithmic field that blends ideas from statistics, computer science and many other disciplines to design algorithms that process data, make predictions, and help make decisions."
- **Deep Learning:** In seeking to provide a unified definition, Zhang, Yang, Lin, Ji, and Gupta (2018) describe deep learning as "a process not only to learn the relation among two or more variables but also the knowledge that governs the relation as well as the knowledge that makes sense of the relation."

2.2 Overview

This chapter consists of several sections containing a review of the research and practice literature for urban life and artificial intelligence (AI) in smart cities, deep learning (DL) and smart cities, machine learning (ML) and smart cities, and ambient learning and smart cities. From this review, a highlighting of key issues, controversies, and problems is provided and a conceptual framework for artificial intelligence in urban life and the ambient in smart cities is formulated and operationalized for use in this chapter. The main research question under exploration in this chapter is reformulated as a proposition for investigation. The methodology for investigation of the proposition in this chapter is described, followed by the presentation of results and a discussion of findings. Limitations of the study underlying this chapter are identified and future directions for research and practice are outlined. The conclusion provides an overview of outcomes, the identification of insights and ideas, development of several exercise questions encouraging further exploration, and identification of the target audience for this chapter.

3. PERSPECTIVES ON ARTIFICIAL INTELLIGENCE AND ASSOCIATED CONCEPTS

This chapter provides a review of the research and practice literature for urban life and evolving understandings of artificial intelligence in smart cities, urban life and deep learning in smart cities, urban life and machine learning in smart cities, and urban life and ambient learning in smart cities.

3.1 Urban Life and Artificial Intelligence (AI) in Smart Cities

Loi (2019) explores the "perceptions, attitudes, thresholds and expectations" of people globally regarding intelligent systems, as in, systems "that leverage the power of artificial intelligence (AI)" in relation to the everyday spaces of homes, cars, and work. Findings by Loi (2019) identify a series of guidelines for designers and developers for intelligent systems futures along with many questions to be explored going forward, from ethics to autonomy and agency, to social and behavioral interactions. Jordan (2019) identifies the challenge associated with "bringing together computers and humans in ways that enhance human life" giving rise to "a new branch of engineering" focusing on "data from and about humans" and is "for the data- and learning-focused fields" requiring "perspectives from the social sciences and humanities." Jordan (2019) claims that "we want computers to trigger new levels of human creativity, not replace human creativity." Cugurullo (2020) refers to urban artificial intelligence as "urban AIs" that are said to enable the transition from automation to autonomy in smart cities as they "act autonomously" and "make decisions in an unsupervised manner, thus displaying a rudimentary form of thinking." As such, Cugurullo (2020) articulates the notion of the autonomous city, cautioning that "a critical understanding of AI in cities" is needed since urban AIs "take actions which can potentially trigger radical changes in the city" giving rise to issues of trust and ethics. Cugurullo (2020) also claims that every urban AI is unique with different impacts, advising that "researchers should avoid treating AI as a homogeneous category of intelligence." Luusua and Ylipulli (2020) seek to formulate an agenda for urban AI through the interdisciplinary research of artificial intelligence, incorporating the three perspectives of engineering, philosophy, and design, in cities. AI is described by Luusua and Ylipulli (2020) as "technology which attempts to mimic human intelligence" or, "as a set of algorithms that are capable of autonomous adaptation and decision-making" and as such, the notion of urban AIs, "in their various forms constitute novel non-human agents that affect human lives in cities" where cities are said to pertain to "urban life, spaces, places, geographies, infrastructures, and practices." Key issues and challenges identified by Luusua and Ylipulli (2020) include city-making ethics, data privacy and surveillance, and the "consequences of AIs on urban life and/or infrastructures." From a behavioral and social perspective, Zhang and He (2020) provide an introduction to a special issue of smart technologies and urban life where artificial intelligence (AI), for example, focusing on areas such as "the application of open source data and deep learning techniques" are considered while encouraging the use of smart technologies to "better understand urban issues and serve societal needs using life-oriented approaches." AI and accessibility are discussed in an interview with Morris (2020) who points to "the importance of considering the socio-technical context in which systems are deployed" while "a human-centered approach considers issues such as the values of end users and the ways in which AI can support, complement, and/or segment human skills (rather than replace them)." Thieme, Cutrell, Morrison, Taylor, and Sellen (2020) argue for the design of AI systems "as a resource for people to extend their own capabilities" where the notion of "interpretability" is "a dynamic of human-AI interaction" and is said to be "evolving through dynamic, situated system interactions." Amabile (2020) encourages study into the potential of creativity and AI to learn more about creative breakthroughs and the associated impact from multiple perspectives – workers, consumers, organizations, and society. Ullah, Al-Turjman, Mostarda and Gagliardi (2020) provide a review of the application of AI in smart cities where, the challenges and future directions for research range from the need for greater accuracy and precision to standardization to security issues. Arrieta, Díaz-Rodríguez, Del Ser, Bennetot, Tabik, Barbado, Garcia, Gil-Lopez, Molina, Benjamins, Chatila, and Herrera (2020) describe the notion of explainable AI (XAI) in relation to methods, concepts, taxonomies, opportunities and challenges, highlighting, for example, the importance of data privacy and responsible AI.

Wu et al. (2021) provide a perspective on AI focusing on creativity while advancing the human-AI co-creation model where "the ability for human and AI to co-live and co-create by playing to each other's strength to achieve more" is articulated for "human society and individuals" and "new opportunities and challenges in technology, society and education." Medaglia, Gil-Garcia, and Pardo (2021) introduce a special issue of artificial intelligence (AI) in government, identifying four areas for future research as governance of AI, trustworthy AI, impact assessment methodologies, and data governance. Ourednik, Mlynář, Mutzner, and Alavi, (2021) discuss the state of AI in relation to social media platforms, described as "failed cities", that can become "better algorithmic cities if the common good becomes their primary focus" and as such, can provide "an effective extension of the city" as in "a civilized space of cohabitation." Jha, Ghimire, Thapa, Jha, and Raj (2021) provide a review of use cases involving AI and

Internet of Things (IoT) for urban planning where such research papers may assist going forward to inform the development of smart cities. It is worth noting that the National Artificial Intelligence Initiative Act (NAIIA) in the United States established the National Artificial Intelligence Initiative (NAII, 2021) and "provides an overarching framework to strengthen and coordinate AI research" and "is organized into six strategic pillars – innovation, advancing trustworthy AI, education and training, infrastructure, applications, and international cooperation." James, Astoria, Castor, Hudspeth, Olstinske, and Ward (2021) make the claim that "[i]ncreasingly, humans will be interacting with computers, robots, cobots, machines, and artificial intelligence rather than solely with other humans" and as such, technologies "are not just tools but are interactants that affect how people make sense of the world." Considering the state of smart cities and looking to the future, Augusto (2021) raises questions pertaining to the range of complex issues and "the extent this will be left to some sort of 'evolution' of the artificial systems 'ecosystem' or will be planned and directed?" To this end, Margetis, Ntos, Antona, and Stephanidis (2021) advance a framework for systematically pursuing human-centered design (HCD) for AI in support of "AI for good" with a "human-in-the-loop approach" and "multidisciplinary collaboration" while being attentive "to issues pertaining to ethics" and the importance of "trustworthy AI." It is worth keeping in mind the one hundred-year, longitudinal study on AI that began in 2014 at Stanford University, involving an evolving array of individuals from many disciplines and sectors around the world (Littman et al., 2021). Five year reports from the study are said to be intended for the general public; industry; local, national, and international governments; and researchers while the purpose of the study "is to provide a collected and connected set of reflections about AI and its influences as the field advances" with "syntheses and assessments that provide expert-informed guidance for directions in AI research, development, and systems design, as well as programs and policies to help ensure that these systems broadly benefit individuals and society" (Littman et al., 2021).

3.2 Urban Life and Deep Learning (DL) in Smart Cities

Meserole (2018) describes deep learning (DL) as flexible where the algorithm is said to learn "on its own" yet, limitations are identified such as the use of probabilistic models. Habibzadeh, Kaptan, Soyata, Kantarci, and Boukerche (2019) explore the application of deep learning in smart cities focusing on one of several planes – the data plane – said to be "where sophisticated processing takes place through data analytic and machine intelligence" and converts raw data to "*useful* information". Atitallah, Driss, Boulila, and Ghézala (2020) provide a review of deep learning (DL) in relation to the Internet of Things (IoT) and big data analytics, where DL is described as "a new area of artificial intelligence (AI)" in support of a range of applications in the development of smart cities such as energy management in the smart home, smart healthcare, and smart transportation. Das and James (2020), highlighting the importance of "openness and transparency", discuss the use of deep learning algorithms "to help understand the impact of social distancing measures on people and vehicle movement within a metropolitan city in real time" using a dashboard with graphic renderings of activity. Concerned with improving solar power prediction capability in smart cities, Hajiabadi, Farhadi, Babaiyan, and Estebsari (2020) propose a method involving deep learning for more accurate forecasting of renewable energy generation such as photovoltaic (PV) generation with results indicating that the "method outperforms the state-of-the-art ML methods for PV energy generation forecasting." Puangpontip and Hewett (2020) address the "excessive usage of deep learning computation" in terms of the potential to "negatively impact the economic benefits and efficiency of smart cities." In response, Puangpontip and Hewett (2020) provide "a formal approach to estimating energy consumption of deep learning" usage in smart cities, focusing in particular on "deep and convolutional neural networks" involving the real world applications of smart drainage, smart transportation, and smart parking systems yielding "promising results." Ullah et al. (2020) identify the need for "more accurate and precise decision-making processes" involving "a large set of training data" in support of improved protocols for deep reinforcement learning (DRL). James et al. (2021) claim that "at the core of smart city platforms is the deep learning in neural networks" that enables processes to "reassess policies and services" based on "large amounts of dynamically accessed data to be used to make decisions on major modifications or adaptations in city activities" citing use cases pertaining to water consumption and conservation, energy management, public health, and economic development. Naoui, Lejdel, Ayad, Amamra, and Kazar (2021) advance "distributed deep learning for smart cities" consisting of a smart city layer, a big data layer, and a deep learning layer where the DL layer is designed for data analysis for energy and the environment. Band, Ardabili, Sookhak, Chronopoulos, Elnaffar, Moslehpour, Csaba, Torok. Pai, and Mosavi (2022) categorize the application of machine learning (ML) methods in smart cities, referring to deep learning (DL) as one of several advanced machine learning methods, finding DL techniques to have "higher accuracy than the hybrid models and ensembles" while requiring "relatively higher computation power."

3.3 Urban Life and Machine Learning (ML) in Smart Cities

Lugmayr, Kening, and Ma (2017) describe the meeting of artificial intelligence (AI) and machine learning (ML) with virtual and augmented realities in the context of semantic ambient media experiences (SAME). Describing machine learning (ML) as recently as 2018, Meserole (2018) claims that "the ability to recognize what is happening in the world with the ability to move and interact with it" demonstrate developments sufficient to "dramatically reshape our world." Habibzadeh et al. (2019) describe machine learning as "sophisticated data analysis" and explore the application of ML in smart cities focusing on one of several planes or viewpoints – the data plane – where, "detecting patterns and trends in past data" algorithms are said to learn and improve. Jordan (2019) argues that "in the public sphere" much "of what is labeled AI today is actually machine learning (ML)." Ullah et al. (2020) describe three categories of machine learning (ML) methods as supervised, unsupervised, and reinforcement learning (RL) in support of smart city applications for transportation, cyber-security, smart girds, and health care. Hurbean, Danaiata, Militaru, Dodea, and Negovan (2021) undertake a systematic review of the literature in the investigation of "open data-based machine learning applications" in smart cities, finding such applications in all six areas of smart cities (e.g., smart governance, economy, mobility, environment, people, living) with "specific ML techniques" found by area, in particular deep learning and supervised learning while "open data platforms represent the most frequently used source of data" and challenges pertaining to open data use are said to range "from quality of data to frequency of data collection, to consistency of data, and data format."

Margetis et al. (2021) use the term "informed machine learning" described as "learning from a hybrid information source that includes data as well as prior knowledge, where the latter is separate from the data and is additionally explicitly integrated into the learning pipeline." Interactive machine learning is also discussed where the key aim "is to directly engage end users in the training of a model" which serves to "assist users to increasingly explore the impact of their own actions" while "empowering them to adapt these actions in the subsequent learning circles in order to address their overall objectives." Federated learning is described by Margetis et al. (2021) as an approach that "promotes decentralized machine learning" by "engaging everyday devices owned and used by everybody for the benefit and improvement of AI" that is "privacy-preserving." Band et al. (2022) provide a systematic review of the research literature on machine learning (ML) methods in smart city applications from energy to healthcare to transportation, to security, and pollution. Band et al. (2022) classify ML algorithms into the four categories of decision trees, support vector machines, artificial neural networks, and advanced machine learning methods such as hybrid methods, ensembles, and deep learning (DL). Categories are assessed in terms of performance pertaining to accuracy, cost, computation power, and processing speed, finding that "differences are negligible." Olmez (2022) describes a project designed "to understand and simulate cities" using machine learning, focusing particularly on factors associated with the social dimension of cities as in "the impact of individual movements and decisions" in relation to how they "play out everyday" in the data context of smart cities, as in "data collected from fixed sensors" for elements such as "traffic counts, air pollution or movements of populations." Wiethof, Roocks, and Bittner (2022) describe the notion of interactive machine learning (iML) where the availability of data is limited providing for challenges to automatic ML so that "human and AI strengths" are combined in a movement "toward so called hybrid intelligence systems (HIS) involving a human-in-the-loop (HITL) approach" to AI "giving the user more control in the development and training of the ML system." As such, Wiethof et al. (2022) advance a prototype for combining gamification and ML" and "a complement to HIS research toward advancing HITL learning."

3.4 Urban Life and Ambient Learning in Smart Cities

Paraskakis (2005) introduces ambient learning as an "innovative e-learning service" in support of "lifelong learners and vocational trainees" where "the key distinguishing features" are identified as content integration, multimodal broadband access, and context management. According to Paraskakis (2005) ambient learning enables "access to high quality e-learning material at a time, place, pace and context that best suits the individual learner." Bick et al. (2007) refer to ambient learning in the context of mobile learning in digital environments, described as "contextualized, personalized knowledge for learners." Graham (2008) argues that "the city sends you a message" claiming that the city of Boston and more particularly Cambridge, sends the message that "you should be smarter." According to Graham (2008) "a city speaks to you mostly by accident – in things you see through windows, in conversations you overhear" such that "It's not something you have to seek out, but something you can't turn off." However, Graham (2008) cautions that "not all cities send a message" claiming that "only those that are centers for some type of ambition do" adding that "it can be hard to tell exactly what message a city sends without living there." It could be argued that the messaging described by Graham (2008) requires an awareness, as in ambient awareness, and that learning to read and receive such a message constitutes a form of ambient learning.

Ravenscroft, Sagar, Baur, and Oriogun (2010) describe ambient pedagogy as "the structure or scaffolding supporting the learning interaction" which "is 'behind the scenes' and yet also implicit in the digital practice that is supported." Exploring ambient and emergent learning in the context of awareness-based emerging technologies, McKenna, Arnone, Kaarst-Brown, McKnight, and Chauncey (2013) found "the need for autonomy and control affordances for people" taking the needs of people into consideration. Sassen (2014) explores the notion of whether the city has speech in terms of "urban capabilities" and "a making of presence" reminiscent perhaps of the notion by Graham (2008) of city messaging. Gams Gu, Härmä, Muñoz, and Tam (2019) note that artificial intelligence is also referred to as machine intelligence

(MI), as in, "the intelligence demonstrated by machines, in contrast to the natural intelligence (NI) displayed by humans and other animals" involving "perceiving, processing input from the environment, and learning." In the context of smart cities, McKenna (2019) developed a conceptual framework for ambient learning, play, and inclusion where the ambient is described "in relation to digital data, education, and computational urbanism in digitally-enabled urban areas and regions" with an emphasis on aware people and awareness-enabling technologies giving rise to the notion of ambient literacies. McKenna (2021) provides a review of approaches to making the invisible visible in smart cities such as computational intelligence (CI) and communication networks (CN). Buseyne, Raes, Rajagopal, Vanneste, Windey, De Croon, Gutiérrez Hernández, Htun, and Verbert, (2022) describe a project – Supporting TEamwork in AMbient Learning Spaces (STEAMS) – involving the learning and teaching of collaborative problem solving (CPS) skills.

3.5 Literature Review Overview

An overview by author and date is provided in this section for the research and practice literature for artificial intelligence (AI), deep learning (DL), machine learning (ML), and ambient learning.

Table 1 provides an overview by author and date of the research and practice literature for urban life and the ambient pertaining to artificial intelligence (AI), spanning the timeframe of 2019 to 2022.

Jordan (2019) identifies challenges associated with AI and bringing about human-computer interactions to enhance human life and trigger new levels of human creativity. Loi (2019) explores intelligent systems (IS) globally in terms of attitudes, expectations, perceptions, and thresholds. Amabile (2020) encourages study into the potential of AI and creativity to learn more about creative breakthroughs and impacts from multiple perspectives. Arrieta et al. (2020) describe explainable AI (XAI) in terms of methods, concepts, and the like while pointing to opportunities and challenges pertaining to data privacy and responsible AI. Cugurullo (2020) describes urban artificial intelligence (AI) as enabling the transition from automation to autonomy, giving rise to issues of trust and ethics in autonomous cities. In formulating an agenda for urban AI, Luusua and Ylipulli (2020) use an interdisciplinary perspective involving engineering, philosophy, and design in cities, giving rise to key issues and challenges associated with ethics and data privacy, to name a few. Morris (2020) discusses AI and accessibility highlighting the importance of the sociotechnical context. Thieme et al. (2020) make the case for the design of AI systems to assist people in developing their own capabilities, highlighting the notion of interpretability. Ullah et al. (2020) provide a review of the application of AI in smart cities where challenges and future directions are found to include accuracy and precision, standardization, and security. Zhang and He (2020) address the area of smart technologies and urban life where the use AI in the form of open source data and deep learning is encouraged to assist in better understanding urban issues to serve the needs of society. Augusto (2021) questions whether the complex issues associated with the artificial systems ecosystem will be left to evolution or will involve planning and direction. James et al., (2021) argue that increasingly people are interacting with technologies (e.g., computers, robots, AI, etc.) rather than other humans which will affect sensemaking of the world. Jha et al. (2021) provide a review of AI and the Internet of Things (IoT) use cases for urban planning in support of smart cities. Littman et al. (2021) describe an interdisciplinary 100-year study on AI around the world with 5-year reports for the public, governments, and researchers offering collected and connected reflections on AI and its influences, including synthesis and assessments for directions going forward to benefit individuals and society. Margetis et al. (2021) advance a framework for human-centered design (HCD) focusing on AI

Table 1. Overview of research for artificial intelligence (AI) in smart cities

Author(s)	Year	Artificial Intelligence (AI)
Jordan	2019	Challenges of HCI to enhance human life, trigger human creativity
Loi	2019	Intelligent systems: perceptions, attitudes, thresholds, expectations globally
Amabile	2020	Study into potential of creativity and AI to learn about creative breakthroughs
Arrieta et al.	2020	Explainable AI (XAI): methods, concepts; data privacy, responsible AI
Cugurullo	2020	Urban AIs: automation to autonomy, rudimentary thinking, trust, ethics
Luusua & Ylipulli	2020	Agenda for urban AI: engineering, philosophy, design; ethics, data privacy
Morris	2020	AI and accessibility, sociotechnical context
Thieme et al.	2020	Design of AI systems: resource to extend human capabilities, interpretability
Ullah et al.	2020	Review of AI application in SCs, accuracy/precision, standardization, security
Zhang & He	2020	Smart technologies & urban life, open source data, deep learning
Augusto	2021	Artificial systems ecosystem complexities, evolution or planned & directed
James et al.	2021	Technologies as interactants affecting how people make sense of the world
Jha et al.	2021	AI & IoT use cases for urban planning and SCs
Littman et al.	2021	Interdisciplinary 100-year longitudinal study on AI, collected/connected reflections
Margetis et al.	2021	Framework for HCD, AI for good, human in the loop, collaboration, ethics, trust
Medaglia et al.	2021	AI in government, research: governance, trustworthy, impact, data governance
NAII	2021	Framework for AI research: innovation, trust, education, infrastructure, appl, IC
Ourednik et al.	2021	AI and social media platforms: failed cities; algorithmic cities, cohabitation
Wu et al.	2021	AI Creativity, human-AI co-creation model: co-live, co-create; tech-society-educ
Shneiderman	2022	AI and HCI integration – amplify, augment, enhance human abilities – creativity

for good, a human-in-the-loop approach, and multidisciplinary collaboration, while being attentive to issues of ethics and trustworthy AI. Medaglia et al. (2021) focus on AI in government identifying the need for research in AI governance, trustworthy AI, impact assessment methodologies, and data governance. The National Artificial Intelligence Initiative (NAII, 2021) provides a framework for AI research based on the pillars of innovation, trustworthy AI, education and training, infrastructure, applications, and international cooperation. Ourednik et al. (2021) describe the state of AI and social media platforms as "failed cities" with the potential to become better algorithmic cities incorporating the common good, enabling civilized cohabitation. Wu et al. (2021) provide a creativity perspective on AI using the human-AI co-creation model in support of AI-human co-living and co-creating to enhance potential giving rise to opportunities and challenges for technology, society, and education. Shneiderman (2022) advances the integrating of AI (artificial intelligence) and HCI (Human-computer interaction) approaches in support of human creativity and the amplification, augmenting and enhancing of human capabilities.

Table 2 provides an overview by author and date of the research and practice literature for urban life and the ambient pertaining to deep learning (DL), spanning the timeframe of 2018 to 2022.

While Meserole (2018) describes the magic and flexibility of deep learning (DL), limitations are also acknowledged pertaining to the use of probabilistic models. Habibzadeh et al. (2019) describe the application of deep learning in smart cities on the data plane or viewpoint where raw data are converted to "*useful* information". Atitallah et al. (2020) provide a review of DL, IoT, and big data analytics in

Table 2. Overview of research for deep learning (DL) in smart cities

Author(s)	Year	Deep Learning (DL)
Meserole	2018	DL as flexible yet with limitations – probabilistic models
Habibzadeh et al.	2019	DL on the data plane in SCs – converting raw data to useful information
Atitallah et al.	2020	DL, IoT, big data analytics: new era of AI: energy, homes, health, transport
Das & James	2020	DL & SC technologies for social distancing – openness, transparency
Hajiabadi et al.	2020	DL method for PV energy generation forecasting in SCs
Puangpontip	2020	Approach to estimating SC DL computation demands – transport, parking ...
Ullah et al.	2020	Decision-making process/SCs: more accurate, precise; training data for DRL
James et al.	2021	SC platforms: DL in neural networks, reassess policies & services
Naoui et al.	2021	DL layer for SC data analysis – energy & environment
Band et al.	2022	DL as advanced ML method outperforms single methods, complexity cost

terms of a new era of AI with an array of applications for smart cities encompassing energy, managing the smart home, smart healthcare and smart transportation. Das and James (2020) describe the use of deep learning algorithms in smart cities for social distancing highlighting the importance of "openness and transparency." Hajiabadi et al. (2020) propose and test a deep learning method for photovoltaic (PV) energy generation forecasting in smart cities (SCs). Puangpontip and Hewett (2020) respond to "excessive usage of deep learning computation" providing an approach to the estimating of DL energy consumption usage in smart cities (SCs), focusing on smart drainage, smart transportation, and smart parking systems. Ullah et al. (2020) argue that decision-making processes in smart cities (SCs) need to be more accurate and precise, involving a large set of training data for deep reinforcement learning (DRL). James et al. (2021) claim that key to smart city (SC) platforms is the deep learning in neural networks where processes are able to reassess policies and services based on large amounts of dynamically accessed data, informing modification and adaptation systems for water, health, energy, and so on. Naoui et al. (2021) describe a deep learning layer for smart cities, designed so far for energy and the environment. Band et al. (2022) describe deep learning (DL) as one of several advanced machine learning (ML) methods, finding DL techniques that "outperform single methods at the cost of higher complexity and processing time."

Table 3 provides an overview by author and date of the research and practice literature for urban life and the ambient pertaining to machine learning (ML) spanning the timeframe of 2017 to 2022.

Lugmayr et al. (2017) describe the meeting of AI and ML with virtual and augmented realities in the context of semantic ambient media experiences (SAME). Meserole (2018) claims the ability of recognition, movement, and interaction in relation to events in everyday life through machine learning have the potential for the dramatic reshaping of the world as we know it. Habibzadeh et al. (2019) describe machine learning (ML) as "sophisticated data analysis" in smart cities on the data plane or data viewpoint where algorithms learn and improve upon patterns and trends detected from past data. Jordan (2019) claims that what passes as AI in the public sphere is machine learning (ML). Ullah et al. (2020) identify several categories of machine learning (ML) methods including supervised, unsupervised, reinforcement learning (RL) for a range of city applications such as transport, cyber-security, smart grids, and health care. Said to be important for future smart city projects, the work of Hurbean et al. (2021)

Table 3. Overview of research for machine learning (ML) in smart cities

Author(s)	Year	Machine Learning (ML)
Lugmayr et al.	2017	AI & ML meet virtual & ARs in semantic ambient media experience context
Meserole	2018	ML recognition + movement + interacting = reshaping of world
Habibzadeh et al.	2019	ML as sophisticated data analysis in SCs on the data plane to learn/improve on trends
Jordan	2019	AI in the public sphere as ML using DL and SL
Ullah et al.	2020	ML methods: supervised, unsupervised, RL: transport, cyber-security, grid, health
Hurbean et al.	2021	Open data-based ML applications in SCs, DL & SL, open data platforms, challenges
Margetis et al.	2021	Interactive; Federated; Informed ML: hybrid info source, data + prior knowledge
Band et al.	2022	Assessment of the application of ML methods in SCs
Olmez	2022	ML to understand & simulate cities – social + smart – traffic, air quality, etc.
Wiethof	2022	Interactive machine learning (iML); gamification, HITL learning

focuses on open data-based machine learning applications in smart cities, finding deep learning (DL) and supervised learning (SL) dominate with open data platforms as the primary source of data and key challenges pertain to quality, frequency of collection, consistency, and format. Margetis et al., (2021) use the term "informed machine learning" as a hybrid information source for learning, involving data and prior knowledge. Interactive ML is also discussed by Margetis et al., (2021), directly involving users in the training of a model, enabling them to explore the impact(s) of their actions and adjust accordingly while federated learning enables decentralized ML through everyday devices to benefit and improve AI while preserving privacy. Band et al. (2022) present a review of the application of machine learning (ML) methods in smart cities (SCs) assessing for accuracy, cost, computation power, and processing speed. Olmez (2022) advances the use of machine learning (ML) to understand and simulate the social and smart dimensions of cities focusing on issues pertaining to traffic volume, air pollution, and the movements of people. Wiethof, Roocks, and Wiethof (2022) highlight iML in advancing the use of gamification in support of human-in-the-loop learning.

Table 4 provides an overview by author and date of the research and practice literature for urban life and the ambient pertaining to ambient learning, spanning the timeframe of 2005 to 2022.

Paraskak (2005) presents ambient learning as an "innovative e-learning service" in support of "access to high quality e-learning material at a time, place, pace and context that best suits the individual learner." Bick et al. (2007) describe the notion of ambient learning in terms of mobile learning in digital environments, characterized as contextualized, personalized knowledge. Graham (2008) claims that a city sends a message which we hear as part of our experience of living in the city such as "be smarter" in the case of Cambridge in the larger city of Boston. Ravenscroft et al. (2010) articulate the notion of ambient pedagogies involving implicit and behind the scenes supports for learning interactions. McKenna et al. (2013) found the need for autonomy and control affordances in the context of awareness-based emerging technologies for ambient and emergent learning. Sassen (2014) speaks of cities having speech in terms of urban capabilities. Gams et al. (2019) note the use of AI for ML for machine intelligence as distinct from natural intelligence (NI) in the perceiving and processing from the environment of input for learning. McKenna (2019) developed a conceptual framework for ambient learning, play, and inclusion involving digital data, education, and computational urbanism (CU), giving rise to the need for ambi-

Table 4. Overview of research for ambient learning in smart cities

Author(s)	Year	Ambient Learning
Paraskak	2005	Innovative e-learning service at a time, place, pace & context suiting the learner
Bick et al.	2007	Mobile learning in digital environments – contextualized, personalized knowledge
Graham	2008	Messages from cities – be smarter
Ravenscroft et al.	2010	Ambient pedagogies – implicit and behind the scenes supports learning interactions
McKenna et al.	2013	Awareness-based emerging tech: autonomy & control affordances
Sassen	2014	Urban capabilities as speech
Gams et al.	2019	AI as ML in machine intelligence in contrast to NI processing environment/learning
McKenna	2019	Ambient learning framework (play & inclusion); data, educ, CU; ambient literacies
McKenna	2021	Making the invisible visible – CI and CN
Buseyne et al.	2022	STEAMS project for teaching & learning CPS skills

ent literacies. McKenna (2021) provides a review of approaches to making the invisible visible such as computational intelligence (CI) and communication networks (CN). Buseyne et al. (2022) introduce the TEamwork in AMbient Learning Spaces (STEAMS) project for learning and teaching collaborative problem solving (CPS) skills.

In terms of patterns and potential relationships, what emerges from this review of the research and practice literature is a comparative overview for ambient learning, artificial intelligence (AI), deep learning (DL), and machine learning (ML) as presented in Table 5.

Elements that emerge across columns of Table 5, include, but are not limited to, accuracy, autonomy, collaboration, computation, data, education, health, precision, privacy, reinforcement learning, and transport.

Table 5. Patterns and potential relationships

Ambient Learning	Artificial Intelligence (AI)	Deep Learning (DL)	Machine Learning (ML)
Ambient literacies	Accuracy	Accurate	Augmented reality
Ambient pedagogy	Autonomy	Computation power	Cyber security
Autonomy affordances	Collaboration	Data (training)	Data plane or viewpoint
Awareness	Creative AI	Data plane or view	Federated learning
City messaging	Data privacy	Decision making	Grid
Collaborative solutions	Education	Energy forecasting	Health
Computational urbanism	Ethics	Flexible	Hybrid – informed ML
Content integration	Explainable AI (XAI)	Health	Interactive
Contextualized	Governance	Homes	Methods
Control affordances	Human capabilities (extend)	Neural networks	Privacy-preserving
Digital data	Human creativity	Openness	Prior knowledge
Digital environments	Innovation	Policies	Public sphere
Education	Interdisciplinary	Precise	Reinforcement learning
Machine intelligence	Precision	Probabilistic models	Sophisticated data analysis
Mobile learning	Security	Reinforcement learning	Supervised
Multimodal	Sensemaking	Services	Transport
Natural intelligence	Trust	Transparency	Unsupervised
Personalized knowledge	Urban planning	Transport	Virtual reality

Table 6. Issues, controversies, and problems

Urban Life	Issues	Controversies	Problems
Artificial Intelligence	• Agency • Autonomy • Enhance human life • Ethics	• Behavioral/social • Data from humans • Human creativity • Urban AI	• Creativity and AI • Data privacy & surveillance • Security • Trust
Deep Learning	• Accuracy/Precision • Flexible • Openness • Transparency	• Decision making • Distributed • DL apps for SCs • Reassess policies	• Computation demands • Improving DRL protocols • Probabilistic models • Reassess policies/services
Machine Learning	• AI/ML confusion • Frequency • Open data • Quality of data	• Data analysis • Informed ML • Reshape the world • Un/Supervised/RL	• Data consistency • Decentralized • Interactive ML • Privacy
Ambient Learning	• Autonomy & Control • Content integration • Context management • Multimodal	• Ambient literacies • City messaging • Digital spaces • Mobile learning	• Ambient pedagogy • Making the invisible visible • Personalized • Collaborative solutions

3.6 Issues, Controversies, Problems

The review of the research and practice literature enables the surfacing of a range of issues, controversies, and problems for urban life and artificial intelligence, deep learning, machine learning, and ambient learning as identified in Table 6.

For artificial intelligence, issues, controversies, and problems pertain to agency; autonomy; ways of enhancing human life through human-computer interactions; ethics; behavioral and social interactions; data from and about humans; triggering new levels of human creativity; urban AI; creativity and AI; data privacy and surveillance; security; and trust.

For deep learning, issues, controversies, and problems pertain to accuracy and precision in decision-making; flexible; openness in relation to algorithms; transparency in relation to algorithms; distributed DL; emerging DL applications for many areas of smart cities (SCs); reassessing of policies and services; computation demands; improved protocols for deep reinforcement learning (DRL); limitations of probabilistic models; and the reassessing of urban policies and services.

For machine learning, issues, controversies, and problems pertain to confusion over what is understood to be AI types of ML employed; frequency of data collection; open data; quality of data; sophisticated data analysis; informed ML; potential for reshaping the world; un/supervised/RL; data consistency; decentralized ML; interactive ML; and privacy.

For ambient learning, issues, controversies, and problems pertain to the need for autonomy and control affordances; content integration; context management; multimodal; ambient literacies; city messaging; digital spaces; mobile learning; ambient pedagogy; approaches to making the invisible visible in smart cities; personalized; and collaborative problem solving.

Of note is the presence of overlap between AI, DL, ML, and ambient learning and across issues, controversies, and problems where data issues and creativity provide a few examples.

Figure 1. Conceptual framework for artificial intelligence in urban life and the ambient in smart cities

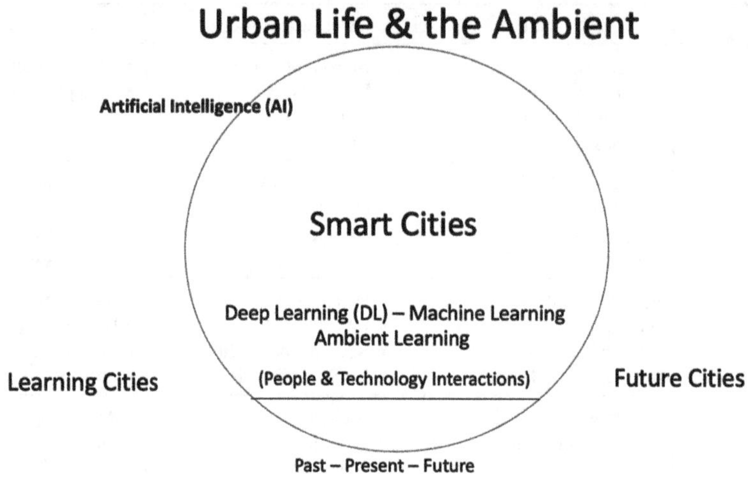

4. CONCEPTUALIZING ARTIFICIAL INTELLIGENCE IN SMART CITIES

Responding to the issues, controversies, and problems emerging from the literature review in the preceding section, a conceptual framework for artificial intelligence in urban life and the ambient in smart cities is formulated in Fig. 1, including deep learning (DL), machine learning (ML), and ambient learning involving people and technology interactions while being attentive to learning cities and future cities over time (past, present, and future).

The research question identified in Section 1 is reformulated as a proposition for exploration in this chapter, as follows:

P7: *Artificial intelligence, including machine learning and deep learning, manifests in urban life and the ambient in smart cities in increasingly pervasive ways in support of ambient learning, mobility, and many other in-the-moment, real time interactivities*

The research design used in exploring the proposition under study in this chapter is described in the Methodology section.

5. METHODOLOGY

Sections 5.1 to 5.3 describe the research design for this chapter in terms of the process used, sources of evidence for data collection, and data analysis.

5.1 Process

In view of the contemporary nature of the inquiry in this chapter, an exploratory case study approach is used (Yin, 2018) and combined with an explanatory correlational design since the two are said to be complementary (Yin, 2018) in that the former may be "needed to examine the underlying processes that might explain a correlation." Using an online space to describe the study and to invite people to participate, interest emerged from individuals in multiple cities and countries (Canada, United States, Greece, Ireland, Israel, etc.) touching several continents (North America, Europe, and the Middle East).

5.2 Sources of Evidence

The online website used for this study enabled the gathering of demographic data from individuals during sign up for the study, including age range, gender, and city. Additionally, people could self-identify in one or more categories (e.g., business, community member, student, educator, etc.). A survey instrument, pre-tested prior to use, was used in this study and contained closed and open-ended questions. Survey questions pertained to a range of items related to smart cities, such as, collaboration, creativity, mobility, and learning. Survey respondents were also invited to participate in an in-depth interview which was guided by an interview protocol that had been pre-tested prior to use in this study.

Additionally, data were gathered systematically through group and individual discussions, conducted in parallel with this study, with individuals from many sectors (e.g., students, faculty, and staff in higher education, business, community members and leaders, and city staff and leaders) in Canadian cities (e.g., Toronto, Vancouver, Victoria).

5.3 Data Analysis

Data analysis in this chapter is guided by the research question posed and the reformulated proposition for exploration, as follows:

P7: *Artificial intelligence, including machine learning and deep learning, manifests in urban life and the ambient in smart cities in increasingly pervasive ways in support of ambient learning, mobility, and many other in-the-moment, real time interactivities.*

For the analysis of quantitative survey data, the use of descriptive statistics pertaining to urban artificial intelligence (AI) along with other associated elements enabled exploration of items such as *collaboration* in relation to *creativity*. The correlation feature of the Real Statistics Resource Pack containing an-add in for Microsoft Excel (Zaiontz, 2022) enabled the conducting of a Spearman Correlation Coefficient for ordinal data in exploring the nature of the relationship between items.

Content analysis was used for qualitative data gathered through open-ended survey questions, interviews, and group and individual discussions so that inductive analysis from data collection was combined with deductive analysis using terminology from the literature review. Overall, an analysis was conducted for n=79 consisting of 42% females and 58% males for people ranging in age from their 20s to 70s.

6. RESULTS AND DISCUSSION

Table 7 provides the results of assessments for urban interactivities pertaining to human and AI creativity using the variables of *collaborating* and *creative opportunities.* In response to the survey question about whether city-focused social media and other aware technologies give rise to possibilities such as *collaborating*, on a 7-point Likert type scale, where 1= Not at all and 7=Absolutely, 17% of respondents

Table 7. Assessments for urban interactivities: collaborating and creative opportunities

Items	Urban Interactivities	Correlation
Collaborating	17% (5); 17% (6); 66% (7)	.81
Creative Opportunities	25% (3); 50% (6); 25% (7)	

selected positions 5 (Sort of) and 6 (Sure) on the scale while 66% responded at the upper end of the scale at position 7. When asked to what extent factors such as *creative opportunities* contribute to the livability of a smart city, 25% of survey respondents selected position 3 (Maybe), 50% responded toward the upper end of the scale at 6 (Sure), and 25% responded at the upper end position 7.

Using the correlation feature of the Real Statistics add-in for Microsoft Excel (Zaiontz, 2022), a Spearman correlation coefficient for ordinal data was conducted on the assessment results of the two items, *collaborating* and *creative opportunities* (Table 7), showing a finding of .81.

Correlations in the .66 – .85 range are said to be "very good" and "good prediction can result from one variable to the other" (Creswell, 2018). Going forward, this finding highlights the importance of collaboration in urban environments as well as spaces, resources, and supports for creative opportunities, which, taken together would seem to have good predictive potential.

Qualitatively, an educator commented that "people who have heard the term" smart city "think of big data and machine learning, often with relevance to sustainability" while "thinking of cities as learning ecologies" adding that "creative thinking must come into play to innovate smarter cities" where "ideas and people need to come first, then we invent the technology to support those ideas and purposes." Another educator and business person spoke of the need for "projects conducted by high-tech companies with collaboration with students and other community centers" incorporating "human values and needs" so as "to include both together, with the IoT, big data, and other tech systems to make the city a better place." Another educator highlighted the need for "collaboration projects between public authorities and community groups, where citizens are considered partners" while a public administration executive emphasized the need for "learning capabilities."

As depicted graphically in Fig. 2, the interconnectivity of artificial intelligence (AI) with deep learning, machine learning, and ambient learning is aided in urban life by ambient awareness in support of the notion and potentials of ambient literacies, ambient media, and ambient pedagogies.

7. LIMITATIONS AND FUTURE DIRECTIONS

Possible limitations of this work are discussed in this section opening the way to potentials for future directions regarding challenges and opportunities for research and practice.

Limitations

Limitations of this chapter pertain possibly to the extent of the literature review, focusing on urban artificial intelligence (AI) in relation to deep learning (DL), machine learning (ML), and ambient learning in smart cities and this is mitigated by the potential to extend such reviews of the literature going forward. Among other limitations of this chapter is that of the small sample size for the underlying study of this

Figure 2. Urban life and the ambient – awareness in smart cities

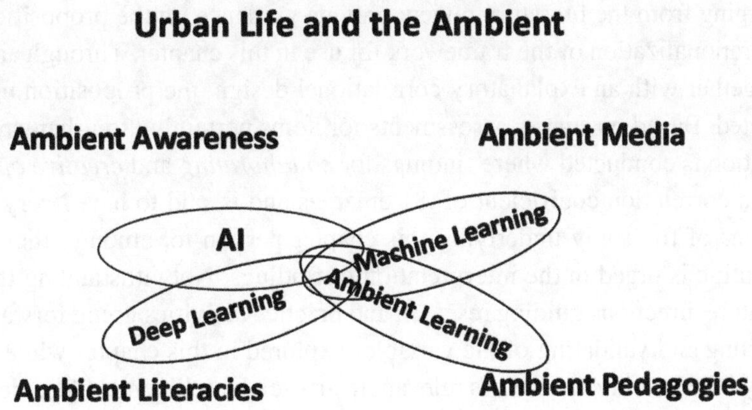

work, so that caution is encouraged in the interpretation of the correlation calculated from the assessment of *collaborating* and *creative opportunities* in smart cities. Going forward, opportunities present for the use of other possible variables, resulting in correlations with similarly strong prediction potentials.

Future Research Directions and Opportunities

For research, future directions emerge in terms of the potential for conducting additional explorations on the variables identified in this work relevant to artificial intelligence (AI). As such, opportunities present for further testing and validation of the variables explored in this chapter while explorations going forward exist for the identification of other similarly strong variables relevant to urban life that foster the notion of ambient learning in the context of urban AI, deep learning (DL), and machine learning (ML) in smart cities, learning cities, and future cities as identified in Table 5. A pathway is also opened for further explorations of ambient awareness, ambient media, ambient literacies, and ambient pedagogies.

Future Practice Directions and Opportunities

Future directions for practice include the potential for pursuing explorations with variables identified in this work relevant to urban AI and the ambient. In addition to deep learning (DL) and machine learning (ML), opportunities going forward for practice include the potential for exploring spaces with other, possibly stronger variables relevant to human creativity and AI creativity that foster real world action and learning in everyday contexts, as in, ambient learning in smart cities, learning cities, and future cities, in addressing the issues, controversies, and problems presented in Table 6.

8. CONCLUSION

Among the contributions of this chapter is that of identifying the challenges and opportunities associated with artificial intelligence (AI) in urban life emerging from the review of the research literature provided from multiple perspective involving machine learning (ML), deep learning (DL), and ambient learning. Another contribution of this chapter is the formulation of a conceptual framework for artificial intelligence in urban life and the ambient in smart cities guided by the literature review. Issues, controversies, and problems emerging from the literature review provide guidance on the proposition explored while contributing to operationalization of the framework for use in this chapter. Through an exploratory case study approach, together with an explanatory correlational design, the proposition under study in this chapter is investigated. Based on survey assessments for items pertaining to urban creativity and interactivities, a correlation is conducted where findings for *collaborating* and *creative opportunities* in the form of a Spearman correlation coefficient of .81 emerges and is said to have "very good prediction" potential. Limitations of the study underlying this chapter pertain to, among other things, the small sample size and caution is urged in the interpretation of findings. Notwithstanding this and other possible limitations, future directions guiding research and practice activities going forward are provided in terms of further testing and validating of the variables explored in this chapter while also encouraging activities to identify other, additional factors relevant to artificial intelligence (AI), creativity, and to the notion of ambient learning in the context of smart, learning, and future cities.

Key insights and ideas emerging from this chapter are provided, along with exercises encouraging further exploration, as follows:

Insights emerging from this chapter include the following:

1. The importance of city-focused social media and other aware technologies as factors giving rise to many possibilities for urban AI in support of more aware people in the form of *collaborating*
2. The importance of factors associated with urban livability and taking the needs of people into consideration in smart cities in the form of *creative opportunities*
3. The potential for many more factors to emerge for understanding urban AI, creativity, and the ambient in smart cities in the form of ambient learning

Ideas emerging from this chapter include the following:

1. Potential opportunities for exploring urban AI and the ambient in relation to a variety of variables (e.g., collaboration, creative opportunities) in the context of smart cities
2. The opportunities potential for the identification of other and additional variables for exploring urban AI and ambient learning in smart environments
3. The use of an interdisciplinary and multi-method approach for the study and practice of urban AI and the ambient in smart, learning, and future cities

Exercises for the reader to consider, encouraging further exploration include the following:

1. Identify one or more ways in which your city is using machine learning (ML) and/or deep learning (DL)
2. Identify one or more ways in which artificial intelligence (AI) would enhance the quality of urban life in your city
3. Describe one or more ways that ambient learning contributes to making your city smarter

The audience for this chapter will be interdisciplinary, including but not limited to community members and leaders, educators, businesses, researchers, urban planners, urban designers, and anyone concerned with artificial intelligence (AI) and creativity in urban life, while being attentive to elements such as deep learning (DL), machine learning (ML), and ambient learning along with associated practice and research opportunities and challenges in smart cities, learning cities, and future cities.

REFERENCES

Amabile, T. M. (2020). Guidepost: Creativity, artificial intelligence, and a world of surprises. *Academy of Management Discoveries*, *6*(3), 351–354. doi:10.5465/amd.2019.0075

Arrieta, A. B., Díaz-Rodríguez, N., Del Ser, J., Bennetot, A., Tabik, S., Barbado, A., Garcia, S., Gil-Lopez, S., Molina, D., Benjamins, R., Chatila, R., & Herrera, F. (2020). Explainable Artificial Intelligence (XAI): Concepts, taxonomies, opportunities and challenges toward responsible AI. *Information Fusion*, *58*, 82–115. doi:10.1016/j.inffus.2019.12.012

Atitallah, S. B., Driss, M., Boulila, W., & Ghézala, H. B. (2020). Leveraging deep learning and IoT big data analytics to support the smart cities development: Review and future directions. *Computer Science Review*, *38*, 100303. doi:10.1016/j.cosrev.2020.100303

Augusto, J. C. (2021). Smart Cities: State of the Art and Future Challenges. In J. C. Augusto (Ed.), *Handbook of Smart Cities*. Springer. doi:10.1007/978-3-030-15145-4_95-2

Band, S. S., Ardabili, S., Sookhak, M., Chronopoulos, A. T., Elnaffar, S., Moslehpour, M., Csaba, M., Torok, B., Pai, H.-T., & Mosavi, A. (2022). When smart cities get smarter via machine learning: An in-depth literature review. *IEEE Access: Practical Innovations, Open Solutions*, *10*, 60985–61015. doi:10.1109/ACCESS.2022.3181718

Bick, M., Kummer, T., Pawlowski, J. M., & Veith, P. (2007). Standards for ambient learning environments. *MMS 2007: Mobilität und mobile Informationssysteme; Proceedings of the 2nd Conference of GI-Fachgruppe MMS*, 103–114.

Buseyne, S., Raes, A., Rajagopal, K., Vanneste, S., Windey, I., De Croon, R., Gutiérrez Hernández, F., Htun, N. N., & Verbert, K. (2022). Supporting TEamwork in AMbient learning Spaces (STEAMS) by Developing Content and Tools. *International Society of the Learning Sciences (ISLS) Annual Meeting Proceedings*.

Cheshire, J., & Uberti, O. (2021). *Atlas of the invisible: Maps and graphics that will change how you see the world.* Norton.

Creswell, J. W. (2018). *Educational research: Planning, conducting, and evaluating quantitative and qualitative research* (6th ed.). Pearson.

Cugurullo, F. (2020). Urban artificial intelligence: From automation to autonomy in the smart city. *Frontiers in Sustainable Cities, 2*(28), 1-14. DOI: doi:10.3389/frsc.2020.00038

Das, R., & James, P. (2020). How smart city technology can be used to measure social distancing. *The Conversation.* Retrieved 3 October 2022 from https://theconversation.com/how-smart-city-technology-can-be-used-to-measure-social-distancing-135139

Gams, M., Gu, I. Y. H., Härmä, A., Muñoz, A., & Tam, V. (2019). Artificial intelligence and ambient intelligence. *Journal of Ambient Intelligence and Smart Environments, 11*(1), 71–86. doi:10.3233/AIS-180508

Graham, P. (2008). Cities and ambition. *Blog*. Retrieved 2 July 2022 from http://www.paulgraham.com/cities.html

Habibzadeh, H., Kaptan, C., Soyata, T., Kantarci, B., & Boukerche, A. (2019). Smart city system design: A comprehensive study of the application and data planes. *ACM Computing Surveys, 52*(2). doi:10.1145/3309545

Hajiabadi, M., Farhadi, M., Babaiyan, V., & Estebsari, A. (2020). Deep learning with loss ensembles for solar power prediction in smart cities. *Smart Cities, 3*(3), 842–852. doi:10.3390martcities3030043

Hurbean, L., Danaiata, D., Militaru, F., Dodea, A.-M., & Negovan, A.-M. (2021). Open data based machine learning applications in smart cities: A systematic literature review. *Electronics (Basel), 10*(23), 2997. doi:10.3390/electronics10232997

James, P., Astoria, R., Castor, T., Hudspeth, C., Olstinske, D., & Ward, J. (2021). Smart cities: Fundamental concepts. In J. C. Augusto (Ed.), *Handbook of Smart Cities*. Springer. doi:10.1007/978-3-030-69698-6_2

Jha, A. K., Ghimire, A., Thapa, S., Jha, A. M., & Raj, R. (2021). A review of AI for urban planning: Towards building sustainable smart cities. *2021 6th International Conference on Inventive Computation Technologies (ICICT)*, 937-944. doi: 10.1109/ICICT50816.2021.9358548

Jordan, M. I. (2019). Artificial intelligence – The revolution hasn't happened yet. *Harvard Data Science Review*. DOI: doi:10.1162/99608f92.f06c6e61

Littman, M. L., Ajunwa, I., Berger, G., Boutilier, C., Currie, M., Doshi-Velez, F., Hadfield, G., Horowitz, M. C., Isbell, C., Kitano, H., Levy, K., Lyons, T., Mitchell, M., Shah, J., Sloman, S., Vallor, S., & Walsh, T. (2021). *Gathering Strength, Gathering Storms: The One Hundred Year Study on Artificial Intelligence (AI100) 2021 Study Panel Report*. Stanford University. Retrieved 27 January 2022 from https://ai100.stanford.edu/gathering-strength-gathering-storms-one-hundred-year-study-artificial-intelligence-ai100-2021-study

Loi, D. (2019). Ten guidelines for intelligent systems futures. *Proceedings of the Future Technologies Conference (FTC) 2018. FTC 2018. Advances in Intelligent Systems and Computing*, 880. 10.1007/978-3-030-02686-8_59

Lugmayr, A., Kening, Z., & Ma, X. (2017). Artificial intelligence MEETS virtual and augmented realities. SIGGRAPH Asia 2017 Workshops, 2a:1–2a:1.

Luusua, A., & Ylipulli, J. (2020). Urban AI: Formulating an agenda for the interdisciplinary research of artificial intelligence in cities (Workshop). *DIS'20, Companion Publication of the 2020 ACM Designing Interactive Systems Conference*. DOI:10.1145/3393914.3395905

Margetis, G., Ntoa, S., Antona, M., & Stephanidis, C. (2021). Human-centered design of artificial intelligence. In Handbook of Human Factors & Ergonomics (5th ed.). Wiley. doi:10.1002/9781119636113.ch42

Marzano, S. (2006). Ambient culture. In True Visions. Springer. doi:10.1007/978-3-540-28974-6_3

McKenna, H. P. (2019). *Ambient Urbanities as the Intersection Between the IoT and IoP in Smart Cities*. IGI Global. doi:10.4018/978-1-5225-7882-6

McKenna, H. P. (2021). *Visibilities and Invisibilities in Smart Cities: Emerging Research and Opportunities*. IGI Global. doi:10.4018/978-1-7998-3850-0

McKenna, H. P., Arnone, M. P., Kaarst-Brown, M. L., McKnight, L. W., & Chauncey, S. A. (2013). Ambient and emergent learning with wireless grid technologies. In *Proceedings of the 5th International Conference on Education and New Learning Technologies (EduLearn13)*. International Association of Technology, Education, & Development (IATED).

Medaglia, R., Gil-Garcia, J. R., & Pardo, T. A. (2021). Artificial intelligence in government: Taking stock and moving forward. *Social Science Computer Review*. Advance online publication. doi:10.1177/08944393211034087

Meserole, C. (2018). *What is machine learning?* The Brookings Institution, Series: A Blueprint for the Future of AI: 2018-2019. Retrieved 2 October 2022 from https://www.brookings.edu/research/what-is-machine-learning/

Miailhe, N., & Hodes, C. (2017). The third age of artificial intelligence. *Field Actions Science Reports. Special Issue on Artificial Intelligence and Robotics in the City.*, *17*, 6–11.

Morris, M. R. (2020). AI and accessibility. *Communications of the ACM*, *63*(6), 35–37. doi:10.1145/3356727

NAII. (2021). *About*. US Government, National Artificial Intelligence Initiative. Retrieved 17 December 2021 from https://www.ai.gov/about/#NAII-NATIONAL-ARTIFICIAL-INTELLIGENCE-INITIATIVE

Naoui, M. A., Lejdel, B., Ayad, M., Amamra, A., & Kazar, O. (2021). Using a distributed deep learning algorithm for analyzing big data in smart cities. *Smart and Sustainable Built Environment*, *10*(1), 90–105. doi:10.1108/SASBE-04-2019-0040

Olmez, S. (2022). *Bringing the social city to the smart city: Using AI and machine learning to understand and simulate cities.* Research Project, The Alan Turing Institute. Retrieved 2 October 2022 from https://www.turing.ac.uk/research/research-projects/bringing-social-city-smart-city

Ourednik, A., Mlynář, J., Mutzner, N., & Alavi, H. (2021). Social media platforms are failed cities. *Interaction*, *28*(6), 62–66. doi:10.1145/3488234

Paraskakis, I. (2005). Ambient learning: A new paradigm for e-Learning. *Proceedings of the 3rd International Conference on Multimedia and ICT in Education (m-ICTE 2005): Recent Research developments in Learning Technologies*.

Puangpontip, S., & Hewett, R. (2020). Energy usage of deep learning in smart cities. *2020 International Conference on Computational Science and Computational Intelligence (CSCI)*, 1143-1148. 10.1109/CSCI51800.2020.00214

Ravenscroft, A., Sagar, M., Baur, E., & Oriogun, P. (2010). Ambient pedagogies, meaningful learning and social software. In S. Dasgupta (Ed.), *Social Computing: Concepts, Methodologies, Tools, and Applications* (pp. 472–489). IGI Global. doi:10.4018/978-1-60566-984-7.ch032

Sassen, S. (2014). Does the city have speech? In T. Haas & K. Olsson (Eds.), *Emergent Urbanism: Urban Planning and Design in Times of Structural and Systemic Change*. Routledge.

Shneiderman, B. (2022). *Human-centered AI: A new synthesis*. Keynote presentation at the Human-Computer Interaction International Conference (HCII2022).

Thieme, A., Cutrell, E., Morrison, C., Taylor, A., & Sellen, A. (2020). Interpretability as a dynamic of human-AI interaction. *Interaction, 27*(5), 40–45. doi:10.1145/3411286

Ullah, Z., Al-Turjman, F., Mostarda, L., & Gagliardi, R. (2020). Applications of artificial intelligence and machine learning in smart cities. *Computer Communications, 154*, 313–323. doi:10.1016/j.comcom.2020.02.069

Wiethof, C., Roocks, T., & Bittner, E. A. C. (2022). Gamifying the human-in-the-loop: Toward increased motivation for training AI in customer service. In H. Degen & S. Ntoa (Eds.), Lecture Notes in Computer Science: Vol. 13336. *Artificial Intelligence in HCI. HCII 2022*. Springer. doi:10.1007/978-3-031-05643-7_7

Wu, Z., Ji, D., Yu, K., Zeng, X., Wu, D., & Shidujaman, M. (2021). AI creativity and the human-AI co-creation model. In M. Kurosu (Ed.), Lecture Notes in Computer Science: Vol. 12762. *Human-Computer Interaction: Theory, Methods and Tools. HCII 2021*. Springer. doi:10.1007/978-3-030-78462-1_13

Yin, R. (2018). *Case study research and applications: Design and methods* (6th ed.). Sage Publications.

Zaiontz, C. (2022). *Real statistics using excel*. Retrieved from www.real-statistics.com

Zhang, J., & He, S. (2020). Smart technologies and urban life: A behavioral and social perspective. *Sustainable Cities and Society, 63*, 102460. doi:10.1016/j.scs.2020.102460

Zhang, W. J., Yang, G., Lin, Y., Ji, C., & Gupta, M. M. (2018). On Definition of Deep Learning. *2018 World Automation Congress (WAC)*, 1-5. DOI: 10.23919/WAC.2018.8430387

ADDITIONAL READING

As, I., & Basu, P. (2022). *Artificial Intelligence in Urban Planning and Design: Technologies, Implementation, and Impacts* (1st ed.). Elsevier.

Bengio, Y., Lecun, Y., & Hinton, G. (2021). Deep learning for AI. *Communications of the ACM, 64*(7), 58–65. doi:10.1145/3448250

Davis, N. (2021). 'Yeah we're spooked': AI starting to have big real-world impact, says expert. *The Guardian*. Retrieved 29 October 2021 from https://www.theguardian.com/technology/2021/oct/29/yeah-were-spooked-ai-starting-to-have-big-real-world-impact-says-expert

ITU. (2021). *AI for good*. Retrieved September 28 from https://aiforgood.itu.int

Joshi, N. (2022). The myriad applications of ambient intelligence in healthcare. *Forbes*. Retrieved January 22, 2022, from https://www.forbes.com/sites/naveenjoshi/2022/01/09/the-myriad-applications-of-ambient-intelligence-in-healthcare/?sh=58c762bb424d

Naughton, J. (2021). Worried about super-intelligent machines? They are already here. *The Guardian*. Retrieved 28 December 2021 from https://www.theguardian.com/commentisfree/2021/dec/25/worried-about-super-intelligent-machines-they-are-already-here

Page, R. (2022). How machine learning could change city design: The creation of more sustainable, liveable cities. *Information Age*. Retrieved 21 February 2022 from https://ia.acs.org.au/article/2022/how-machine-learning-could-change-city-design.html

Pretz, K. (2021). Stop calling everything AI, machine-learning pioneer says. *IEEE Spectrum*. Retrieved March 12, 2022, from https://spectrum.ieee.org/stop-calling-everything-ai-machinelearning-pioneer-says

Stuart, S. C. (2021). Your smart city is boring: How to preserve its soul when the AI sensors move in. *PCmag*. Retrieved 28 December 2021 from https://www.pcmag.com/news/your-smart-city-is-boring-how-to-preserve-its-soul-when-the-ai-sensors

KEY TERMS AND DEFINITIONS

Ambient Collaboration: Ambient collaboration in a smart city context refers to working together in a dynamic way, involving interactions with technology and with each other to create or improve one or more aspects of life in the city.

Collaboration: Collaboration in a smart city context refers to working together to create or improve some aspect of life in the city.

Creative Opportunities: Creative opportunities in the context of smart environments refer to spaces, resources, supports, and the like enabling the use of imaginative and original ideas.

Chapter 8
Smart Technologies and the Augmented and Virtual in Smart Cities:
Urban Life With Smart Homes, Electric Vehicles, Robotics, and Extended Realities

ABSTRACT

The purpose of this chapter is to explore urban life and the ambient in relation to smart technologies, the augmented and virtual, electric vehicles, robotics, and extended realities. A review of the research literature is conducted in the context of smart, learning, and future cities for smart homes, electric vehicles, robotics, and extended realities. Issues, controversies, and problems emerging from the literature review are highlighted, contributing to formulation of a conceptual framework for smart technologies, the augmented, and virtual in urban life and the ambient in smart cities. Using an exploratory case study approach combined with an explanatory correlational design, variables pertaining to the exploration are identified, and the nature of their relationship is assessed. Through the lens of smart technologies in urban life, ranging from smart homes to extended realities, applications are explored in smart cities while informing directions for future research and practice.

1. INTRODUCTION

Zhang and Shenjing (2020) provide an introduction to a special issue of smart technologies and urban life that aims "to generate a comprehensive understanding of the interrelations" impacting humanity, nature, sustainability, and society where smart technologies are defined broadly and are said to include "smartphone apps, apps in other mobile devices, wearable devices such as smart watches, autonomous vehicles, intelligent transport systems, drones, robots, smart houses, and new types of data generated by smart technologies." Cugurullo (2020) employs the notion of the autonomous city in the context of

DOI: 10.4018/978-1-6684-4096-4.ch008

smart cities to "stress the presence of urban artificial intelligence capable of thinking and acting in an unsupervised manner" involving "autonomous cars, robots, and city brains" that "are increasingly performing tasks and taking on roles which have traditionally been the domain of humans." Stanney, Nye, Haddad, Hale, Padron, and Cohn (2021) articulate a "virtuality continuum" consisting of the real world and extending to augmented reality to mixed reality and to virtual reality. This chapter is significant in that it builds on the research literature for such smart technologies giving rise to the motivation for an exploration of smart homes, electric vehicles, robotics, and extended realities in the context of urban life and the ambient in smart cities, learning cities, and future cities.

Objectives: The primary objectives of this chapter are to a) explore urban life and the ambient in relation to smart homes, electric vehicles (EV), and robotics in the context of smart cities; b) formulate a conceptual framework for smart technologies, the augmented, and the virtual in urban life and the ambient in smart cities; and c) explore the nature of the relationship between extended realities and smart cities in learning more about the ambient. These objectives give rise to the main research question under exploration in this chapter – *How do smart homes, electric vehicles, extended realities, and robotics contribute to urban life and the ambient in smart cities?*

2. BACKGROUND

A smart home concept is described by Marzano (2006) in terms of a system of modules that would assist people while being unobtrusive, as in, "more or less invisible, either hidden completely out of sight or incorporated into objects that have always belonged in the home" such as tables, chairs, cupboards, lamps, and the like. Soar, Livingstone, and Wang, (2009) describe the notion of ambient living in the context of a health care model of wellness in Australia. Robotics findings by Studley and Little (2021) identify "the necessity of benchmarking" to discern "the state-of-the-art in Robotics" in order to continue learning about "robot-human interactions and co-working." Celani, Falcone, and d'Alessandro (2019) describe the notion of the augmented city "where citizens and communities are the main actors of urban intelligence" and "they are encouraged to create and co-design their services" based on "the real needs of people in the city."

2.1 Definitions

Definitions from the research literature are provided for key terms used in this chapter.

- **Augmented Reality:** Hopkins, Bae, Uhr, Banić, Zheng, and Do (2021) describe augmented reality (AR) as interfaces such as "screens, glasses, head-mounted displays, projection" that "superimpose digital 3D content on the user's view of the physical world using image recognition or location tracking to overlay digital content in a real-world location in real-time."
- **Extended Reality:** Stanney et al. (2021) describe eXtended reality (XR) as "a technical capability that offers a range of computer-generated immersive experiences that mirror reality to varying degrees" and is said to have become "a common-place tool for training and education, collaborating and partnering, networking and entertaining, and other applications which have yet to be envisioned and realized."

- **Mixed Reality (MR):** Stanney et al. (2021) describe mixed reality as an "augmenting of the real world with virtualized content" that "allows the virtual content to be aware of and interact with the real world; or vice versa with real objects in a virtual world" as in "Augmented Virtuality (AV)" said to be "another form of MR."
- **Virtual Reality (VR):** Stanney et al. (2021) describe virtual reality as "an entirely simulated virtual world."

2.2 Overview

This chapter provides a review research and practice literature for urban life and eXtended reality, including augmented, mixed, and virtual reality, as well as smart homes, electric vehicles, and robotics in smart cities. This review enables the highlighting of key issues, controversies, and problems and the formulation of a conceptual framework for smart technologies, the augmented, and the virtual in urban life and the ambient in smart cities that is then operationalized for use in this chapter. The main research question under exploration in this chapter is reformulated as a proposition for investigation and the methodology used is described, followed by the presentation of results and a discussion of findings. Key limitations of the study underlying this chapter are identified and future directions for research and practice are outlined. The conclusion provides an overview of findings, the identification of insights and ideas, development of several exercise questions encouraging further exploration, and identification of the target audience for this chapter.

3. PERSPECTIVES ON URBAN LIFE AND THE AMBIENT: EVERYDAY APPLICATIONS

This chapter provides a review of the of the research and practice literature for urban life and the ambient in relation to evolving everyday applications such as smart homes, electric vehicles, robotics, and eXtended reality, including augmented, mixed, and virtual reality in smart cities.

3.1 Smart Homes in Smart Cities

Solaimani, Keijzer-Broers, and Bouwman (2015) provide a systematic analysis of the smart home literature focusing on technological as well as "non-technological aspects" such as "social-organizational, economical, organizational, law/legislation, and entrepreneurial from a strategic and an operational perspective" while preferring the term "smart living" as less limiting to that of "smart home". Wilson, Hargreaves, and Hauxwell-Baldwin (2015) provide a systematic review of the literature for smart homes and their users, highlighting three perspectives – functional, instrumental, and socio-technical – along with the concerns of privacy and control as well as access to data, trust, reliability, and transparency. Mechant, All, and De Marez (2018) develop a framework "to explore how technology enables and constrains agency and engagement in smart homes" as a methodology for exploring user experience. Baig, Himarish, Panaya, and Ahmed (2018) speak in terms of "ultra-aware cities" for the future and the need for "a 3 tier cognitive architecture" for smart homes consisting of perception, attention-memory, and decision layers, articulated through a framework based on use of the Internet of Things (IoT) and Information and Communication Technology (ICT). Sahni, Cao, and Shen (2018) use the example of smart homes to explore the challenges and opportunities of designing smart spaces advancing the need for a more user-centric approach. Loi (2019) provides a reflection on "the role that intelligent systems and ambient computing may play in future homes and cities" focusing on the 65+ population, advocating for participatory design approaches with an emphasis on developing "with (instead of for)" people.

From a sociological perspective, Kylili, Afxentiou, Georgiou, Panteli, Morsink-Georgalli, Panayidou, Papouis, and Fokaides (2020) explore the environmental impact of remote working in Europe during the COVID-19 global pandemic claiming that "working from home, or the reduced time and distance traveled to a coworking space" based on findings "contributes to the tackling of alarming environmental issues and the transition toward the establishment of smart cities and communities" wherein "within a very short amount of time, the shift of a great number of everyday activities from the industry and office environments to the homes" occurred where "the potential of the role of teleworking in the future smart cities has surfaced." As such, Kylili et al. (2020) "established impact indicators that demonstrate the contribution of remote working models in tackling energy and environmental challenges for the transition of European cities to smart energy regions" and thus, "demonstrate the degree the adoption of remote working models can help to combat energy and environmental issues" and in doing so "contribute to the development of smart cities." According to Kylili et al. (2020), challenges associated with smart homes are said to be associated with "technological solutions for support and security." Sessa (2020) explores the smart home from a sociological perspective, while identifying and describing the features of a smart home – comfort, solidity, flexibility, security, interconnection, sustainability, low cost, and inclusion. Kim et al. (2021) provide a review of the research literature for smart energy conservation in relation to smart homes and smart sustainable cities. Bisson (2022) claims that "ambient computing has arrived" in the form of a mixing of "the Internet of Things (IoT), machine learning, and hyperscale compute cloud" rendering computing as ignorable in that "it's there, but it's in the background, doing the job we've built it to" as in "a computer you use without knowing that you're using it" using the example of "a smart speaker" as "the interface to a smart home" or simply walking into a room and the lights come on based on your presence. Emekci (2022) points to "the potential contribution of smart homes to smart city concepts" in relation to "energy efficiency and GHG emissions with the help of the bottom up approach" where GHG refers to greenhouse gas. It is worth noting the discussion by Loi (2022), from a design perspective, on the distinction between a home and a house where a house is said to be "a space one buys or rents" while home is "the place that one lives in." In this sense, a home is described by Loi (2022) as "a much more complex artifact to design" involving "surfaces, materials, colors, and volume" along with "a multisensorial nature" that is "defined by intangible trajectories, overlapping flows, invisible routines accumulated over time, and a rather complex web of interstitial spaces." Loi (2022) argues that three ingredients are needed to transform a house into a home, namely participation, continuous evolution, and intangible aspects (e.g., memories, etc.). As such, it would seem that such ingredients will be important to consider in the context of the smart home.

3.2 Electric Vehicles in Smart Cities

In an interview (Sisson, 2016), Ratti speaks of ambient technology enabled through "real time information" in support of "modes of transportation", giving rise to the notion of "ambient mobility" in smart cities. Potdar, Batool, and Krishna (2018) identify the risks and challenges associated with the adoption of electric vehicles (EV) in smart cities, ranging from the battery capacity of EVs to the availability of "efficient charging infrastructure" to "the consistent supply of raw materials needed for EVs", also stated as "unreliable supply of rare earth battery materials", to "waste management of used EV batteries", to name a few. Toh, Sanguesa, Cano, and Martinez (2020) address the area of transport in smart cities, exploring the notion of "smart roads" for future cities, particularly "electrified roads" enabling the automatic charging of electric vehicles (EVs), using an initiative in the city of Stockholm in Sweden as an example. Yeo and Lin (2020) focus on "the notions of sensation, action, and encounter" to propose "a reconceptualization of autonomous vehicles that better accounts for human agency and sentience in an increasing automated urban world" and in doing so, undertake "to illuminate the emergent and creative potential of urban life to unsettle and complicate the taken-for-granted power of smart and autonomous technologies in enacting urban transformations." Jain, Mani, and Siddiqui (2021) address concerns with design challenges for electric vehicles (EVs) in smart cities in terms of efficient integration and associated impacts pertaining to costs, power grid demand, and sustainable mobility. From a business, urban design, and innovation perspective, Jones (2021) discusses how the rapid move to electric mobility including electric vehicles (EVs), e-bikes, e-scooters, electric taxis and the like is contributing to an electric ecosystem that is reshaping cities in terms of the "broader implications for how we might reimagine our urban environments" including streets and buildings while "EVs will become technology platforms" giving rise to "data and energy wallets" with, for example, "the option to sell the excess energy in our EV batteries." Ruggieri, Ruggeri, Vinci, and Poponi (2021) address the challenge of clean energy consumption in smart cities pointing to the importance of "the decarbonization of transport" in the form of "electric mobility" in support of efforts to "reduce polluting emissions on the road" exploring the mobility transition strategies of six European cities. Sanguesa, Torres-Sanz, Garrido, Martinez, and Marquez-Barja (2021) provide a review of the research literature focusing on advances in electric vehicle (EV) developments pertaining to battery technology trends (lead-acid to lithium-ion); charging methods, standards, and sources; along with "new research challenges and open opportunities" and "future prospects." Battery developments are said to be critical for factors such as EV autonomy, standards, as well as for sustainability and environmental impact, in terms of materials used and their recyclability (Sanguesa et al., 2021). From an urban planning and sociology of technology perspective, Freudendal-Pedersen (2022) explores "the role and significance of communities and emotions in mobile everyday lives" encouraging thinking beyond the car and "electrifying the car fleet" to other possible types and alternatives. Skjølsvold, Henriksen, and Ryghaug (2022) explore electric vehicle charging infrastructures at "the intersection of the smart grid and smart city" finding spaces for engagement, discussion, and action while noting that "smart can be interpreted both as outcome and as a driving force."

3.3 Robotics in Smart Cities

Dainow (2017) includes robotics as one of "the essential functional characteristics of the smart city" contributing to an environment with "both humans and machines moving within the same spaces" and this is said to "require adjustments in human behavior." Macrorie, Marvin, and While (2021) explore robotics and automation systems in smart cities in terms of how "they are being layered upon existing urban digital networks, extending the capabilities and capacities of human agency" on the one hand and "infrastructure networks" on the other, while "reshaping the city and citizen's everyday experiences." Myung and Wang (2021) introduce a special issue on robotic sensing and smart cities, focusing on areas such as construction automation and construction informatics in support of "sensing, analysis, and design" for sustainable built environments involving "intensive collaborations" for applications such as autonomous vehicles, human-robot interaction systems, inspection robots, and mobile robots. Song, Song, and Zhang (2021) explore swarm robotics and coordinated navigation in a smart city using RoboMaze as a simulator space to demonstrate various real world applications, whether for autonomous vehicles and the optimizing of routes and urban traffic, or emergency situations where people need to be evacuated from a building in a coordinated way. RoboMaze is said to provide an environment for "the development and test[ing] of new algorithms" and additionally "serves as a wonderful education and outreach tool for smart city awareness" (Song, Song, and Zhang, 2021). Studley and Little (2021) argue that "robots could have an enormous impact on the way that cities are designed and operate" through an initiative – the Robots in Smart Cities Challenge – designed to "promote public discussion" adding that "the closer integration of such 'autonomous' machines into our living, working public spaces will similarly change the way we interact with the city and with each other." Among the findings (Studley and Little, 2021) were "how accepting the inhabitants of Milton Keynes are to the gathering and use of data within their city, and also the extent to which they would trust robots to perform the everyday tasks which they had seen them attempt in the episodes" and having "delivered value to all the stakeholder groups." Bardaro, Daga, Carvalho, Chiatti, and Motta (2022) use a robotic competition venue to describe service robots in smart cities, said to be "a surprisingly under-explored" area. Seeking to "stimulate advances on the integration between robots and smart cities", Bardaro et al. (2022) focus on "activities specifically designed to require cooperation between robots" and "the smart city data infrastructure" of Milton Keynes, where applications pertain to museum guides, automated delivery drones, and so on, such that robots contribute to both the data acquisition and the data generation process. Lightman (2022) claims that "for a city to truly be smart, it needs to go beyond data and technology" placing a "focus on addressing the root causes of inequities and improving quality of life for all its residents." As such, Lightman (2022) welcomes the $1T Investment and Jobs Act recently signed by the United States government as having "the potential to enable the equitable deployment of smart cities" and the use of emerging technologies such as robotics "enabling communities to be connected, equitable, and sustainable." Savage (2022) addresses the different ways that robotic technologies are being improved for use in nursing home and other care settings "to augment how people care for people."

3.4 Extended Reality in Smart Cities

Celani et al. (2019) use the example of a 3-D city model enabling augmented experiences through a virtual reality application providing interactive information. Concerned with opportunities "for more people to interact with the built environment in a contextual, intuitive, and meaningful way", Harkins and Heard (2020) advance the notion of mixed reality as "a form of open data" as in, "interfacing the city" drawing on "building information modelling (BIM)" and gaming technologies in an effort to increase "health, wellbeing, and liveability of our cities for the betterment of everyone." Hu, Nanjappan, and Georgiev (2021) claim that extended reality technologies "can help remove the barriers in the mind by directly addressing underlying mental mechanisms" enabling such technologies to "more strongly support creativity and innovation if their full capabilities can be realized." Stanney et al. (2021) identify issues associated with eXtended reality (XR) systems "affecting the utility and use of these systems" in relation to "health, safety, and social concerns" with guidance on the management of such issues. From a technology, design, and application perspective, Stanney et al. (2021) provide an update on the advances of XR approaches making it "more accessible for education and training, collaborating and partnering, networking and entertaining" and for many more uses and potential uses. Hopkins, Bae, Uhr, Banić, Zheng, and Do (2021) refer to extended reality (XR) as "a blanket term used to describe the spectrum, or reality-virtuality continuum", drawing on the work of Milgram and Kishino (1994), "of the physical world versus digital content that is displayed in the user's field of view using augmented reality (AR) and virtual reality (VR) interfaces." Hopkins et al. (2021) describe various uses of XR beyond gaming, such as urban planning and design. Al Faleh (2021) describes an application that "will use mobile augmented reality as a tool and as a lens that allows the process of revealing the invisible" through, for example, "an experiential urban walk" while "bringing forth a hidden dimension of the city" although privacy and security issues are identified as challenges. Egliston and Carter (2022) explore mixed reality (MR) as an 'interface of the future' and imagined, speculative ways of adopting such technologies as "surveillant futures" with a call "not to reject MR" but instead "renegotiate the terms on which we are asked to accept MR into our lives" while extending imaginings of what is possible in terms of "technological advance." Ma (2022) describes challenges associated with evolving renderings of the metaverse from design to adoption, suggesting that centering the metaverse around augmented reality (AR) would enable an intersecting of the virtual world with the physical world as we know it, resulting possibly in "mediated, magical worlds everywhere" accessible via smartphones. Rzeszewski and Naji (2022) explore virtual geographical environments (VGEs) in terms of the "unique affordances and potentialities" of the application of extended reality or cross-reality (XR) technologies incorporating narrative immersion and literary placemaking to "evoke a stronger sense of place and a feeling of presence and belonging in users" in support of urban planning processes.

3.5 Literature Review Overview

An overview by author and date is provided in this section for the research and practice literature for smart homes, electric vehicles, robotics, and extended reality.

Table 1 provides an overview by author and date of the research and practice literature for urban life and the ambient pertaining to smart homes, spanning the timeframe of 2006 to 2022.

Marzano (2006) describes a smart home concept as an invisible, unobtrusive system embedded in objects in the home such as chairs, lamps, and so on, that would assist people. Soar et al. (2009) address the notion of ambient living in the city in relation to a health care model for wellness. Solaimani et al. (2015) provide a systematic review of the smart home literature from several perspectives, claiming the smart living term to be more expansive. Wilson et al. (2015) also provide a systematic review of the smart home literature finding concerns with access to data, control, privacy, reliability, and trust. Baig et al. (2018) describe ultra-aware cities with smart homes featuring three layers, namely – perception, attention-memory, and decision. Mechant et al. (2018) provide a framework as a methodology for exploring how agency and engagement are enabled or constrained by technology. Sahni et al. (2018) address the challenges and opportunities of designing user-centric smart spaces focusing on smart homes. Loi (2019) provides a reflection on the role of intelligent systems and ambient computing in homes of the future, working with, not for, the 65+ population, using participatory design. Kylili et al. (2020) identify challenges associated with smart homes in relation to technological solutions pertaining to support and security. Sessa (2020) identifies eight features of a smart home and Kim et al. (2021) focus on energy conservation in relation to smart homes and smart sustainable cities. Bisson (2022) describes ambient computing in the context of the smart home where computing recedes into the background delivering experiences and information to people in everyday life based on their interactions and preferences. Emekci (2022) addresses the potential of smart homes in contributing to the energy dimension of smart cities. Loi

Table 1. Overview of research for smart homes in smart cities

Author(s)	Year	Smart Homes
Marzano	2006	Unobtrusive system embedded in objects in the home to assist people
Soar et al.	2009	Ambient living
Solaimani et al.	2015	Systematic review of the smart home literature; Smart living
Wilson et al.	2015	Access to data, control, privacy, reliability, trust
Baig et al.	2018	Ultra-aware cities; smart homes with perception, attention-memory, decision
Mechant et al.	2018	Tech as enabler & constrainer of agency & engagement in smart homes
Sahni et al.	2018	Challenges and opportunities designing smart spaces as user-centric
Loi	2019	Reflection on role of IS & AC in future homes, with 65+, participatory design
Kylili et al.	2020	Technological solutions identified as challenges – support and security
Sessa	2020	Features
Kim et al.	2021	Smart homes to sustainable smart cities, energy conservation system
Bisson	2022	Ambient computing for experience and information delivery
Emekci	2022	Potentials of smart homes for smart cities – energy
Loi	2022	Making a house a home: participation, continuously evolving, intangible

(2022) identifies three ingredients for making a house a home in the form of participation, continuously evolving, and the intangible and these would seem to be important considerations for the smart home.

Table 2 provides an overview by author and date of the research and practice literature for urban life and the ambient pertaining to electric vehicles, spanning the timeframe of 2016 to 2022.

Ratti describes (Sisson, 2016) ambient technology and ambient mobility in relation to real time information for modes of transport. Potdar et al. (2018) discuss a range of risks and challenges associated with EVs from battery capacity to efficient charging infrastructure, unreliable supply of rare earth battery materials, and waste management. Toh et al. (2020) describe the "smart roads" concept, enabling the dynamic charging of electric vehicles (EVs). Yeo and Lin (2020) emphasize the importance of human

Table 2. Overview of research for electric vehicles in smart cities

Author(s)	Year	Electric Vehicles
Sisson	2016	Ambient technology, ambient mobility
Potdar et al.	2018	Risks & challenges: battery, charging, materials, waste management of used EVs
Toh et al.	2020	Smart roads – electrified roads for EV charging
Yeo & Lin	2020	Human agency in urban life and AVs – sensation, action, encounter
Jain et al.	2021	EV design challenges, SC integration for sustainable mobility
Jones	2021	Electric mobility – reshaping & reimagining cities – data & energy
Ruggieri et al.	2021	EVs & Europe – effectiveness of mobility transition in reducing pollution
Sanguesa et al.	2021	EV review of developments & future prospects – batteries, standards, materials
Freudendal-Pedersen	2022	Alternative visions for mobile everyday life beyond the car & electrified cars
Skjølsvold et al.	2022	Socio-technical implications of transport electrification, local engagement

Table 3. Overview of research for robotics in smart cities

Author(s)	Year	Robotics
Dainow	2017	Functional characteristic of SCs involving humans & machines
Macrorie et al.	2019	Research agenda for robotics & automation in SCs – responsible innovation
Cugurullo	2020	Autonomous city, urban AI, unsupervised thinking & acting
Zhang & Shenjing	2020	Smart technology with impacts on humanity, nature, sustainability, society
Myung & Wang	2021	Robotic sensing applications & SCs
Song et al.	2021	Swarm robotics & coordinated navigation – dev, testing, education
Studley & Little	2021	Design of SCs, robot-human interaction, public discussion, trust, value
Bardaro et al.	2022	Service robots in SCs, integration – urban data access & generation
Lightman	2022	Tech for SCs – robotics: connected, equitable & sustainable communities
Savage	2022	Caring robots in hospitals and nursing homes to augment care

agency in urban life focusing on sensation, action, and encounter in relation to autonomous vehicles (AVs). Jain et al. (2021) explore challenges, opportunities, limitations, and benefits associated with electric vehicle (EV) integration in smart cities (SCs) for sustainable mobility. Jones (2021) provides a discussion of electric mobility pointing to the broader implications for reimaging and reshaping urban environments and advancing EVs as technology platforms for data and energy. Ruggieri et al. (2021) show the effectiveness of policies in six European cities in support of transitioning to electric mobility in reducing pollution emissions. Sanguesa et al. (2021) highlight the critical factors for electric vehicles (EVs) associated with batteries in terms of autonomy, standards, and sustainability and environmental impact. Freudendal-Pedersen (2022) encourages the rethinking of mobile everyday living for alternatives to the car and electrified cars. Skjølsvold et al. (2022) explore the sociotechnical implications of transport electrification at the local level in Norway.

Table 3 provides an overview by author and date of the research and practice literature for urban life and the ambient pertaining to robotics, spanning the timeframe of 2017 to 2022.

Dainow (2017) describes robotics as a functional characteristic of smart cities involving the movement of humans and machines in the same spaces. Macrorie et al. (2019) advance a research agenda for robotics and automation in smart cities with the "potential for responsible urban innovation." Cugurullo (2020) uses the concepts of autonomous city and urban artificial intelligence (AI) in describing the presence of technologies that are increasingly capable of unsupervised thinking and acting. Zhang and Shenjing (2020) include robotics as a form of smart technology having impacts on humanity, nature, sustainability, and society. Myung and Wang (2021) provide an overview of a special issue on robotic sensing and smart cities, where a range of applications are described for autonomous vehicles, human-robot interaction systems, inspection robots, and mobile robots. Song et al. (2021) describe RoboMaze as a simulation space to explore and demonstrate swarm robotics applications and coordinated navigation in smart city contexts while serving as a tool for algorithm development, education, and outreach. Studley and Little (2021) discuss the study of robot-human interaction involving public discussion and the potential that robots will have an enormous impact on the design of smart cities while considering issues such as trust and value. Bardaro et al. (2022) explore service robots in smart cities, in terms of advancing their integration while focusing on how robots contribute to both the data acquisition and the data generation process. Lightman (2022) highlights the importance of upgrading urban infrastructure and the use of technologies (including robotics) in supporting communities to be connected, equitable, and sustainable in addressing "real-world problems that are often complicated, messy, and unattractive." Savage (2022) addresses the different ways that robotic technologies are being improved and provides an update on the notion of caring robots for use in nursing homes and other care settings to augment care.

Table 4 provides an overview by author and date of the research and practice literature for urban life and the ambient pertaining to extended reality, spanning the timeframe of 1994 to 2022.

Milgram and Kishino (1994) describe physical and digital content appearing for the user through interfaces for augmented reality (AR) and virtual reality. Celani et al. (2019) discuss the notion of an augmented city involving citizens and communities while describing augmented experiences enabled through a virtual reality application. Harkins and Heard (2020) discuss opportunities to increase people interaction with built environments in contextual, intuitive, and meaningful ways through mixed reality, drawing on open data and gaming technologies to improve the health, livability, and wellbeing of cities for everyone. Al Faleh (2021) describes a mobile augmented reality (AR) application as a tool and lens for revealing the invisible or hidden dimension of the city through an experimental urban walk while noting privacy and security issues. Hopkins et al. (2021) describe extended reality (XR) as the reality-virtuality spectrum or continuum, identifying uses such as urban planning and design. Hu et al. (2021) argue that human creativity can be enhanced with XR technologies thus breaking cognitive barriers. Stanney et al. (2021) identify issues with eXtended reality (XR) systems pertaining to health, safety, and social concerns while highlighting advances pertaining to collaboration, education, and networking, to name a few. Egliston and Carter (2022) explore mixed reality (MR) as an 'interface of the future' calling for renegotiated terms of acceptance and an extending of imaginings of what is possible in terms of "technological advance." Ma (2022) describes challenges ranging from adoption to design associated with the metaverse and in response proposes that the initiative be centered on augmented reality (AR) to complement and extend experiences of everyday life using devices such as smartphones. Rzeszewski and Naji (2022) discuss virtual geographical environments (VGEs) and the application of extended reality or cross-reality (XR) technologies incorporating narrative immersion and literary placemaking tools in support of user engagement based on "a stronger sense of place and a feeling of presence and belonging" for urban planning processes (UPP).

In terms of patterns and potential relationships, what emerges from this review of the research and practice literature is a comparative overview for smart homes, electric vehicles, robotics, and extended reality as presented in Table 5.

Elements that emerge across columns of Table 5, include, but are not limited to, privacy, security, and trust.

3.6 Issues, controversies, problems

The review of the research and practice literature enables the surfacing of a range of issues, controversies, and problems as identified in Table 6.

For smart homes, issues, controversies, and problems pertain to functional; instrumental; socio-technical; support; access to data; agency; privacy and control; transparency; engagement and participation; reliability; security; and trust.

For electric vehicles, issues, controversies, and problems pertain to battery capacity; smart roads; consistent supply of raw materials; waste management of used EV batteries; human agency and sentience;

Table 4. Overview of research for extended reality in smart cities

Author(s)	Year	Extended Reality
Milgram & Kishino	1994	Physical & digital content, viewable by AR & VR interfaces
Celani et al.	2019	Augmented city: create, co-design; augmented experiences, virtual reality
Harkins & Heard	2020	Contextual, intuitive, meaningful interactions; mixed reality, open data
Al Faleh	2021	Mobile AR tool: experimental urban walk; privacy & security issues
Hopkins et al.	2021	Reality-virtuality continuum; urban planning & design
Hu et al.	2021	Enhancing creativity with XR technologies breaking human mind barriers
Stanney et al.	2021	XR cons: health, safety, social; XR pros: collaborate, education, network
Egliston & Carter	2022	MR as an 'interface of the future' needing negotiations and imaginings
Ma	2022	Metaverse as centered on AR to overcome design & adoption challenges
Rzeszewski & Naji	2022	VGEs & XR tech – narrative immersion & literary placemaking – UPP

availability of charging infrastructure; real time information; sensation – action – encounter; costs; design challenges and integration; materials; and power grid demands.

For robotics, issues, controversies, and problems pertain to intensive collaborations; design impact on cities; swarm; trust; extending the capabilities and capacities of human agency; key functional characteristic of smart cities (SCs); need for human behavior adjustments in spaces of robot-human interactions; value; human-machine interactions; reshaping citizen's experiences; reshaping the city; and sustainable built environments.

Table 5. Patterns and potential relationships

Smart Homes	Electric Vehicles	Robotics	Extended Reality
Access to data	Ambient mobility	Autonomous city	Augmented city
Adaptive living	Ambient technology	Collaboration – intensive	- co-design
Control	Autonomy	Education/outreach tool	- create
Energy conservation	Charging methods	Functional characteristic	Augmented experiences
Participatory design	Charging standards	Human agency extended	Collaboration
Privacy	Costs	Human behavior	Content, physical/digital
Reflection, role of	Design challenges	Mobile robots	- AR/VR interfaces
- intelligent systems	Engagement	Public discussion	Contextual
- ambient computing	Electrified roads	Reshaping experiences	Creativity
Reliability	Environmental impact	Reshaping the city	Education
Security	EV risks/challenges	Robot-human interaction	Experimental urban walks
Smart living	- batteries	Sensing, analysis, design	Health
Support	- charging infrastructure	Smart technology impact	Intuitive
Tech: constrainer/enabler	- rare earth materials	- humanity	Meaningful interactions
- agency	- waste management	- nature	Mixed reality
- engagement	Human agency	- society	Mobile AR tool
Trust	Human sentience	- sustainability	Open data
Ultra-aware cities	Power grid demands	Swarm robotics	Privacy
- attention-memory	Smart driving force	Trust	Reality-virtuality continuum
- decision	Smart outcomes	Unsupervised	Safety
- perception	Smart roads	- acting	Security
Unobtrusive systems	Sustainable mobility	- thinking	Social
- assist people	Technology platforms	Urban AI	Urban planning & design
User-centric	Urban transformation	Value	Virtual reality

Table 6. Issues, controversies, and problems

	Issues	Controversies	Problems
Smart Homes	• Functional • Instrumental • Socio-technical • Support	• Access to data • Agency • Privacy and control • Transparency	• Engagement & participation • Reliability • Security • Trust
Electric Vehicles	• Battery capacity • Raw materials • Smart roads • Used batteries	• Human agency & sentience • Infrastructure • Real time information • Sensation, action, encounter	• Costs • Design challenges & integration • Materials • Power grid demands
Robotics	• Collaborations • Design impact • Swarm • Trust	• Extending human agency • Functional characteristic/SCs • Human behavior adjustments • Value	• Human-machine interactions • Reshaping citizen's experiences • Reshaping the city • Sustainable built environments
Extended Reality	• Health & Safety • Livability • Privacy/Security • Well-being	• Accessibility • Contextual, intuitive … • Creativity & innovation • Surveillant futures	• Collaborating & partnering • Education & training • Social concerns • Urban planning & designing

For extended reality, issues, controversies, and problems pertain to health and safety; livability; privacy and security; well-being; accessibility; contextual – intuitive – meaningful interactions; creativity and innovation capabilities; surveillant futures; collaborating and partnering; education and training; social concerns; and urban planning and designing.

Of note is the presence of overlap between smart homes, EVs, robotics, and extended reality and across issues, controversies, and problems where collaboration and trust provide a few examples.

4. SMART TECHNOLOGY, THE AUGMENTED, AND THE VIRTUAL IN URBAN LIFE AND THE AMBIENT IN SMART CITIES

Responding to the issues, controversies, and problems emerging from the literature review in the preceding section, a conceptual framework for smart technologies, the augmented, and the virtual in urban life and the ambient in smart cities is formulated in relation to smart homes, electric vehicles (EVs), robotics, and extended realities as depicted in Fig. 1. Building upon elements in preceding chapters, people and technology interactions figure strongly over the continuum of the past, present, and future in learning, smart, and future cities.

The research question identified in Section 1 is reformulated as a proposition for exploration in this chapter, as follows:

P8: *Smart homes, electric vehicles, extended realities, and robotics contribute to urban life and the ambient in smart cities in support of sustainability and livability on the planet*

The research design used in exploring the proposition under study in this chapter is described in the Methodology section.

Figure 1. Conceptual framework for smart technologies, the augmented, and the virtual in urban life and the ambient in smart cities

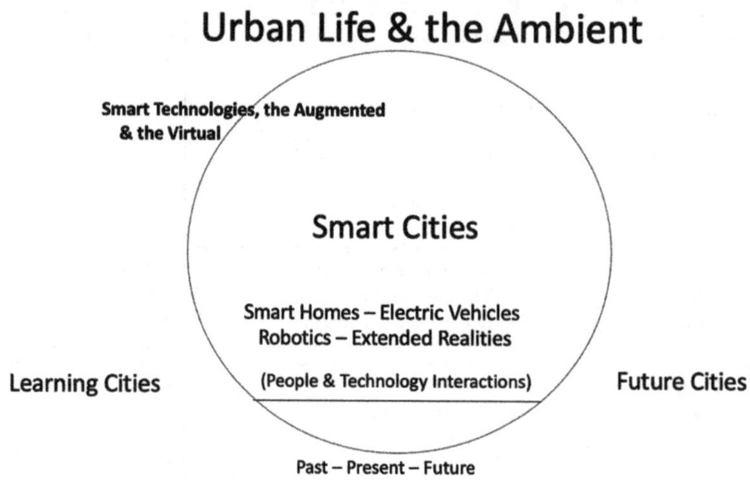

5. METHODOLOGY

Sections 5.1 to 5.3 describe the research design for this chapter in terms of the process used, sources of evidence for data collection, and data analysis.

5.1 Process

In view of the contemporary nature of the inquiry in this chapter, an exploratory case study approach is used (Yin, 2018) and combined with an explanatory correlational design since the two are said to be complementary (Yin, 2018) in that the former may be "needed to examine the underlying processes that might explain a correlation." Using an online space to describe the study and invite people to participate, interest emerged from individuals in a variety of cities in multiple countries (Canada, United States, Greece, Ireland, Israel, etc.) reaching several continents.

5.2 Sources of Evidence

Use of an online website enabled the gathering of demographic data from individuals during sign up for the study, which included age range, gender, and city. Additionally, people could self-identify in one or more categories (e.g., business, community member, student, educator, etc.). A survey instrument, pre-tested prior to use, was used in this study and contained closed and open-ended questions. Survey questions pertained to a range of items related to smart cities such as electric vehicles, extended realities, and the like. Survey respondents were also invited to participate in an in-depth interview which was guided by an interview protocol that had been pre-tested prior to use in this study.

Additionally, data were also gathered systematically through group and individual discussions, conducted in parallel with this study, with individuals in many sectors (e.g., students, faculty, and staff in higher education, business, community members and leaders, and city staff and leaders) in several Canadian cities (e.g., Toronto, Vancouver, Victoria).

Table 7. Assessments for connecting and trust

Items	Aware Technologies in SCs	Correlation
Connecting	17% (5); 83% (7)	.56
Trust (aware technologies)	17% (5); 33% (6); 50% (7)	

5.3 Data Analysis

Data analysis in this chapter is guided by the research question posed that is reformulated into a proposition for exploration, as follows:

P8: *Smart homes, electric vehicles, extended realities, and robotics contribute to urban life and the ambient in smart cities in support of sustainability and livability on the planet.*

For the analysis of quantitative survey data, the use of descriptive statistics pertaining to *access to public data* and *trust* along with other associated elements enabled exploration in relation to *livability* and *data value.* The correlation feature of the Real Statistics Resource Pack containing an-add in for Microsoft Excel (Zaiontz, 2022) enabled the conducting of a Spearman Correlation Coefficient for ordinal data in exploring the nature of relationships between items pertaining to electric vehicles, extended realities, and the like.

Content analysis was used for qualitative data gathered through open-ended survey questions, interviews, and group and individual discussions so that inductive analysis from data collection was combined with deductive analysis using terminology from the literature review. Overall, an analysis was conducted for n=79 consisting of 42% females and 58% males for people ranging in age from their 20s to 70s.

6. RESULTS AND DISCUSSION

Table 7 provides the results of assessments in response to a survey question about the extent to which social media and other aware technologies give rise to possibilities for factors such as *connecting* and *trust* in smart city. On a 7-point Likert type scale, where 1= Not at all and 7=Absolutely, for *connecting*, 17% of survey respondents selected position 5 (Sort of) on the scale, and 83% responded at the upper end position 7. For *trust*, 17% of survey respondents selected position 5 (Sort of) on the scale, 33% selected position 6 (Sure) and 50% responded at the upper end position of 7. Using the correlation feature of the Real Statistics add-in for Microsoft Excel (Zaiontz, 2022), a Spearman correlation coefficient for ordinal data was conducted on the assessment results of the two items, *connecting* and *trust* (Table 7), showing a finding of .56.

Exploring further, as shown in Table 8, when asked to assess to what extent elements such as *access to public data* contribute to the success of a smart city project, 50% responded at position 6 on the scale and 50% at position 7. When asked to assess to what extent factors such as *trust* contribute to increased value for data in smart cities, 25% responded at position 6 on the scale and 75% responded at position 7.

Table 8. Assessments for access to public data (for success) and trust

Items	Data and Success SCs	Correlation
Access to Public Data (success)	50% (6); 50% (7)	.57
Trust (data value)	25% (6); 75% (7)	

Using the correlation feature of the Real Statistics add-in for Microsoft Excel (Zaiontz, 2022), a Spearman correlation coefficient for ordinal data was conducted on the assessment results of the two items, *access to public* and *trust* (Table 8), showing a finding of .57.

Exploring further, as shown in Table 9, when asked to assess to what extent elements such as *access to public data* contribute to the livability of a smart city, 25% responded at position 4 on the scale, 25% at position 5, 25% at position 6, and 25% at position 7. When asked to assess to what extent factors such as *trust* contribute to increased value for data in smart cities, 25% responded at position 6 on the scale and 75% responded at position 7.

Using the correlation feature of the Real Statistics add-in for Microsoft Excel (Zaiontz, 2022), a Spearman correlation coefficient for ordinal data was conducted on the assessment results of the two items, *access to public* and *trust* (Table 9), showing a finding of .77. It is worth noting that correlations in the range of .66 – .85 are said to be "very good" and "good prediction can result from one variable to the other" (Creswell, 2018).

Two different survey questions were used to explore *access to public data* in smart cities, one pertaining to success and the other pertaining to livability. It is worth noting that when correlated with trust (associated with data value), access to public data associated with livability emerges much more strongly (with good predictability) than that associated with smart city success. It is also worth noting that trust associated with social media and aware technology, when correlated with connecting in smart cities, shows a correlation at .56 with much more room to strengthen, possibly indicative of the need going forward for social media platforms and aware-enabling technologies to take people and their needs into greater consideration.

Qualitatively, findings of interest pertaining to social media include comments from an educator who indicated that "people are the reason for cities" and "technology is only playing a bit part" in support of "ideas and purposes." A community leader commented that regarding the affective aspects of social media and aware technologies on the experience of the city, "the biggest benefit they can offer is to assist in citizen/visitor education and awareness" as in, "focused around people and their needs." Another educator commented that "social media has a huge influence" and "it helps people to connect" while the "adoption of technologies, community weaving, trust relationships between the authorities and the public make all the difference."

Table 9. Assessments for access to public data (for livability) and trust

Items	Data and Livability in SCs	Correlation
Access to Public Data (livability)	25% (4); 25% (5); 25% (6); 25% (7)	.77
Trust (data value)	25% (6); 75% (7)	

Figure 2. Urban life and the ambient – smart technologies in smart cities

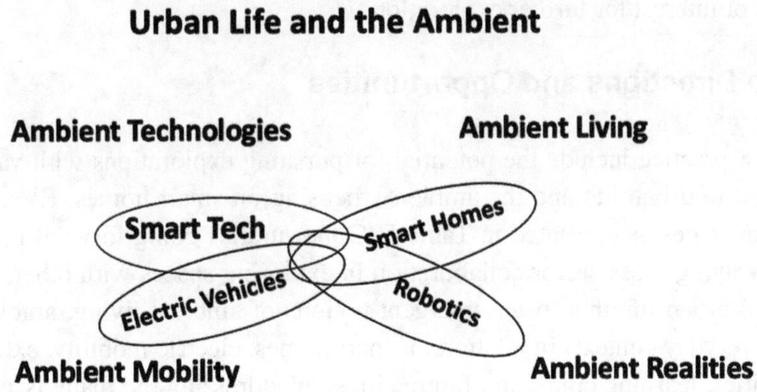

As depicted graphically in Fig. 2, the interconnectivity of smart technologies with electric vehicles, smart homes, and robotics is aided in urban life by ambient technologies giving way to the notion and potentials of ambient living, ambient mobility, and ambient realities.

7. LIMITATIONS AND FUTURE DIRECTIONS

Possible limitations of this work are discussed in this section opening the path for future challenges and opportunities in terms of directions for research and practice.

Limitations

Limitations of this chapter pertain possibly to the extent of the literature review on urban life and the ambient in relation to smart homes, electric vehicles, extended reality (including augmented, mixed, and virtual reality), and robotics in smart cities. Among other limitations of this chapter is that of the small sample size for the underlying study of this work, and as such, caution is encouraged in the interpretation of correlations calculated from the assessment of variables used. Going forward opportunities may emerge for the use and exploration of additional variables, resulting in possibly stronger correlations with greater prediction potentials.

Future Research Directions and Opportunities

For research, future directions emerge in terms of the potential for conducting additional exploration, testing, and validation of the variables identified in this work relevant to urban life and the ambient in relation to smart homes, electric vehicles, extended reality, and robotics. Additionally, this chapter creates a space for explorations going forward in terms of the identification of other and possibly stronger variables relevant to smart homes, EVs, extended reality, and robotics for urban life in the context of smart cities, learning cities, and future cities as identified in Table 5. Going forward, an opening emerges for further explorations of ambient technologies, ambient living, ambient mobility, and ambient realities.

Of note in Table 9 is the presence of a neutral response for *access to public data (livability)* identifying a potential area of interest for further exploration.

Future Practice Directions and Opportunities

Future directions for practice include the potential for pursuing explorations with variables identified in this work relevant to urban life and the ambient, focusing on smart homes, EVs, extended reality, and robotics in smart cities as presented in Table 5. Opportunities going forward for practice include the potential for involving cross-sector collaboration in exploring spaces with other, possibly stronger variables relevant to urban life that foster emergent notions of ambient living, ambient mobility, and ambient reality in everyday contexts in relation to smart homes, electric mobility, extended reality, and robotics in smart cities, learning cities, and future cities, in addressing the many issues, controversies, and problems presented in Table 6.

8. CONCLUSION

A key contribution of this chapter is identification of the challenges and opportunities associated with urban life and the ambient emerging from the review of the research literature provided from multiple perspectives involving smart homes, electric vehicles, extended reality (including augmented, mixed, and virtual reality), and robotics. Another contribution of this chapter is the formulation of a conceptual framework for smart technologies, the augmented, and virtual in urban life and the ambient in smart cities encompassing smart homes, electric vehicles (EVs), extended reality, and robotics. Based on the literature review, emerging issues, controversies, and problems provide guidance on the proposition explored while contributing to operationalization of the framework for use in this chapter. Through an exploratory case study approach, together with an explanatory correlational design, the proposition under study in this chapter is investigated, based on survey data assessments. Correlations of assessment data are conducted, such as that for *connecting* and *trust* where, among the findings is that of a Spearman correlation coefficient of .56, with useful although limited prediction potential while a Spearman correlation coefficient of .77, with good prediction potential was found between *access to public data* (in relation to livability) and *trust* (associated with data value). Limitations of the study underlying this chapter pertain to, among other things, the small sample size and caution is urged in the interpretation of findings, giving rise possibly to opportunities for further research. Notwithstanding this and other possible limitations, future directions guiding research and practice activities going forward are provided in terms of further validating the findings in this chapter while also encouraging activities to identify other, additional factors relevant to urban life and the ambient and to the notions of ambient living, ambient mobility, ambient reality, and ambient technologies, in the context of smart, learning, and future cities.

Key insights and ideas emerging from this chapter are provided, along with exercises encouraging further exploration, as follows:

Insights emerging from this chapter include the following:

1. The importance of the relationship between the value of urban data in the form of *access to public data* as a factor contributing to livability when correlated with *trust* in the making of a smart city
2. The importance of factors such as *trust* when associated with *connecting* in smart cities

3. The potential for many more factors to emerge for understanding *smart homes, electric vehicles (EVs), robotics,* and *extended reality* in smart cities

Ideas emerging from this chapter include the following:

1. Potential opportunities for exploring urban life and the ambient in relation to a variety of variables pertaining to smart homes, electric vehicles (EVs), and robotics (e.g., access to public data, connecting, and trust) in the context of smart cities
2. The opportunities potential for the identification of other and additional variables for exploring urban life and the ambient in smart environments in relation to mobility and extended reality (e.g., augmented, mixed, and virtual)
3. The use of an interdisciplinary and multi-method approach for the study and practice of urban life and access to public data, connecting, and trust in smart, learning, and future cities

Exercises for the reader to consider, encouraging further exploration include the following:

1. Identify one or more ways in your city that smart homes or electric vehicles (EVs) matter
2. Identify one or more ways in which extended reality would enhance the quality of life in your city
3. Describe one or more ways in which robotics would enhance the quality of life in your city

The audience for this chapter is intended to be interdisciplinary in nature, including but not limited to community members and leaders, educators, businesses, researchers, urban planners, urban designers, and anyone concerned with smart homes, electric vehicles (EVs), and robotics in urban life, while being attentive to extended reality elements such as augmented, mixed, and virtual reality along with associated practice and research opportunities and challenges in smart cities, learning cities, and future cities.

REFERENCES

Al Faleh, O. (2021). Crowdsourced urban annotations and augmented reality as design thinking tools to navigate and interact with urban data. In N. Streitz & S. Konomi (Eds.), Lecture Notes in Computer Science: Vol. 12782. *Distributed, Ambient and Pervasive Interactions. HCII 2021*. Springer. doi:10.1007/978-3-030-77015-0_2

Baig, M. N., Himarish, M. N., Panaya, Y. C., & Ahmed, M. R. (2018). Cognitive architecture based smart homes for smart cities. *Proceedings of the 2nd International Conference on Trends in Electronics and Informatics (ICOEI)*, 461-465. 10.1109/ICOEI.2018.8553774

Bardaro, G., Daga, E., Carvalho, J., Chiatti, A., & Motta, E. (2022). Introducing a smart city component in a robotic competition: A field report. *Frontiers in Robotics and AI*, *9*, 728628. Advance online publication. doi:10.3389/frobt.2022.728628 PMID:35252360

Bisson, S. (2022). Ambient computing has arrived: Here's what it looks like, in my house. *ZDNET*. Retrieved 16 September 2022 from https://www.zdnet.com/article/ambient-computing-has-arrived-heres-what-it-looks-like-in-my-house/

Celani, P., Falcone, R., & d'Alessandro, E. (2019). The challenge of augmented city: New opportunities of common spaces. In *New Metropolitan Perspectives. ISHT 2018. Smart Innovation, Systems and Technologies* (vol. 100). Springer. https://link.springer.com/chapter/10.1007/978-3-319-92099-3_18

Cheshire, J., & Uberti, O. (2021). *Atlas of the invisible: Maps and graphics that will change how you see the world.* Norton.

Creswell, J. W. (2018). *Educational research: Planning, conducting, and evaluating quantitative and qualitative research* (6th ed.). Pearson.

Cugurullo, F. (2020). Urban artificial intelligence: From automation to autonomy in the smart city. *Frontiers in Sustainable Cities, 2*(28), 1-14. DOI: doi:10.3389/frsc.2020.00038

Dainow, B. (2017). Smart city transcendent: Understanding the smart city by transcending ontology. *Orbit (Amsterdam, Netherlands), 1*, 1–15.

Egliston, B., & Carter, M. (2022). 'The interface of the future': Mixed reality, intimate data and imagined temporalities. *Big Data & Society*, *9*(1). Advance online publication. doi:10.1177/20539517211063689

Emekci, S. (2022). From smart homes to smart cities: How smart homes contribute to the sustainable development goals. In A. Pego (Ed.), *Smart Cities, Citizen Welfare, and the Implementation of Sustainable Development Goals* (pp. 321–337). IGI Global. doi:10.4018/978-1-7998-7785-1.ch018

Freudendal-Pedersen, M. (2022). *Making Mobilities Matter*. Routledge. doi:10.4324/9781003100515

Harkins, J., & Heard, C. (2020). Interfacing the city: Mixed reality as a form of open data. In S. Hawken, H. Han, & C. Pettit (Eds.), *Open Cities | Open Data*. doi:10.1007/978-981-13-6605-5_11

Hopkins, T., Bae, S. S., Uhr, J., Banić, A., Zheng, C., & Do, E. Y. L. (2021). User interfaces in smart cities. In J. C. Augusto (Ed.), *Handbook of Smart Cities*. Springer. doi:10.1007/978-3-030-15145-4_94-1

Hu, X., Nanjappan, V., & Georgiev, G. (2021). Breaking the block in the human mind: Enhancing creativity with extended reality technologies. *Interaction, 38*(3). Advance online publication. doi:10.1145/3460114

Jain, A., Mani, A., & Siddiqui, A. S. (2021). Design challenges for efficient integration of electric vehicles in smart cities. In Smart Energy Management Systems and Renewable Energy Resources. AIP Publishing. doi:10.1063/9780735422827_009

Jones, D. (2021). 4 ways electric mobility will reshape our cities. *Gensler Blog*. Retrieved 6 October 2022 from https://www.gensler.com/blog/4-ways-electric-mobility-will-reshape-our-cities

Kim, H., Choi, H., Kang, H., An, J., Yeom, S., & Hong, T. (2021). A systematic review of the smart energy conservation system: From smart homes to sustainable smart cities. *Renewable & Sustainable Energy Reviews, 140*(110755). Advance online publication. doi:10.1016/j.rser.2021.110755

Kylili, A., Afxentiou, N., Georgiou, L., Panteli, C., Morsink-Georgalli, P.-Z., Panayidou, A., Papouis, C., & Fokaides, P. A. (2020). The role of remote working in smart cities: Lessons learnt from COVID-19 pandemic. *Energy Sources. Part A, Recovery, Utilization, and Environmental Effects*, 1–16. Advance online publication. doi:10.1080/15567036.2020.1831108

Lightman, K. (2022). In focus: Redefining the smart in smart cities – Emerging robotics and AI technologies can help meet critical goals of connectivity, equity, and sustainability. *Area Development*. Retrieved 5 October 2022 from https://www.areadevelopment.com/sustainable-development/q1-2022/redefining-the-smart-in-smart-cities.shtml

Loi, D. (2019). HOME: Exploring the role of ambient computing for older adults. In M. Antona & C. Stephanidis (Eds.), Lecture Notes in Computer Science: Vol. 11573. *Universal Access in Human-Computer Interaction. Multimodality and Assistive Environments. HCII 2019*. Springer. doi:10.1007/978-3-030-23563-5_39

Loi, D. (2022, May-June). Designing the unfinished: A home is not a house. *Interaction*, *29*(3), 16–18. doi:10.1145/3529163

Ma, A. (2022). What is the metaverse and what can we do there? *The Conversation*. Retrieved 10 July 2022 from https://theconversation.com/what-is-the-metaverse-and-what-can-we-do-there-179200

Macrorie, R., Marvin, S., & While, A. (2021). Robotics and automation in the city: A research agenda. *Urban Geography*, *42*(2), 197–217. doi:10.1080/02723638.2019.1698868

Marzano, S. (2006). Ambient culture. In True Visions. Springer. doi:10.1007/978-3-540-28974-6_3

Mechant, P., All, A., & De Marez, L. (2018). Evaluating user experience in smart home contexts: A methodological framework. In N. Streitz & S. Konomi (Eds.), Lecture Notes in Computer Science: Vol. 10921. *Distributed, Ambient and Pervasive Interactions: Understanding Humans. DAPI 2018*. Springer. doi:10.1007/978-3-319-91125-0_7

Myung, H., & Wang, Y. (2021). Robotic sensing and systems for smart cities. *Sensors*, *21*(9). doi:10.3390/s21092963

Potdar, V., Batool, S., & Krishna, A. (2018). Risks and challenges of adopting electric vehicles in smart cities. In Z. Mahmood (Ed.), *Smart Cities. Computer Communications and Networks*. Springer. doi:10.1007/978-3-319-76669-0_9

Ruggieri, R., Ruggeri, M., Vinci, G., & Poponi, S. (2021). Electric mobility in a smart city: European overview. *Energies*, *14*(2), 315. doi:10.3390/en14020315

Rzeszewski, M., & Naji, J. (2022). Literary placemaking and narrative immersion in extended reality virtual geographic environments. *International Journal of Digital Earth*, *15*(1), 853–867. doi:10.1080/17538947.2022.2061619

Sahni, Y., Cao, J., & Shen, J. (2018). Challenges and Opportunities in Designing Smart Spaces. In B. Di Martino, K. C. Li, L. Yang, & A. Esposito (Eds.), *Internet of Everything. Internet of Things*. Springer. doi:10.1007/978-981-10-5861-5_6

Sanguesa, J. A., Torres-Sanz, V., Garrido, P., Martinez, F. J., & Marquez-Barja, J. M. (2021). A review on electric vehicles: Technologies and challenges. *Smart Cities*, *4*(1), 372–404. doi:10.3390martcities4010022

Savage, N. (2022). The slow rise of the caring robot. *Nature*, 601.

Sessa, M. (2020). Home smart home: A sociology of living in the age of reflective materialism. In R. Iannone & R. Gurashi (Eds.), *Smart Society: A Sociological Perspective on Smart Living*. Routledge.

Sisson, P. (2016). The future of smart city technology, from an MIT Professor: Carlo Ratti's the city of tomorrow examines how tech will shift and reshape the urban landscape. *Curbed*. Retrieved 8 February 2022 from https://archive.curbed.com/2016/5/24/11761274/smart-city-carlo-ratti-city-of-tomorrow

Skjølsvold, T. M., Henriksen, I. M., & Ryghaug, M. (2022). Beyond the car: How electric vehicles may enable new forms of material politics at the intersection of the smart grid and smart city. *Urban Geography*, 1–18. Advance online publication. doi:10.1080/02723638.2022.2044692

Soar, J., Livingstone, A., & Wang, S. Y. (2009). A case study of an ambient living and wellness management health care model in Australia. In Ambient Assistive Health and Wellness Management in the Heart of the City. ICOST 2009. Lecture Notes in Computer Science (vol. 5597). Springer. doi:10.1007/978-3-642-02868-7_7

Solaimani, S., Keijzer-Broers, W., & Bouwman, H. (2015). What we do – and don't – know about the smart home: An analysis of the smart home literature. *Indoor and Built Environment*, *24*(3), 370–383. doi:10.1177/1420326X13516350

Song, J., Song, A., & Zhang, F. (2021). RoboMaze: Swarm robotics and coordinated navigation in smart city. *2021 IEEE 23rd Int Conf on High Performance Computing & Communications; 7th Int Conf on Data Science & Systems; 19th Int Conf on Smart City; 7th Int Conf on Dependability in Sensor, Cloud & Big Data Systems & Application (HPCC/DSS/SmartCity/DependSys)*, 1881-1885. doi: 10.1109/HPCC-DSS-SmartCity-DependSys53884.2021.00279

Stanney, K. M., Nye, H., Haddad, S., Hale, K. S., Padron, C. K., & Cohn, J. V. (2021). Extended reality environments. In Handbook of Human Factors & Ergonomics (5th ed.). Wiley. doi:10.1002/9781119636113.ch30

Studley, M. E., & Little, H. (2021). Robots in smart cities. In How Smart Is Your City? Intelligent Systems, Control and Automation: Science and Engineering (vol. 98). Springer. doi:10.1007/978-3-030-56926-6_7

Toh, C. K., Sanguesa, J. A., Cano, J. C., & Martinez, F. J. (2020). Advances in smart roads for future smart cities. *Proceedings of the Royal Society of London. Series A*, *476*(2233). doi:10.1098/rspa.2019.0439 PMID:32082053

Wilson, C., Hargreaves, T., & Hauxwell-Baldwin, R. (2015). Smart homes and their users: A systematic analysis and key challenges. *Personal and Ubiquitous Computing*, *19*(2), 463–476. doi:10.100700779-014-0813-0

Yeo, S. J. I., & Lin, W. (2020). Autonomous vehicles, human agency and the potential of urban life. *Geography Compass*, *14*(10), e12531. doi:10.1111/gec3.12531

Yin, R. (2018). *Case study research and applications: Design and methods* (6th ed.). Sage Publications.

Zaiontz, C. (2022). *Real statistics using excel*. Retrieved from www.real-statistics.com

Zhang, J., & Shenjing He, S. (2020). Smart technologies and urban life: A behavioral and social perspective. *Sustainable Cities and Society*, *63*, 102460. doi:10.1016/j.scs.2020.102460

ADDITIONAL READING

Alvarez, R. (2017). The relevance of informational infrastructures in future cities. *Field Actions Science Reports, 17*, 12–15.

Bauer, R. (2015). *How ambient intelligence will change our lives.* World Economic Forum. Retrieved 19 October 2021 from https://www.weforum.org/agenda/2015/10/how-ambient-intelligence-will-change-our-lives/

Campbell, I. C. (2020). Amazon is building the anti-metaverse with 'ambient intelligence'. *Inverse*. Retrieved 4 October 2022 from https://www.inverse.com/gear/amazon-ambient-intelligence-metaverse-alternative

Chopra, S., & Addam, O. (2021). Towards an ambient smart city: Using augmented reality to geospatially filter the right Airbnb via artificial intelligence. In H. Degen & S. Ntoa (Eds.), Lecture Notes in Computer Science: Vol. 12797. *Artificial Intelligence in HCI. HCII 2021*. Springer. doi:10.1007/978-3-030-77772-2_31

Hampson, M. (2021). How smart homes alter human behavior: Shouldn't intelligent appliances require less fiddling to get the settings just right? *IEEE Spectrum*. Retrieved February 6, 2022, from https://spectrum.ieee.org/smart-homes-alter-human-behavior

Joshi, N. (2022). 2 lesser-known domestic robotics applications in smart cities. *Forbes*. Retrieved May 15, 2022, from https://www.forbes.com/sites/naveenjoshi/2022/01/20/3-lesser-known-domestic-robotics-applications-in-smart-cities/?sh=88868dc28cba

Laudante, E. (2021). Smart cities and robotic technologies for a model of integrated growth. *Revista Iberoamericana, Académico Científica de Humanidades, Arte Y Cultura*, 9.

Lecomte, P. (2021). Get ready for the invasion of smart building technologies following COVID-19. *The Conversation*. Retrieved 3 February 2022 from https://theconversation.com/get-ready-for-the-invasion-of-smart-building-technologies-following-covid-19-168646

Pira, S. (2021). The social issues of smart home: A review of four European cities' experiences. *European Journal Futures Resaerch, 9*(3), 3. Advance online publication. doi:10.118640309-021-00173-4

Tengler, S. (2021). Why aren't more people talking about solar cars? *Forbes*. Retrieved February 8, 2022, from https://www.forbes.com/sites/stevetengler/2021/07/01/why-arent-more-people-talking-about-solar-cars/?sh=43c24b9349a9

von Gravrock, E. (2021). Evaluating smart home challenges to improve future smart cities. *Forbes*. Retrieved May 15, 2022, from https://www.forbes.com/sites/forbesbusinesscouncil/2021/04/27/evaluating-smart-home-challenges-to-improve-future-smart-cities/?sh=7df409fd6109

KEY TERMS AND DEFINITIONS

Access to Public Data: Access to public data in a smart city context refers to the ability to freely use data that is in the public domain.

Connecting: Connecting in a smart city context refers to activities, resources, or services in support of bridging, joining, and the fostering of relationships and interactions.

Livability: Livability in a smart city context refers to the quality of life afforded by spaces and environments, taking into consideration a range of factors such as cultural, social, economic, environmental, political, and emotional.

Success: Success in a smart city context refers to the achieving of a desired outcome or the completing of one or more tasks in fulfillment of an objective.

Trust: Trust in a smart city context refers to the ability to rely upon people, technologies, services, and governance.

Section 4

Theories, Approaches, and Metrics in Guiding Understandings of Urban Life and the Ambient and Emergent in Support of Directions for Smart, Learning, and Future Cities

This section introduces the reader to theories and approaches for urban life and the ambient involving greater adaptability, awareness, and creativity. Metrics, indices, and standards are explored along with the use of visualizations and simulations for imagining improvements for governance, livability, and sustainability in navigating a way forward involving human flourishing with the belief that we can do better.

Chapter 9
Theories, Approaches, and Metrics in Smart, Learning, and Future Cities:
Understanding Urban Life and the Ambient

ABSTRACT

The purpose of this chapter is to explore understandings of urban life and the ambient in smart, learning, and future cities in relation to urban theories, approaches, and metrics. As such, a review of the research literature is conducted in the context of smart cities and regions pertaining to urban theories, methodologies, and metrics. Issues, controversies, and problems emerging from the literature review are highlighted, contributing to formulation of a conceptual framework for theories, approaches, and metrics in relation to urban life and the ambient in smart cities. Using an exploratory case study approach combined with an explanatory correlational design, variables pertaining to the exploration are identified, and the nature of their relationship is assessed. Through the lens of urban theories, approaches, and metrics, urban life and the ambient is explored in smart cities while informing directions for future research and practice.

1. INTRODUCTION

Harrison and Donnelly (2011) point to the instrumentation of smart cities "as a key enabler for new theories of cities." Schmitt (2015) identifies methodologies for smart cities and future cities such as simulation and experiment. Batty (2020) articulates the notion of high and low frequency cities where change occurs "over very fine time intervals" as in "seconds minutes, hours, days" for the former and "over years, decades, centuries" for the latter with implications for the adapting of theory, approaches, and tools more generally for the understanding of cities and more particularly, smart cities. In response, this chapter is significant in that it seeks to explore urban life and the ambient in smart cities through the

DOI: 10.4018/978-1-6684-4096-4.ch009

lens of evolving theory, approaches, and metrics giving rise to the motivation for this work as described in the objectives below.

Objectives: The main objectives of this chapter are to: a) provide a review of the research literature for urban life and the ambient in the context of smart cities pertaining to theory, approaches, and metrics; b) formulate a conceptual framework for urban life and the ambient in smart cities in relation to theory, approaches, and metrics; and c) explore the relationship between variables in smart cities such as collaborating and data visualizations as factors shedding insight on the evolving of theory, approaches, and metrics in smart environments.

These objectives give rise to the main research question under exploration in this chapter – *Why are tools such as theory needed for understanding urban life and the ambient in smart, learning, and future cities?*

2. BACKGROUND

The notion of innovating metrics for smart cities is advanced by McKenna (2019) as is the proposing of ambient theory for smart cities (McKenna, 2021) in accommodating the aware people and awareness-enabling technologies dimensions of smartness in urban environments and regions. Considerable work has been undertaken by many, such as, the United Nations, European cities, and the International Standards Organization (ISO) on metrics for smart cities as described by the OECD (2020) in the context of measuring the performance of smart cities. Lopes and Rodrigues (2020) explore methodologies for the design, development, implementation, and evaluation of smart cities, proposing a five-phase approach consisting of context analysis, capability assessment, strategy, action plan, and evaluation. Qaed (2020) describes design thinking as an approach to smart cities "to strengthen the human dimension." Backhouse (2020) develops a taxonomy for smart cities based on the identification of four types of measures "indicator standards, models, composite indices, and other measures" in coming to an understanding of measures, their uses, differences, and potentials.

2.1 Definitions

Definitions from the practice and research literature are provided for key terms used in this chapter.

- **Metrics:** In the context of an online course on smart cities, The Open University (2020) describes metrics as "measurements that are based on a standardised method."
- **Theory:** Gregor (2006) describes theory as that which is able "to encompass conjectures, models, frameworks, or body of knowledge" adding more recently (2017) that "theory can serve multiple purposes, from sense-making to predicting."

2.2 Overview

This chapter provides a review of the research and practice literature for urban life and the ambient pertaining to theory, approaches, and metrics, in understanding evolving renderings of smart cities. This review enables the highlighting of key issues, controversies, and problems and the formulation of a conceptual framework for theories, approaches, and metrics in relation to urban life and the ambient in smart cities that is then operationalized for use in this chapter. The main research question under exploration in this chapter is reformulated as a proposition for investigation and the methodology used is described, followed by the presentation of results and a discussion of findings. Key limitations of the study underlying this chapter are identified and future directions for research and practice are outlined. The conclusion provides an overview of findings, the identification of insights and ideas, development of several exercise questions encouraging further exploration, and identification of the target audience for this chapter.

3. PERSPECTIVES ON THEORY, APPROACHES AND METRICS FOR SMART CITIES

This chapter provides a review of the of the research and practice literature for urban life and the ambient pertaining to urban theory, approaches, and metrics for smart cities. A tabular overview of the literature review is then presented.

3.1 Urban Life and Theory for Smart Cities

McKenna and Chauncey (2014a) proposed use of the frictionless learning environment (FLE) innovation combined with activity theory (AT) in formulation of the FLEAT model in support of the notion of "smart teams" for engagement with "collaborative real-world problem solving in smart cities" as "learning cities in action." Going further, McKenna and Chauncey (2014b) propose employing the FLEAT model in developing an "interpretation of digital inequalities" using the 'frictions' concept "as an analytic lens" for the learning cities component of smart cities. Petrucco and Ferranti (2017) demonstrate the use of activity theory in the design of applications for smart city interventions from a knowledge management and e-learning perspective, finding that the smart city model used – Cultural Historical Activity Theory (CHAT) – enabled understanding of "the interaction" and the solving of "contradictions between different stakeholders and systems." Chauncey and Simpson (2020) propose use of a combination of self-determination theory along with "conceptual frameworks for communities of networked expertise and activity theory" in exploring "the role of learning city 'smart teams' in promoting, supporting, and extending the community school model" with multiple goals, one of which is "reimagining and helping students to actualize new identities as learners and contributors to society." McKenna (2021) advances ambient theory for smart cities (ATSC) in an effort to further develop "definitions and understandings of smart cities and regions" while contributing to "a theoretical foundation for the ambient in smart environments" and "opening spaces for dialogue and practice." Nardi (2021) discusses the importance of "appropriating theory" for theory development using the example of activity theory, arguing that "many existing theories are underexploited." According to Nardi (2021), "interacting with theory is especially important to the human sciences for the obvious but usually overlooked reason that human life is so complex, indeterminate, and historical" such that "for the foreseeable future, we will probably be engaging multiple theories, and we will need to be conversant with more than one to continue to develop theoretically." Nardi cites the work of Erickson (Nardi and Redmiles, 2002) in identifying the importance of theory – to "promote dialog, create community, and illumine the world through conceptual richness."

3.2 Urban Life and Smart City Approaches

According to Schmitt (2015), "theory, experiment, and simulation need to work hand-in-hand" to aid the understanding of city, where theory "entails research on the reality, the planning, and the implementation of the city" while experiment involves conducting "Design Research Studios, with the city as a living laboratory" and simulation serves "to make the invisible visible and to test and visualise future scenarios." Schmitt (2017) identifies simulation as an important approach to "understand, shape, plan, design, build, manage, and continually adapt a city" in the context of smart and future cities. Garau and Pavan (2018) formulate "a methodology to measure the quality of urban life based on investigative checklists" together with "objective and subjective indicators" that are "aggregated to develop an index to evaluate a city's level of smart urban quality" for application in an Italian city context. Bibri and Krogstie (2019) undertake to "analyze, investigate, and develop a novel model for smart sustainable cities of the future using backcasting as a scholarly approach." Backcasting is a "scenario methodology" for model building, serving "as a planning tool for facilitating urban sustainability." As such, backcasting scenarios are said to enable the exploration of "future uncertainties, create opportunities, build capabilities, and improve decision-making processes", integrating the compact city model and the eco-city model with the data-driven city model (Bibri and Krogstie, 2019). Timeus, Vinaixa, and Pardo-Bosch (2020) propose a 'city business model' for smart cities; describe use of the Business Model Canvas (BMC) as a practical framework for analysis of the model in identifying value and as a methodology for smart cities to guide the "design, delivery, and assessment of smart services." Qaed (2020) employs a design thinking approach to "add value to smart city design" focusing on the four areas of humanizing design, engagement, sustainability, and innovation in relation to "the five phases of design thinking – empathise, define, ideate, prototype, and test." Eneqvist and Karvonen (2021) explore experimental approaches to governance in smart cities where "laboratories, testbeds, platforms, and innovation districts" are said to be used "to address a wide range of complex sustainability challenges." An analytic framework is developed by Eneqvist and Karvonen (2021) "of the most common strategic functions of municipalities in experimental governance" with findings that "reveal how the strategic functions of visioning, facilitating, supporting, amplifying, and guarding are producing new opportunities and challenges to urban planning practices in twenty-first century cities." Follesa, Yao, Liang, and Zhou (2021) explore the use of "narrative design strategies for urban spaces" in smart cities and propose "the 'Perception-Interaction-Make sense' narrative design model" together with "data as a tool for urban narrative" in support of the importance of culture and identity as dynamic components. Sun, Cao, and Mao (2020) claim that developing a smart city "is inseparable from the application of data visualization technology" and as such, "the development of Virtual Reality technology" is said to enable a "new way for the 3D visualization of city models and the study of user behavior." Ammara, Rasheed, Mansoor, Al-Fuqaha, and Qadir (2022) present an approach to understanding smart cities as a system of sub-systems (SoS) accommodating the complexity, adaptability, and dynamic nature of cities while noting the "dependency and interaction between the various stakeholders, components, and subsystems."

3.3 Urban Life and Metrics for Smart Cities

Marans and Stimson (2011) describe the evolving of approaches to the study of the quality of life (QOL) and the quality of urban life (QOUL) using objective indicators, subjective indicators, and the use of geographic information systems (GIS) to integrate the two. Taking into consideration the culture of sustainability, Marans (2015) calls for the integrating of indicators for sustainability with the quality of urban life (QOUL) including objective, subjective, and behavioral indicators, in support of city prosperity indexes. Bell, Benatti, Edwards, Laney, Morse, Piccolo and Zanetti (2018) address the issue of "making metrics meaningful (MMM)" or "M^3" in smart cities using a "rapid participatory method" namely, "imagine" over the time-period of five months focusing on "information system innovation" pertaining to transport, involving participants who are given "a voice in the data collection landscape" and "a role in app design." McKenna (2019) introduces an innovation framework for emergent and evolving ambient metrics for smart cities as "more aware, adaptive, and dynamic" focusing on awareness, engagement, learning, and openness. The OECD (2020) describes smart cities in relation to the use of digital technologies to enhance well-being and increase inclusiveness, sustainability, and resilience of a city while identifying the challenges and threats of digitalization in smart cities associated with privacy risks, regulation, and widening inequalities. In determining how smart cities are performing and who is benefitting, the OECD (2020) provides a review of existing measurement frameworks, associated complexities, and proposes a measurement framework for implementation. According to the OECD (2020), indicators may be organized within the three categories of input indicators (e.g., "a measure of the effort that is devoted to pursuing a policy"), output indicators (e.g., "measure quantities produced by a policy" as in, percentage of households with a smart energy meter), and outcome indicators (e.g., "monitor the effectiveness of a policy in achieving its objectives" as in "reduction in commuting time, satisfaction with life or the city, or energy savings"). Albert and Pandey (2021) provide an overview of ways in which communities are able to "self-assess and benchmark their progress along a continuum of smart, intelligent, and sustainable" focusing on metrics for sustainability and livability such as trust, engagement, and circular economy. Dashkevych and Portnov (2022) provide a systematic review of the research literature in an effort to empirically identify metrics in support of smart city identification. Key categories emerge including "smart digital technology" and "ecological sustainability" and Dashkevych and Portnov (2022) note the latter is used less often with few criteria pertaining to "urban residents' needs" giving rise to the need for developing a "balanced SC assessment" in the form of a "ranking system based on the nine most important metrics" focusing on "the improvement of the quality of life of local residents" and then applied globally in major cities to demonstrate "use and usefulness." De Sanctis, Iovino, Rossi, and Wimmer (2022) present an automatic assessment modeling framework for smart city key performance indicators (KPIs) which is shown to be open, extensible, and usable.

3.5 Literature Review Overview

An overview by author and date is provided in this section for the research and practice literature for urban life and the ambient pertaining to theory, smart city approaches, and metrics for smart cities.

Table 1 provides an overview by author and date of the research and practice literature for urban life and the ambient pertaining to theory, spanning the timeframe of 2006 to 2021.

Gregor (2006) describes theory in terms what it encompasses, such as, a body of knowledge, conjectures, models, and frameworks. Harrison and Donnelly (2011) point to the instrumentation of smart cities as enabling new theories of cities. McKenna and Chauncey (2014a) propose use of the frictionless learning environment (FLE) innovation combined with activity theory (AT) enabling smart teams to engage and collaborate as learning cities (LCs) in action. The FLEAT model is further advanced by McKenna and Chauncey (2014b) using the frictions concept as an analytic lens for the interpretation of digital inequalities. Gregor (2017) articulates the multi-purpose nature of theory, ranging from sense-

Table 1. Overview of research for urban life and theory for smart cities

Author(s)	Year	Urban Life and Theory
Gregor	2006	Theory encompassing body of knowledge, conjectures, models, frameworks
Harrison & Donnelly	2011	Instrumentation of SCs, new theories of cities
McKenna & Chauncey	2014a	FLE+AT=FLEAT: smart teams, engagement, collaboration, LCs in action
McKenna & Chauncey	2014b	FLEAT model with frictions concept as analytic lens digital inequalities
Gregor	2017	Multi-purpose: sense-making to predicting
Petrucco & Ferranti	2017	AT design of apps for SC interventions; interaction, contradictions
Batty	2020	High & low frequency cities with implications for the adapting of theory
Chauncey & Simpson	2020	Self-determination theory, communities of expertise, AT: contributors
McKenna	2021	Ambient theory for smart cities: define, understand, dialogue, practice
Nardi	2021	Appropriating theory – dialog, create community, illumine the world

making to predicting. Petrucco and Ferranti (2017) employ AT in the design of applications for smart city interventions, helpful for understanding interactions and solving contradictions. High and low frequency cities are described by Batty (2020) where, in the former case, change occurs "in seconds, minutes, hours, days" while in the latter case, the time interval involves "years, decades, centuries" and as such this has implications for the adapting of theory for smart cities. Chauncey and Simpson (2020) combine the use of self-determination theory, communities of networked expertise and AT to explore learning city (LC) smart teams in supporting and extending the community school model in assisting students, for example, to become contributors to society. McKenna (2021) advances ambient theory for smart cities to assist in defining and understanding smart cities while encouraging dialogue and testing through use in practice. Nardi (2021) articulates the notion of appropriating theory for theory development and, inspired by Erickson (Nardi and Redmiles, 2002), Nardi (2021) describes the role of theory to encourage dialog, create community, and illumine the world.

Table 2 provides an overview by author and date of the research and practice literature for urban life and the ambient pertaining to smart city approaches, spanning the timeframe of 2015 to 2022.

Schmitt (2015) describes the interplay of theory, experiment, and simulation in designing cities and future city scenarios. Schmitt (2017) describes how simulation is important for the design and many other aspects of building and continually adapting smart cities and future cities. Garau and Pavan (2018) describe the quality of urban life methodology designed to measure smart urban quality. Bibri and Krogstie (2019) employ a backcasting scenario approach to visioning smart sustainable cities of the future. Qaed (2020) discusses design thinking in relation to smart cities in support of engagement, humanizing design, innovation, and sustainability. Sun et al. (2020) discuss the importance of data visualization for smart cities where virtual reality (VR) technology makes way for 3D visualization of city models and the study of human behavior. Timeus et al. (2020) advance the notion of a city business model, employing the Business Model Canvas (BMC) in identifying value on the one hand and on the other as a methodology to guide smart city (SC) development. Eneqvist and Karvonen (2021) describe experimental approaches to smart city governance in addressing complex sustainability challenges revealing the importance of activities such as amplifying, visioning, and the like. Follesa et al. (2021) advance the Perception-Interaction-Make sense narrative design model in combination with data as an

Table 2. Overview of research for urban life and smart city approaches

Author(s)	Year	Urban Life and Smart City Approaches
Schmitt	2015	Theory, experiment & simulation in designing future city scenarios
Schmitt	2017	Simulation of SCs, continually adapting cities
Garau & Pavan	2018	Quality of urban life methodology to measure smart urban quality
Bibri & Krogstie	2019	Backcasting approach for future smart sustainable cities
Qaed	2020	Design thinking: engagement, humanizing design, innovation, sustainability
Sun et al.	2020	Data visualization, VR, 3D visualization of city models study human behavior
Timeus et al.	2020	City business model for SCs; BMC: analysis (value) & methodology (guide)
Eneqvist & Karvonen	2021	Experimental - governance; complex sustainability: amplifying, visioning
Follesa et al.	2021	Perception-Interaction-Make sense narrative design model + data; culture
Ammara et al.	2022	System of sub-systems (SoS) – complexities, dependencies & interactions

urban narrative tool where culture and identity are dynamic components. Ammara et al. (2022) present a system of sub-systems (SoS) approach to understanding the complexities, dependencies, and interactions of smart cities.

Table 3 provides an overview by author and date of the research and practice literature for urban life and the ambient pertaining to metrics, spanning the timeframe of 2011 to 2022.

Table 3. Overview of research for urban life and metrics for smart cities

Author(s)	Year	Urban Life and Metrics
Marans & Stimson	2011	QOL, QOUL, GIS
Marans	2015	QOUL, sustainability
Bell et al.	2018	Making metrics meaningful – Imagine – rapid participatory method
McKenna	2019	Ambient metrics: awareness, engagement, learning, openness
Backhouse	2020	Taxonomy for SCs: composite indices, indicator standards, models, other
OECD	2020	Input, output, outcome indicators: inclusion, resilience, sustainability, well-being
Open University	2020	Measurements based on a standardised method
Albert & Pandey	2021	Performance metrics & sustainable cities – trust, engagement, circular economy
Dashkevych & Portnov	2022	Assessing smartness; human-centric considerations for quality of life
DeSanctis et al.	2022	MIKADO – automatic assessment modeling framework for SC KPIs

Marans and Stimson (2011) describe the study of quality of life (QOL), quality of urban life (QOUL), and the use of geographic information systems (GIS) as an integrator. Marans (2015) calls for the integrating of sustainability indicators for the QOUL. Bell et al. (2018) describe the making metrics meaningful (MMM) project, using the 'imagine' rapid participatory method to engage participants in information system innovation of transport data collection and app design. McKenna (2019) advances a framework for ambient metrics in smart cities involving awareness, engagement, learning, and openness. Backhouse

Table 4. Patterns and potential relationships

Approaches	Metrics	Theory
City business model for SCs	Ambient metrics	Activity theory
- Business model canvas	Awareness	- design of apps for SC interventions
- guide	Circular economy	Ambient theory for smart cities
- value	Engagement	Body of knowledge
Continually adapting SCs	Geographic Info Systems	Collaboration
Data visualization	Indicators	Communities of networked expertise
- 3D visualization of city models	- Input / Output	Conjectures
- study human behavior	- Outcome	Contradictions
- virtual reality	Inclusion	Contributors to society
Design thinking	Learning	Dialogue
- engagement	Livability	Engagement
- humanizing design	Meaningful metrics	FLEAT
- innovation	Openness	Frameworks
- sustainability	Quality of life	High & low frequency cities
Designing cities & SCs	Quality of urban life	- adapting of theory
- experiment	Rapid participatory ….	Instrumentation of SCs
- simulation	Resilience	- new theories of cities
- theory	Standardization method	Interaction
Experimental governance	Sustainability	Learning cities in action
- amplifying	Taxonomy for SCs	Models
- visioning	- composite indices	Multi-purpose
Perception-Interaction-Make sense	- indicator standards	- predicting
- narrative design model + data	- models	- sensemaking
- culture & identity	- other measures	Practice
Quality of urban life methodology	Trust	Self-determination theory
- measure smart urban quality	Well-being	Smart teams

(2020) provides a taxonomy for smart cities with four types of measures including composite indices, indicator standards, models, and other measures. The OECD (2020) identifies the three categories of input indicators, output indicators, and outcome indicators in measuring elements such as inclusion, resilience, sustainability, and well-being in determining how smart cities are performing and who may be benefitting. Albert and Pandey (2021) provide an overview of performance metrics for smart, intelligent, and sustainable cities focusing on sustainability and livability in terms of trust, engagement, and circular economy. The Open University (2020) provides a definition for metrics as measurements based on a standardized method. Dashkevych and Portnov (2022) develop and apply a balanced set of criteria for assessing the movement toward smartness in cities taking people into consideration and the importance of quality of life. De Sanctis et al. (2022) advance an automatic assessment modeling framework for smart city KPIs.

In terms of patterns and potential relationships, what emerges from this review of the research and practice literature is a comparative overview for theory, approaches, and metrics as presented in Table 4. Elements that emerge across columns of Table 4, include, but are not limited to, engagement, quality of life, quality of urban life, and sustainability.

3.6 Issues, Controversies, Problems

As shown in Table 5, an overview of the research and practice literature review is provided in terms of the issues, controversies, and problems for theory, approaches, and metrics for urban life and the ambient in smart cities.

Table 5. Issues, controversies, and problems

	Issues	Controversies	Problems
Theory	• Collaboration • Conceptual richness • Engagement • Frictionless learning	• Appropriating theory for smart cities • Learning cities in action • Reimagining • Smart teams	• Community • Dialogue • Frictions • Understanding
Approaches	• Backcasting scenarios • Compact-eco-data driven • Engagement • Simulation	• Humanizing urban design • Quality of urban life • Smart urban quality, level • Sustainability	• BMC • Design • Narrative • Visualization
Metrics	• Learning • Make metrics meaningful • Openness • Quality of urban life	• Ambient metrics – aware, adaptive … • Rapid participatory method – Imagine • Smart, intelligent, sustainable • Sustainability	• Inclusiveness • Prosperity • Resilience • Well-being

For urban theory, issues, controversies, and problems pertain to collaboration; conceptual richness; engagement; frictionless learning; appropriating theory for smart cities; learning cities in action; reimagining; smart teams; community; dialogue; frictions; and understanding.

For urban approaches, issues, controversies, and problems pertain to backcasting scenarios; compact-eco-data driven; engagement; simulation; humanizing urban design; quality of urban life; level of smart urban quality; sustainability; business model canvas (BMC); design thinking; narrative design model; and visualization.

For metrics, issues, controversies, and problems pertain to learning; making metrics meaningful; openness; quality of life in urban spaces and regions; ambient metrics as aware – adaptive – dynamic; the rapid participatory method – imagine; smart-intelligent-sustainable continuum; sustainability; inclusiveness; prosperity; resilience; and well-being.

Of note is the presence of overlap between theory, approaches, and metrics and across issues, controversies, and problems where learning and sustainability provide a few examples.

4. CONCEPTUALIZING THEORY, APPROACHES AND METRICS FOR SMART CITIES

Responding to the issues, controversies, and problems emerging from the literature review in the preceding section, a conceptual framework for theories, approaches, and metrics in relation to urban life and the ambient in smart cities is formulated in Fig. 1, highlighting design thinking, experimentation, simulation, and visualization, building upon elements in preceding chapters pertaining to people and technology interactions on a past – present – future timeline in learning, smart, and future cities.

The research question identified in Section 1 is reformulated as a proposition for exploration in this chapter, as follows:

P9: *Tools such as theory are needed for understanding urban life and the ambient in smart, learning, and future cities to identify variables important for the selection of approaches and metrics for researchers and practitioners*

Figure 1. Framework for theories, approaches, and metrics: urban life and the ambient in smart cities

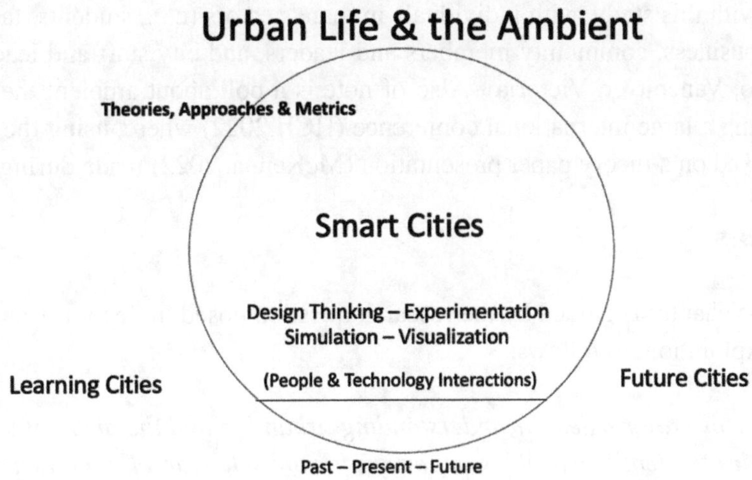

The research design used in exploring the proposition under study in this chapter is described in the Methodology section.

5. METHODOLOGY

Sections 5.1 to 5.3 describe the research design for this chapter in terms of the process used, sources of evidence for data collection, and data analysis.

5.1 Process

Use of an exploratory case study approach in this chapter is supported in view of the contemporary nature of the inquiry (Yin, 2018) and is combined with an explanatory correlational design since the two are said to be complementary (Yin, 2018) and as such, the former may be "needed to examine the underlying processes that might explain a correlation." An online space was used to describe the study and to invite people to participate. Interest emerged from individuals in many cities in multiple countries (Canada, United States, Greece, Ireland, Israel, etc.) reaching several continents.

5.2 Sources of Evidence

Use of an online website enabled the gathering of demographic data from individuals during sign up for the study, which included age range, gender, and city. In addition, people could self-identify in one or more categories (e.g., business, community member, student, educator, etc.). A survey instrument, pre-tested prior to use, was used in this study and contained closed and open-ended questions. Survey questions pertained to a range of items related to smart cities such as collaboration and the visualizing of data. Survey respondents were also invited to participate in an in-depth interview which was guided by an interview protocol that had been pre-tested prior to use in this study.

In addition, data were also gathered systematically through group and individual discussions, conducted in parallel with this study, with individuals in many sectors (e.g., students, faculty, and staff in higher education, business, community members and leaders, and city staff and leaders) in Canadian cities (e.g., Toronto, Vancouver, Victoria). Also of note is a poll about ambient theory for smart cities, conducted during a large international conference (HCII 2022) where, using the Whova platform, feedback was received on a theory paper presentation (McKenna, 2022) made during this conference.

5.3 Data Analysis

Data analysis in this chapter is guided by the research question posed in Section 1 and reformulated as a proposition for exploration, as follows:

P9: *Tools such as theory are needed for understanding urban life and the ambient in smart, learning, and future cities to identify variables important for the selection of approaches and metrics for researchers and practitioners.*

For the analysis of quantitative survey data, the use of descriptive statistics pertaining to theory, approaches, and metrics along with other associated elements enabled exploration of items such as *collaboration* in relation to *the visualizing of data.* The correlation feature of the Real Statistics Resource Pack containing an-add in for Microsoft Excel (Zaiontz, 2022) enabled the conducting of a Spearman Correlation Coefficient for ordinal data in exploring the nature of the relationship between items assessed.

Content analysis was used for qualitative data gathered through open-ended survey questions, interviews, and group and individual discussions so that inductive analysis from data collection was combined with deductive analysis using terminology from the literature review. Overall, an analysis was conducted for n=79 consisting of 42% females and 58% males for people ranging in age from their 20s to 70s.

6. RESULTS AND DISCUSSION

Table 6 provides the results of assessments pertaining to approaches in smart cities in response to a survey question about the extent to which social media and other aware technologies give rise to possibilities for factors such as *collaborating* in a smart city. On a 7-point Likert type scale, where 1= Not at all and 7=Absolutely, for *collaborating*, 17% of survey respondents selected position 4 (Neutral), 17% selected position 5 (Sort of), and 66% responded at the upper end position 7. When asked to assess the extent to which factors such as *visualizations of data that educate, inform, and inspire in real time* contribute to increased value for data in smart cities, 75% of respondents selected position 6 on the scale, and 25% selected position 7.

Using the correlation feature of the Real Statistics add-in for Microsoft Excel (Zaiontz, 2022), a Spearman correlation coefficient for ordinal data was conducted on the assessment results of the two items, *collaborating* and *visualizations of data that educate, inform, and inspire in real time* (Table 6), showing a finding of .33.

Creswell (2018) cautions that with correlations in the .20 to .35 range "there is only a slight relationship" and as such, "may be valuable to explore the interconnection of variables but of little value in prediction studies." As new approaches to visualization emerge in smart cities (Sun et al., 2020), it

Table 6. Approaches in smart cities: assessments for collaborating and visualizations of data

Items	Approaches in Smart Cities	Correlation
Collaborating	17% (4); 17% (5); 66% (7)	.33
Visualizations of data	75% (6); 25% (7)	

may be that correlations with other variables could improve going forward. It is also worth noting that this low correlation involves *collaborating* associated with social media and aware technology, possibly indicative of the need going forward for social media platforms and aware-enabling technologies to take people and their needs into greater consideration.

Qualitatively, findings of interest pertaining to social media and aware technologies include comments from a community leader that "the biggest benefit they can offer is to assist in citizen/visitor education and awareness" and be "focused around people and their needs" while an educator indicated that "people are the reason for cities" and "issues for citizen quality of life have remained fairly constant throughout recent history" including "housing, jobs, health, education, transport, and local economy" and the importance of "thinking about them as an integrated whole, utilizing technologies in easily accessible and useful ways for all citizens."

Also of note, using the Whova networking tool during a virtual international conference of HCII 2022, a Poll was conducted where conference attendees could respond to the following question which was accompanied by 5 checkbox options (items a-e):

How would ambient theory for smart cities (ATSC) be usable in your city?

a. ATSC could guide urban researchers
b. ATSC could guide urban practitioners
c. ATSC could help us think differently about urban design
d. I have no idea
e. Other

Over the six-day duration of the conference, 93 responses were received from conference participants with the following results:

Option a was selected by 23% of responders
Option b was selected by 20% of responders
Option c was selected by 46% of responders
Option d was selected by 32% of responders
Option e was selected by 4% of responders

It should be noted that a paper was delivered at this conference on the usefulness and usability of ambient theory for smart cities (McKenna, 2022) and poll responses would seem to show an encouraging potential for the usability of the theory for urban researchers (23%), urban practitioners (20%), for those concerned with thinking differently about urban design (46%), and for those who are envisioning other usability options (4%). As a relatively new theory (McKenna, 2021), it is perhaps not surprising that 32% of responders selected option d.

Qualitatively, at this same conference (HCII2022) when asked in a session question, *Why do smart cities matter?* it is worth noting that the response of a session attendee included the importance of design – "They are an opportunity to rethink the city and to lower levels of automobile traffic and promote better, more healthy and environmentally friendly spaces. Whether it becomes this is dependent on how well IoT folks approach their designs." In response to a theory paper presentation on the usability and usefulness of ambient theory for smart cities at this same conference, a participant commented that "the attributes of usability in this ambient theoretical scoping is actually a very, very interesting way of thinking about usability factors in these contexts" while observing that "when you see a theory like this it does promote quite inventive thinking in other directions."

As depicted graphically in Fig. 2, the interconnectivity of theory with approaches, and metrics is aided in urban life by understandings of design thinking, experimentation and simulation giving way to the notion and potentials of ambient approaches, ambient collaboration, ambient metrics, and ambient theory.

Figure 2. Urban life and the ambient – theory in smart cities

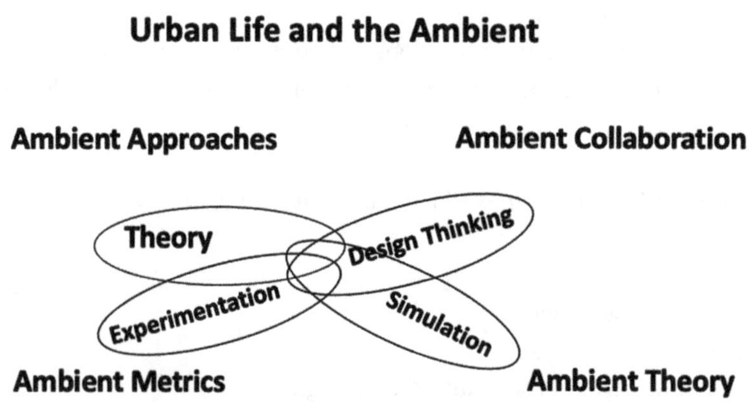

7. LIMITATIONS AND FUTURE DIRECTIONS

Possible limitations of this work are discussed in this section along with challenges and opportunities for research and practice going forward.

Limitations

Limitations of this chapter pertain possibly to the extent of the literature review on urban life and the ambient in relation to theory, approaches, and metrics in smart cities and this is mitigated by the potential to extend the reviews of the research literature going forward. Among other limitations of this chapter is that of the small sample size for the underlying study of this work, and as such, caution is encouraged in the interpretation of the correlation calculated from the assessment of variables used for approaches in smart cities. Going forward, as theory, approaches, and metrics evolve, opportunities may emerge for the use of other variables, resulting in possibly stronger correlations with greater prediction potentials.

Future Research Directions and Opportunities

For research, future directions emerge in terms of the potential for conducting additional exploration, testing, and validation of the variables identified in this work relevant to approaches to urban life and the ambient in smart cities. Additionally, opportunities for explorations going forward may emerge for the identification of other and possibly stronger variables relevant to theory, approaches, and metrics for urban life in the context of smart cities, learning cities, and future cities as identified in Table 4. An opening emerges going forward for further explorations of ambient approaches, ambient collaboration, ambient metrics, and ambient theory.

Of note in Table 6 is the presence of a neutral response for *collaborating,* identifying a potential area of interest for further exploration.

Future Practice Directions and Opportunities

Future directions for practice include the potential for pursuing explorations with variables identified in this work relevant to urban life and the ambient, focusing on theory, approaches, and metrics in smart cities, such as well-being (Table 4). Opportunities going forward for practice include the potential for exploring spaces with other, possibly stronger variables relevant to urban life that foster emergent notions of ambient theory and ambient metrics in everyday contexts in smart cities, learning cities, and future cities, in addressing the many issues, controversies, and problems presented in Table 5.

8. CONCLUSION

Challenges and opportunities involving theory, approaches, and metrics associated with urban life and the ambient emerge from the review of the research literature provided in this chapter from multiple perspective. A key contribution of this chapter is the formulation of a conceptual framework for urban life and the ambient in relation to theory, approaches, and metrics in smart cities based on the literature review where emerging issues, controversies, and problems contribute to operationalization of the framework for use in this chapter. Through an exploratory case study approach, together with an explanatory correlational design, the proposition under study in this chapter is explored. Based on survey data, a correlation is conducted for *collaborating* and *visualizations of data that educate, inform, and inspire in real time* with the finding of a Spearman correlation coefficient of .33, with "only a slight relationship" having "little value for prediction." Limitations of the study underlying this chapter pertain to, among other things, the small sample size and caution is urged in the interpretation of correlation findings, providing opportunities going forward for further research. Future directions guiding research and practice activities going forward are provided in terms of further validating the findings in this chapter while also encouraging activities to identify other, additional variables relevant to urban life and the ambient in relation to the notions of ambient theory, ambient approaches, ambient metrics, and ambient collaboration, in the context of smart, learning, and future cities.

Key insights and ideas emerging from this chapter are provided, along with exercises encouraging further exploration, as follows:

Insights emerging from this chapter include the following:

1. The importance of the emerging relationship between *collaborating* and *visualizations of data that educate, inform, and inspire in real time* in smart cities
2. The importance of factors pertaining to people and their needs in smart cities
3. The potential for many more factors to emerge for understanding *theory, approaches,* and *metrics* in smart cities such as capabilities associated with the emergent, involving design thinking

Ideas emerging from this chapter include the following:

1. Potential opportunities for exploring urban life and the ambient in relation to a variety of variables pertaining to theory, approaches, and metrics in the context of smart cities
2. The opportunities potential for the identification of other and additional variables for exploring urban life and the ambient in smart environments in relation to ambient theory or ambient metrics (e.g., more dynamic, adaptive, real-time, etc.)
3. The use of an interdisciplinary and multi-method approach for the study and practice of urban life and data visualizations and collaborating in smart, learning, and future cities

Exercises for the reader to consider, encouraging further exploration include the following:

1. Identify one or more ways in your city that urban theory matters
2. Identify one or more ways in which new urban approaches would enhance the quality of life in your city
3. Describe one or more ways in which urban metrics would enhance the quality of life in your city

The audience for this chapter is intended to be interdisciplinary in nature, including but not limited to community members and leaders, educators, businesses, researchers, urban planners, urban designers, and anyone concerned with theory, approaches, and metrics in urban life and the ambient in smart cities, while being attentive to emergent ambient theory and ambient metrics along with associated practice and research opportunities and challenges in smart cities, learning cities, and future cities.

REFERENCES

Albert, S., & Pandey, M. (2021). *Performance Metrics for Sustainable Cities* (1st ed.). Routledge. doi:10.4324/9781003096566

Ammara, U., Rasheed, K., Mansoor, A., Al-Fuqaha, A., & Qadir, J. (2022). Smart cities from the perspective of systems. *Systems, 10*(3), 77. doi:10.3390ystems10030077

Backhouse, J. (2020). A taxonomy of measures for smart cities. In *Proceedings of the 13th International Conference on Theory and Practice of Electronic Governance (ICEGOV2020)*. Association for Computing Machinery. 10.1145/3428502.3428593

Batty, M. (2018). *Inventing future cities.* MIT Press. doi:10.7551/mitpress/11923.001.0001

Batty, M. (2020). Defining smart cities: High and low frequency cities, big data and urban theory. In K. S. Willis & A. Aurigi (Eds.), *The Routledge companion to smart cities* (pp. 51–60). Routledge. doi:10.4324/9781315178387-5

Bell, S., Benatti, F., Edwards, N. R., Laney, R., Morse, D. R., Piccolo, L., & Zanetti, O. (2018). Smart cities and M^3: Rapid research, meaningful metrics and co-design. *Systemic Practice and Action Research, 31*(1), 27–53. doi:10.100711213-017-9415-x

Bibri, S. E., & Krogstie, J. (2019). Generating a vision for smart sustainable cities of the future: A scholarly backcasting approach. *European Journal of Futures Research, 7*(5), 5. Advance online publication. doi:10.118640309-019-0157-0

Chauncey, S. A., & Simpson, G. I. (2020). The role of learning city "smart teams" in promoting, supporting, and extending the community school model. In Lecture notes in computer science: Vol. 12425. *HCI International 2020—Late breaking papers: Cognition, learning and games. HCII 2020*. Springer. doi:10.1007/978-3-030-60128-7_25

Cheshire, J., & Uberti, O. (2021). *Atlas of the invisible: Maps and graphics that will change how you see the world*. Norton.

Creswell, J. W. (2018). *Educational research: Planning, conducting, and evaluating quantitative and qualitative research* (6th ed.). Pearson.

Dashkevych, O., & Portnov, B. A. (2022). Criteria for smart city identification: A systematic literature review. *Sustainability, 14*(8), 4448. doi:10.3390u14084448

De Sanctis, M., Iovino, L., Rossi, M. T., & Wimmer, M. (2022). MIKADO: A smart city KPIs assessment modeling framework. *Software & Systems Modeling, 21*(1), 281–309. doi:10.100710270-021-00907-9

Eneqvist, E., & Karvonen, A. (2021). Experimental governance and urban planning futures: Five strategic functions for municipalities in local innovation. *Urban Planning, 6*(1), 183–194. doi:10.17645/up.v6i1.3396

Follesa, S., Yao, P., Liang, S., & Zhou, M. (2021). Cultural identity of the cities—The use of narrative design in urban spaces. In J. Charytonowicz, A. Maciejko, & C. S. Falcão (Eds.), *Advances in Human Factors in Architecture, Sustainable Urban Planning and Infrastructure. AHFE 2021. Lecture Notes in Networks and Systems* (Vol. 272). Springer. doi:10.1007/978-3-030-80710-8_19

Garau, C., & Pavan, V. M. (2018). Evaluating urban quality: Indicators and assessment tools for smart sustainable cities. *Sustainability, 10*(3), 575. doi:10.3390u10030575

Harrison, C., & Donnelly, I. A. (2011). A theory of smart cities. In *Proceedings of the 55th Annual Meeting of the International Society for the Systems Sciences*. University of Hull Business School.

Lopes, N. V. M., & Rodrigues, J. (2020). Smart methodologies for smart cities: A comparative analysis. In *Smart Governance for Cities: Perspectives and Experiences*. EAI/Springer Innovations in Communication and Computing. Springer. https://link.springer.com/chapter/10.1007/978-3-030-22070-9_1

Marans, R. W. (2015). Quality of urban life & environmental sustainability studies: Future linkage opportunities. *Habitat International, 45*(Part 1), 47–52. doi:10.1016/j.habitatint.2014.06.019

Marans, R. W., & Stimson, R. (2011). An overview of quality of urban life. In R. Marans & R. Stimson (Eds.), Social Indicators Research Series: Vol. 45. *Investigating Quality of Urban Life*. Springer. doi:10.1007/978-94-007-1742-8_1

McKenna, H. P. (2019). Innovating metrics for smarter, responsive cities. *Data*, *4*(1), 25. doi:10.3390/data4010025

McKenna, H. P. (2021). The importance of theory for understanding smart cities: Making a case for ambient theory. In N. Streitz & S. Konomi (Eds.), Lecture Notes in Computer Science: Vol. 12782. *Distributed, Ambient and Pervasive Interactions. HCII 2021*. Springer. doi:10.1007/978-3-030-77015-0_4

McKenna, H. P. (2022). Exploring the Usefulness and Usability of Ambient Theory for Smart Cities. In N. A. Streitz & S. Konomi (Eds.), Lecture Notes in Computer Science: Vol. 13325. *Distributed, Ambient and Pervasive Interactions. Smart Environments, Ecosystems, and Cities. HCII 2022*. Springer. doi:10.1007/978-3-031-05463-1_12

McKenna, H. P., & Chauncey, S. A. (2014a). Taking learning to the city: An exploration of the frictionless learning environment innovation. *Proceedings of the 6th International Conference on Education and New Learning Technologies (EduLearn14),* 6324-6334.

McKenna, H. P., & Chauncey, S. A. (2014b). Frictionless learning environments for 21st century education and learning cities: A response to digital inequalities. *Proceedings of the 7th International Conference of Education, Research and Innovation (iCERi2014),* 2505-2515.

Nardi, B. (2021). Appropriating theory. In D. H. Sonnenwald (Ed.), *Theory Development in the Information Sciences* (pp. 204–221). University of Texas Press. doi:10.7560/308240-014

Nardi, B., & Redmiles, D. (Eds.). (2002). Activity Theory and the Practice of Design. The Journal of Computer-supported Cooperative Work, 11(1-2).

OECD. (2020). Measuring smart cities' performance: Do smart cities benefit everyone? 2nd *OECD Roundtable on Smart Cities and Inclusive Growth.*

Petrucco, C., & Ferranti, C. (2017). Design smart city apps using activity theory. *Knowledge Management & E-Learning: An International Journal*, *9*(4), 499–511.

Qaed, F. (2020). The value of design thinking for smart cities. *3rd Smart Cities Symposium (SCS 2020)*, 640-645. doi: 10.1049/icp.2021.0934

Schmitt, G. (2015). *Information Cities*. Apple Books. https://books.apple.com/ca/book/information-cities/id970529491

Schmitt, G. (2017). *Future Cities.* MOOC (Massive Open Online Course). ETHzurich. Retrieved 11 December 2017 from https://online.ethz.ch/courses/course-v1:ETH+FC-01x+2019_T1/about

Sun, T., Cao, Y., & Mao, B. (2020). 3D City Cadaster data visualization and viewer behavior analysis in virtual reality environment. *2020 IEEE/WIC/ACM International Joint Conference on Web Intelligence and Intelligent Agent Technology (WI-IAT)*, 936-939. 10.1109/WIIAT50758.2020.00143

The Open University. (2020). *Smart city measurement: Metrics and indicators.* Course on Smart Cities, OpenLearn. Retrieved 21 February 2020 from https://www.open.edu/openlearn/mod/oucontent/view.php?id=67889&section=1

Timeus, K., Vinaixa, J., & Pardo-Bosch, F. (2020). Creating business models for smart cities: A practical framework. *Public Management Review*, *22*(5), 726–745. doi:10.1080/14719037.2020.1718187

Yin, R. (2018). *Case study research and applications: Design and methods* (6th ed.). Sage Publications.

Zaiontz, C. (2022). *Real statistics using excel.* Retrieved from www.real-statistics.com

ADDITIONAL READING

Burns, R., Fast, V., Levenda, A., & Miller, B. (2021). Smart cities: Between worlding and provincializing. *Urban Studies (Edinburgh, Scotland)*, *58*(3), 461–470. doi:10.1177/0042098020975982

Darisi, T., & Watt-Kapitain, H. (2018). *Theory of change for smart cities: Envisioning a circular food economy*. Retrieved 15 May 2022 from https://guelph.ca/wp-content/uploads/Theory-of-Change-for-Smart-Cities.pdf

Doors to Switzerland. (2021). *Smart cities and creativity: Towards a new intersection of urban and culture theory – A challenge after the pandemic.* Symposium. Retrieved 15 May 2022 from https://doorstoswitzerland.com/en/events/smart-cities

Lom, M., & Pribyl, O. (2020). Smart city model based on systems theory. *International Journal of Information Management*. Advance online publication. doi:10.1016/j.ijinfomgt.2020.102092

Mora, L., Deakin, M., Zhang, X., Batty, M., de Jong, M., Santi, P., & Appio, F. P. (2021). Assembling sustainable smart city transitions: An interdisciplinary theoretical perspective. *Journal of Urban Technology*, *28*(1-2), 1–27. doi:10.1080/10630732.2020.1834831

Pira, M. (2021). A novel taxonomy of smart sustainable city indicators. *Humanities and Social Sciences Communications*, *8*(1), 197. doi:10.105741599-021-00879-7

Ruddick, S., Peake, L., Tanyildiz, G. S., & Patrick, D. (2018). Planetary urbanization: An urban theory for our time? *Environment and Planning. D, Society & Space*, *36*(3), 387–404. doi:10.1177/0263775817721489

SEEC. (2019). *Partnering nudge theory and SMART city technology.* Schulich Executive Education Centre (SEEC), School of Business. Retrieved 15 May 2022 from https://seec.schulich.yorku.ca/news/partnering-nudge-theory-and-smart-city-technology/

Smith, H., Medero, G. M., Crane De Narváez, S., & Castro Mera, W. (2022). Exploring the relevance of 'smart city' approaches to low-income communities in Medellín, Colombia. *GeoJournal*. Advance online publication. doi:10.100710708-022-10574-y

KEY TERMS AND DEFINITIONS

Ambient Approaches: Ambient approaches to smart cities refer to more dynamic and adaptive ways of working together in designing solutions for complex challenges.

Ambient Collaboration: Ambient collaboration in an urban context refers to working together in a dynamic way, possibly involving interactions with technology to create or improve one or more aspects of life in the city.

Ambient Metrics: Ambient metrics in the context of smart cities refer to more dynamic and adaptive ways of measuring any aspect of urban environments and spaces in-the-moment and rendering such measurements in meaningful ways.

Ambient Theory: Ambient theory in the context of smart, learning, and future cities refers to theory that takes into consideration elements such as awareness, adaptive, dynamic, emergent, interactive, and pervasive as well as meaningfulness for people.

Collaboration: Collaboration in an urban context refers to working together to create or improve some aspect of life in the city.

Chapter 10
Governance, Livability, and Sustainability in Smart, Learning, and Future Cities: Enriching Urban Life and the Ambient

ABSTRACT

The purpose of this chapter is to provide a review of the research literature for urban governance, livability, and sustainability in smart cities and regions. Issues, controversies, and problems emerging from the literature review are highlighted in formulation of a conceptual framework for enriching urban life and the ambient in smart cities aided by urban governance, livability, and sustainability. Using an exploratory case study approach combined with an explanatory correlational design, variables pertaining to the exploration are identified, such as housing affordability and useful community functions and purposes, and the nature of their relationship is assessed. Through the lens of urban governance, livability, and sustainability, potentials for enriching urban life and the ambient are explored in smart cities, attentive to health and well-being and quality of life, while informing directions for future research and practice such as human flourishing.

1. INTRODUCTION

Timmer and Seymoar (2005) refer to the "achieving and sustaining" of a livable city as "a living dynamic experiment" where livability both in cities and on the planet more generally is described as "a central challenge." Batty (2018) describes a "great transition from a world with few cities to a world of all cities" and yet, what if the earth is unable to sustain such a scenario as "a world of all cities"? It is worth noting that during the global COVID-19 Pandemic, reports (Thomson Reuters, 2022) indicate that an "urban exodus" is occurring for example, from larger Canadian cities such as Montreal and Toronto to smaller cities and communities. Also of note perhaps is the theme of the 2014 Annual Challenges of Government Conference (CGC, 2014) which was "flourishing cities" where, after four days of "thought-

DOI: 10.4018/978-1-6684-4096-4.ch010

provoking discussions" involving many people around the world, Dean Ngaire Woods noted that "four factors emerged as essential for building and developing cities that 'flourish': people, sustainable planning, investment, and good governance." This chapter is significant in that it undertakes an exploration of governance, livability, and sustainability in relation to urban life and the ambient in smart cities giving rise to the motivation of this work as described in the objectives that follow.

Objectives: The main objectives of this chapter are to: a) enrich the notion of urban life and the ambient in the context of smart cities in relation to governance, livability, and sustainability; b) develop a conceptual framework for enriching urban life and the ambient in smart cities aided by governance, livability, and sustainability; and c) explore the relationship between variables in smart cities such as *livability* and *housing affordability* as factors relevant to human flourishing that may provide insight into navigating the way forward in smart environments. These objectives give rise to the main research question under exploration in this chapter – *How do governance, livability, and sustainability contribute to the enriching of urban life and the ambient in smart, learning, and future cities?*

2. BACKGROUND

Joss (2018) seeks to "reinvigorate the place of the public in the future city" arguing "for strengthening institutional frameworks" while "advancing a more active role for citizens to become involved in actualizing and scrutinizing future cities" where the complexity and challenge of public governance is said to involve "institutional arrangements, conceptual discourses, and on-the-ground practices." From an urban sustainability and planning perspective, Garau and Pavan (2018) develop an index "to evaluate a city's level of smart urban quality" where urban quality of life is said to refer to "all factors that, either directly or indirectly, affect citizens' quality of life" contributing to a smart sustainable city. Higgs et al. (2019) advance a livability index that takes health and well-being into consideration and the need for integrated urban governance across sectors to inform urban planning policies. According to Abrahams (Voices, 2020), "how well local government can navigate" the "contestations, demands, and expectations" is key for "realizing [a] sustainable future for urban dwellers" identifying "material outcomes" as important, such that, "it is the material impact on urban dwellers that ultimately evaluates how well we have governed."

2.1 Definitions

Definitions from the research and practice literature are provided for key terms used in this chapter.

- **Governance (urban):** Abrahams (Voice, 2020) states that "urban governance is about the management of both urban demands *and* expectations all while self-managing a local bureaucratic infrastructure that is at the mercy of political tides and under constant scrutiny by the Treasury."
- **Livability:** Higgs et al. (2019) cite the work of Lowe et al. (2015) in providing a definition of liveability as "safe, attractive, socially cohesive and inclusive, and environmentally sustainable" communities that have "affordable and diverse housing linked by convenient public transport, walking and cycling infrastructure to employment, education, public open space, local shops, health and community services, and leisure and cultural opportunities."

- **Smart Sustainable Cities:** A joint definition for smart sustainable cities is provided by the UNECE (United Nations Economic Commission for Europe) and the ITU (International Telecommunication Union) as "an innovative city that uses ICTs" as in, Information and Communication Technologies, "and other means to improve quality of life, efficiency of urban operation and services, and competitiveness, while ensuring that it meets the needs of present and future generations with respect to economic, social, environmental as well as cultural aspects" (UNECE, 2022).

2.2 Overview

This chapter provides a review of the research and practice literature for governance, livability, and sustainability, in relation to evolving understandings of urban life and the ambient in smart cities. The literature review enables the highlighting of key issues, controversies, and problems and the formulation of a conceptual framework for enriching urban life and the ambient aided by governance, livability, and sustainability that is then operationalized for use in this chapter. The main research question under exploration in this chapter is reformulated as a proposition for investigation and the methodology used is described, followed by the presentation of results and a discussion of findings. Key limitations of the study underlying this chapter are identified and future directions for research and practice are outlined. The conclusion provides an overview of findings, the identification of insights and ideas, development of several exercise questions encouraging further exploration, and identification of the target audience for this chapter.

3. PERSPECTIVES ON ENRICHING URBAN LIFE AND THE AMBIENT IN SMART CITIES

This chapter provides a review of the of the research and practice literature for enriching urban life and the ambient in relation to governance, livability, and sustainability for smart cities.

3.1 Urban Life and Governance for Smart Cities

Norman (2018) identifies "urban governance" as a research priority for cities in an effort to provide "a better understanding of the multiple ways the developing science of cities can be incorporated into the planning, design and management of cities" utilizing a "systems approach to the urban future." Jiang, Geertman, and Witte (2020) propose a framework for the notion of "smart urban governance" involving the "three intertwined components" of the "spatial, institutional, and technological." Testing the framework in Singapore and Helsinki, Jiang et al. (2020) found that "smart urban governance varies remarkably" such that "as urban issues differ in different contexts, the governance modes and relevant ICT functionalities applied also differ considerably." As such, context matters according to Jiang et al. (2020) and a "sociotechnical way of governing cities" is promoted. Timeus, Vinaixa, and Pardo-Bosch (2020) propose the use of business model logic to the application of smart city governance and in doing so, provide a definition for the 'smart city business model' as follows: "the way in which a city government organizes its services to create and deliver value for its citizens" adding that such a way be "economically viable, socially inclusive and environmentally sustainable." To this end, the city model canvas (CMC) is advanced as a practical tool to guide the design and implementation of smart city services with analysis capabilities in support of governance by city councils and other stakeholders.

Reflecting on "governance challenges on the way to a sustainable urban world" with One Earth (Voices, 2020), Abrahams claims that "the challenge of urban governance is almost always local" as in "keeping systems running and dealing with everyday demands" while emphasizing the importance of "functioning as the interface between the state and its people." Abrahams (Voices, 2020) adds that this interface "must navigate all of the contestation that plays out in urban space, all of the demands that affect everyday urban living" along with "all of the imagined structure that must function as a conducive environment to facilitate a good quality of life." Drapalova and Wegrich (2020) speculate "that smart city strategies and their impacts will inevitably vary across contexts" so using a combination of "existing research on the governance of smart cities" together "with theories of business-state relations from political economy" what emerges is "an analytical framework and typology that captures varieties (and similarities) of smart city policies and governance" for application to the study of smart cities. Grossi, Meijer, and Sargiacomo (2020) contribute a public management perspective on smart cities, noting that the "smart urban governance" literature "has exponentially increased since 2016" with "three sets of papers" categorized as prescriptive, critical, and empirical, expressing, "optimism, pessimism, and realism" respectively and, where empirical studies "analyze the emerging patterns of technology-mediated urban governance" which "highlights the need to study rules, actors, and contexts." Shadowen, Lodato, and Loi (2020) explore the role of automated decision-making systems in smart cities from a participatory governance and participatory design perspective, involving citizens, while cautioning that such undertakings may be "messy, contextual, and frictional." Fernandez, Sanchez, Velez, and Moreno (2020) propose an enterprise architecture framework (EAF) designed for smart cities that is "supported by a hybrid model based on a geographic information system (GIS) and associated graph-database (GDB)" in support of a range of governance challenges, from quality of life to digital transformation. Eneqvist and Karvonen (2021) explore the area of "experimental governance" involving "laboratories, testbeds, platforms, and innovation districts" designed "to address a wide range of complex sustainability challenges." An analytic framework is developed by Eneqvist and Karvonen (2021) for "the most common strategic functions of municipalities in experimental governance" namely, "visioning, facilitating, supporting, amplifying, and guarding" and then applied to the city of Stockholm in order to learn how this approach is "producing new opportunities and challenges" for urban planning practices. Among the many challenges highlighted by Eneqvist and Karvonen (2021) are those of "learning, scaling, and implementation." Allam, Sharifi, Bibri, and Chabaud (2022) conduct a bibliometric analysis of the research literature providing a mapping of trends and knowledge structures for smart urban governance using VOSviewer, pointing to the "need for citizen participation in urban policies" and the need for "sustained innovation for e-participation, e-governance, and policy frameworks." Gasco-Hernandez, Gil-Garcia, and Luna-Reyes (2022) provide an exploration of the four areas of technology, leadership, governance, and collaborative capacities in the context of digital government.

3.2 Urban Life and Livability for Smart Cities

Patorniti and Stevens (2017) show that data associated with street ambience elements enables measurement "to manage light and noise pollution, and traffic volumes, and to enhance liveability, safety and air quality" and as such, "ambience is an important component of the social, perceptual quality and environment functional purposes." Braem, Panzica La Manna, Devis-Clavijo, Lefever, Lievens, and

Marquez-Barja (2019) call for the transformation of cities toward more livable spaces, as in "lively, safe, and healthy" with "an active community that is engaged" in this endeavor employing the use of testbeds and "environmental wireless sensors" for "air quality and noise monitoring" in developing "future applications for a more humanized city." Trencher and Karvonen (2019) explore the potential to stretch the agenda for smart cities beyond that of "environmental protection and economic development to encompass social issues and lifestyle" in addressing "social problems and realise more sustainable, equitable and liveable cities." Findings by Trencher and Karvonen (2019) "contribute to an emerging discourse on the possibilities to utilize digital technologies and smart planning strategies to improve the human dimensions of urban life" focusing on health, well-being, and longevity. Tan, Lim, Zhang, and Tan (2020) respond to the interest in livable, smart cities with a global index providing ranking analysis, simulation, and policy evaluation. Bickley, Macintyre, and Torgler (2021) explore safety with "urban analytics and digital technologies" in relation to crime, traffic, and public health while arguing for the importance of "the 'human factor' when using smart technologies to design a safe and livable smart city." Martino, Girling, and Lu (2021) provide a socioeconomic perspective on livability taking into consideration indicators such as accessibility, social diversity, affordability, and economic vitality as well as built environment indicators such as intensity, centrality, and diversity in the metropolitan region of Vancouver, Canada. It is worth noting that the Architecture, Media, Politics Society (AMPS) is featuring an upcoming conference on livable cities (AMPS, 2023) focusing on architecture, design, city planning, health, technology, urban economics, and social policy encompassing many other areas such as teaching, learning, education, to be followed by another conference on livable cities in 2024 in London.

3.4 Urban Life and Sustainability for Smart Cities

Peris-Ortiz, Bennett, and Pérez-Bustamante Yábar (2017) introduce a book providing guidance on the movement from industrial cities to sustainable smart cities focusing on elements such as citizen participation, information and communication technologies (ICTs), quality of life, and cooperation for public and private initiatives. Norman (2018) argues that the notion of "autonomous cities" may apply locally "where cities will increasingly do many things very well at the urban level and through city networks" however, it is argued that "collectively we will develop a much more sustainable urban future when we are working together from the local to the global scale." Sengers, Späth, and Raven (2018) focus on the notion of "sustainability transitions" using a socio-technical perspective where the smart city is interpreted "not as a fixed narrative or static artifact" but rather, "as an emerging practice" involving "a range of public and private actors" who "experiment in various ways" with "constructions-in-action" as "discursive, institutional and material." Evans, (Guest Editors / Authors), Karvonen, Luque-Ayala, Martin, McCormick, Raven, and Palgan (2019) introduce a special journal issue addressing the gap of "pipedreams and practicalities" in smart cities in relation to "social equity and environmental sustainability" that need to be created and also "nurtured and maintained as they materialise in particular places" in support of future possibilities.

Regarding issues such as urban density, Pafka (2020) argues for the need to learn from the COVID-19 global pandemic "to rethink urban resilience from a broad and nuanced public health perspective" in figuring out how to "design a more adaptable city that maximizes capacities for change in both its architecture and urban design" while undoing the "rigidities of the 'master plan'." From the perspective of the field of law, Giuffrida (2021) challenges the "sustainability claims" pertaining to smart cities, drawing attention to issues of accountability such as the privatization of data collection on a large scale,

Table 1. Overview of research for governance in smart cities

Author(s)	Year	Governance
CGC	2014	Flourishing cities: good governance, investment, people, sustainable planning
Joss	2018	Public governance, involve citizens actively, institutional frameworks
Norman	2018	Urban governance, research priority: planning, design, management
Drapalova & Wegrich	2020	Analytic framework & typology of SC governance & policies
Fernandez et al.	2020	EAF for SCs; GIS, graphic database; quality of life to digital transformation
Grossi et al.	2020	Smart urban governance: prescriptive, critical, empirical; actor, context, rule
Jiang et al.	2020	Smart urban governance framework: spatial, institutional, technological
Shadowen et al.	2020	ADM in SCs; participatory governance & design; contextual-frictional-messy
Timeus et al.	2020	Business model logic: economic viability, socially inclusive, sustainable
Voices	2020	Challenges: local; state to people interface; quality of life; material outcome
Eneqvist & Karvonen	2021	Experimental governance: learning, scaling, implementation
Allam et al.	2022	Emerging trends & knowledge structures for smart urban governance
Gasco-Hernandez et al.	2022	Technology, leadership, governance, and collaborative capacities

personal data exposure, and "the datafication of citizens" with implications for smart city governance. Focusing on urban design, transport, and health, Giles-Corti et al. (2022) advance the "urgent need for evidence-informed city planning policies and standards" aligned with a detailed framework "for healthy and sustainable cities." Where smart growth has more to do with principles and policies concerned with ensuring good planning for cities and regions (e.g., walkable neighborhoods, variety of transportation choices, community and stakeholder collaboration in decision making) as articulated in an edited work by Knaap, Lewis, Chakraborty, and June-Friesen (2022), Goodspeed (2022) argues for the leveraging of smart cities capabilities in support of smart growth and sustainable urbanization through the use of mapping technologies for land use, data sharing, and urban sensing.

3.5 Literature Review Overview

An overview by author and date is provided in this section for the research and practice literature for governance, livability, and sustainability in smart cities.

Table 1 provides an overview by author and date of the research and practice literature for urban life and the ambient in smart cities pertaining to governance, spanning the timeframe of 2014 to 2022.

The notion of flourishing cities emerged from a Challenges of Government Conference (CGC, 2014) where key factors identified included good governance, investment, people, and sustainable planning.

Joss (2018) argues for more active public involvement in governance involving citizens while also strengthening institutional frameworks. Norman (2018) argues for urban governance to be a research priority for cities, in support of a systems approach to planning, design, and management. Drapalova and Wegrich (2020) employ an analytic framework and typology of smart cities governance and policies, showing varieties and similarities. An enterprise architecture framework (EAF) is advanced for smart cities (SCs) by Fernandez et al. (2020) using GIS and graphic database technologies, in addressing a range of governance challenges from quality of life to digital transformation. Grossi et al. (2020)

Table 2. Overview of research for livability in smart cities

Author(s)	Year	Livability
Timmer & Seymoar	2005	Central challenge within cities and the planet – living, dynamic experiment
Lowe et al.	2015	Defining livability
Patorniti & Stevens	2017	Street ambience elements, data; liveability, safety; social, perceptual
Braem et al.	2019	Lively, safe, healthy spaces: Testbeds – air quality & noise monitoring
Higgs et al.	2019	Livability index, health, well-being; integrated urban governance
Trencher & Karvonen	2019	Digital tech + smart planning: health, longevity, well-being; social issues
Tan et al.	2020	Global Liveable & Smart Cities Index (GLSCI)
Bickley et al.	2021	Safety in livable SCs – urban analytics & digital tech & human factor
Martino et al.	2021	Socioeconomic livability & livability
AMPS	2023	Livable Cities Conferences – New York & London

categorize the growing smart urban governance literature as prescriptive, critical, and empirical where the empirical studies focusing on technology-mediated urban governance patterns show the need for studying actors, contexts, and rules. Jiang et al. (2020) advance a smart urban governance framework with the three interwoven components of the spatial, institutional, and technological for a sociotechnical approach to governing. Shadowen et al. (2020) describe the role of automated decision-making (ADM) in smart cities using participatory governance and participatory design, involving citizens which is described as contextual, frictional, and messy. Timeus et al. (2020) advance the use of business model logic in smart city governance where services to citizens are economically viable, socially inclusive, and environmentally sustainable. Governance challenges are said (Voices, 2020) to be, for the most part, local where the interface between people and state matters in support of good quality of life and material outcomes serve to indicate how well governance has been conducted. Eneqvist and Karvonen (2021) explore challenges for experimental governance highlighting those associated with learning, scaling, and implementation. Allam et al. (2022) identify emerging trends and knowledge structures pertaining to smart urban governance. Gasco-Hernandez et al. (2022) explore of the role of technology, leadership, governance, and collaborative capacities in the context of digital government.

Table 2 provides an overview by author and date of the research and practice literature for urban life and the ambient in smart cities pertaining to livability, spanning the timeframe of 2005 to 2023.

Timmer and Seymoar (2005) describe livability in cities and the planet as a living, dynamic experiment posing a central challenge. The work of Lowe et al. (2015) is cited by Higgs et al. (2019) in providing a definition of livability. Patorniti and Stevens (2017) show the importance of street ambience elements that enable the generating of data pertaining to noise, traffic and the like that can be used to improve livability and safety and as such, is important for social, perceptual, and environmental purposes. Braem et al. (2019) describe liveable spaces as lively, safe, healthy and active using testbeds to monitor air quality and noise. Higgs et al. (2019) push for integrated urban governance as well as a livability index, focusing on health and well-being. Trencher and Karvonen (2019) advance the use of digital technologies and smart planning to improve health, longevity, and well-being while encompassing social issues as part of the agenda for smart cities. Tan et al. (2020) provide an updated rendering of the global liveable and smart cities index (GLSCI) with ranking analysis, simulation, and policy evaluation. Bickley et al.

(2021) place the focus on safety in livable smart cities using "urban analytics and digital technologies" and the human factor to address issues of crime, traffic, and public health. Martino et al. (2021) consider the intensity, centrality, and diversity of urban form and livability in relation to socioeconomic indicators such as accessibility, social diversity, affordability, and economic vitality. Of note are the upcoming Architecture, Media, Politics Society conferences on livable cities in New York (AMPS, 2023) and then in London in 2024.

Table 3 provides an overview by author and date of the research and practice literature for urban life and the ambient in smart cities pertaining to sustainability, spanning the timeframe of 2017 to 2022.

Peris-Ortiz et al. (2018) articulate the movement from industrial cities to sustainable smart cities where critical elements include but are not limited to citizen participation, information and communication technologies (ICTs), quality of life, and cooperation for public and private initiatives. Garau and Pavan (2018) advance an index for smart urban quality where urban quality of life affects the quality of life of citizens and in turn smart sustainable cities. Norman (2018) advances working together collectively from the local to global in support of more sustainable urban futures. From a socio-technical perspective, Sengers et al. (2018) discuss sustainability transitions where the city is an emergent practice with public and private actors experimenting. Evans et al. (2019) introduce an emerging agenda of possibilities for smart sustainable cities focusing on the making, nurturing, and maintaining of social equity and environmental sustainability. Pafka (2020) pushes for a rethinking of urban resilience from a health perspective, enabling more adaptive cities, greatly increasing capacities for change. Giuffrida (2021) explores accountability, sustainability, and data from the perspective of law, highlighting challenges for smart city governance. Giles-Corti et al. (2022) develop a framework to guide policies and standards for human and planetary health. Goodspeed (2022) advances the idea of smart growth in relation to smart cities and sustainable urbanization. Sustainable cities are defined (UNECE, 2022) as using ICTs (Information and Communication Technology) and other methods to improve quality of life, efficiency, and competitiveness, now and in the future, encompassing the economic, social, environmental, and cultural (esec).

In terms of patterns and potential relationships, what emerges from this review of the research and practice literature is a comparative overview for governance, livability, and sustainability as presented in Table 4.

Elements that emerge across columns of Table 4, include, but are not limited to, design, experimental, health, planning, quality of life, and social.

3.6 Issues, Controversies, and Problems

The review of the research and practice literature for urban life and the ambient in terms of issues, controversies, and problems pertaining to governance, livability, and sustainability is shown in Table 5.

For governance, issues, controversies, and problems pertain to the institutional, local/day-to-day, spatial, technological, business model logic, context, digital transformation, sociotechnical, experimental governance, participatory design, participatory governance, and quality of life.

For livability, issues, controversies, and problems pertain to data analysis, healthy, lively, safe, global index, humanized city, social aspects, street ambience elements, accessibility, active and engaged community, health / well-being / longevity, and social.

For sustainability, issues, controversies, and problems pertain to adaptability, ICTs, quality of life, urban density, collective initiatives, environmental sustainability, resilience, social equity, accountabil-

Table 3. Overview of research for sustainability in smart cities

Author(s)	Year	Sustainability
Peris-Ortiz et al.	2017	From industrial cities to sustainable smart cities
Garau & Pavan	2018	Smart urban quality index, urban quality of life, citizen QOL, city sustainability
Norman	2018	Local to global collective work – sustainable urban futures
Sengers et al.	2018	Transitions; socio-technical; emergent; public, private actors, experimental
Evans et al.	2019	Agenda of possibilities for smart sustainable cities – social & environmental
Pafka	2020	Urban resilience rethink, health perspective, adaptability, capacities for change
Giuffrida	2021	Accountability, sustainability & data – challenges for smart city governance
Giles-Corti et al.	2022	Urban design, transport, health policies and standards for human/planetary health
Goodspeed	2022	Smart growth leveraging technology capabilities of smart cities
UNECE	2022	Sustainable cities definition, improve QOL, efficiency, competitiveness, encompassing the esec

ity and data, citizen participation, cooperation across sectors, and the sociotechnical – the discursive, institutional, material.

It is worth noting that some elements emerge across the governance, livability, and sustainability categories such as health, quality of life, and the social.

Table 4. Patterns and potential relationships

Governance	Livability	Sustainability
Analytic framework & typology	Accessibility	Adaptability
Business logic model	Affordability	Capacities for change
Citizens involved actively	Air quality	Citizen participation
Context	Architecture	Citizen quality of life
Design	Data	City sustainability
Digital transformation	Design	Competitiveness
e-governance	Digital technologies	Cooperation
e-participation	Economic vitality	Cultural
Experimental governance	Engaged community	Economic
Flourishing cities	Global index	Efficiency, operations & services
GIS	Health	Emergent
Graphic database	Humanized city; Human factor	Environmental
Investment	Integrated urban governance	Experimental
Good governance	Light pollution	Health perspective
Institutional frameworks	Livability index	ICTs
Management	Longevity	Local to global collective work
Material outcome	Noise pollution	Private actors
Participatory governance & design	Perceptual	Public actors
People	Policy evaluation	Quality of life
Planning	Safety	Smart growth
Policies	Smart planning	Smart urban quality index
Public governance	Social	Social
Quality of life	Social issues	Socio-technical perspective
Spatial	Street ambience elements	Sustainable urban futures
Sustainability	Traffic volume	Transitions
Technological	Urban analytics	Urban quality of life
Urban governance	Well-being	Urban resilience rethink

Table 5. Issues, controversies, and problems

	Issues	Controversies	Problems
Governance	• Institutional • Local, day-to-day • Spatial • Technological	• Business model logic • Context • Digital transformation • Sociotechnical	• Experimental governance • Participatory design • Participatory governance • Quality of life
Livability	• Data analysis • Healthy • Lively • Safe	• Global index • Humanized city • Social aspects • Street ambience elements	• Accessibility • Active & engaged community • Health, well-being, longevity • Social
Sustainability	• Adaptability • ICTs • Quality of life • Urban density	• Collective initiatives • Environmental sustainability • Resilience • Social equity	• Accountability & data • Citizen participation • Cooperation • Sociotechnical, institutional

4. CONCEPTUALIZING GOVERNANCE, LIVABILITY AND SUSTAINABILITY FOR SMART CITIES

Responding to the issues, controversies, and problems emerging from the literature review in the preceding section, a conceptual framework for enriching urban life and the ambient in smart cities aided by governance, livability, and sustainability is formulated in Fig. 1, while building upon elements in preceding chapters pertaining to people and technology interactions over a time continuum (past – present – future) also involving learning cities and future cities.

The research question identified in Section 1 is reformulated as a proposition for exploration in this chapter, as follows:

P10: *Governance, livability, and sustainability contribute to the enriching of urban life and the ambient in smart, learning, and future cities influencing health and well-being, quality of life, and resilience*

Figure 1. Framework for enriching urban life and the ambient in smart cities aided by governance, livability, and sustainability

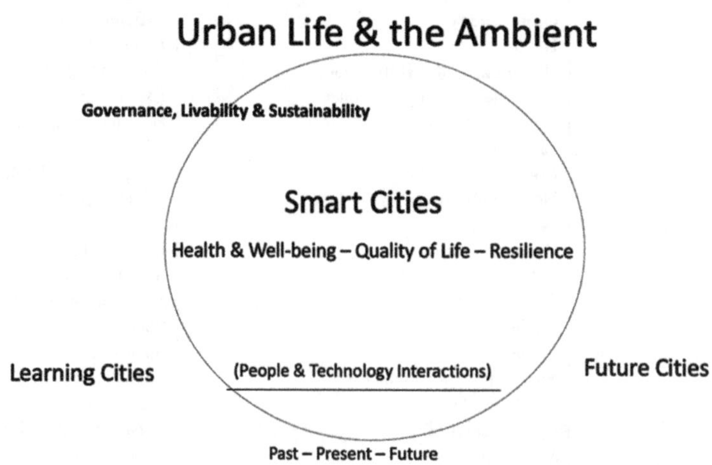

The research design used in exploring the proposition under study in this chapter is described in the Methodology section.

5. METHODOLOGY

Sections 5.1 to 5.3 describe the research design for this chapter in terms of the process used, sources of evidence for data collection, and data analysis.

5.1 Process

Use of an exploratory case study approach in this chapter is supported in view of the contemporary nature of the inquiry (Yin, 2018) and is combined with an explanatory correlational design since the two are said to be complementary (Yin, 2018) and as such, the former may be "needed to examine the underlying processes that might explain a correlation." An online space was used to describe the study and to invite people to participate. Interest emerged from individuals in many cities in multiple countries (Canada, United States, Greece, Ireland, Israel, etc.) reaching several continents.

5.2 Sources of Evidence

The online website enabled the gathering of demographic data from individuals during sign up for the study, which included age range, gender, and city. In addition, people could self-identify in one or more categories (e.g., business, community member, student, educator, etc.). A survey instrument, pre-tested prior to use, was used in this study and contained closed and open-ended questions. Survey questions pertained to a range of items related to smart cities such as governance, livability, and sustainability. Survey respondents were also invited to participate in an in-depth interview which was guided by an interview protocol that had been pre-tested prior to use in this study.

In addition, data were also gathered systematically through group and individual discussions, conducted in parallel with this study, with individuals in many sectors (e.g., students, faculty, and staff in higher education, business, community members and leaders, and city staff and leaders) in several Canadian cities (e.g., Toronto, Vancouver, Victoria).

5.3 Data Analysis

Data analysis in this chapter is guided by the research question posed in Section 1, reformulated as a proposition for exploration, as follows:

P10: *Governance, livability, and sustainability contribute to the enriching of urban life and the ambient in smart, learning, and future cities influencing health and well-being, quality of life, and resilience*

For the analysis of quantitative survey data, the use of descriptive statistics pertaining to governance, livability, and sustainability are explored in relation to elements such as *housing affordability* and *useful community functions and purposes.* The correlation feature of the Real Statistics Resource Pack containing an-add in for Microsoft Excel (Zaiontz, 2022) enabled the conducting of a Spearman Correla-

Table 6. Assessments for livability and housing affordability

Items	Livability and Sustainability in SCs	Correlation
Livability	17% (5); 17% (6); 66% (7)	.81
Housing affordability	25% (4); 25% (6); 50% (7)	

tion Coefficient for ordinal data in exploring the nature of the relationship between items pertaining to governance, livability, and sustainability.

Content analysis was used for qualitative data gathered through open-ended survey questions, interviews, and group and individual discussions so that inductive analysis from data collection was combined with deductive analysis using terminology from the literature review. Overall, an analysis was conducted for n=79 consisting of 42% females and 58% males for people ranging in age from their 20s to 70s.

6. RESULTS AND DISCUSSION

Table 6 provides the results of assessments in response to a survey question about the extent to which elements such as livability are associated with smart cities. On a 7-point Likert type scale, where 1= Not at all and 7=Absolutely, for *livability*, 17% of survey respondents selected position 5 (Sort of), 17% selected position 6 (Sure), and 66% responded at the upper end position of 7. When asked to assess the extent to which factors such as *housing affordability* contribute to the livability of a smart city, 25% of respondents selected position 4 (Neutral), 25% selected position 6 on the scale, and 50% selected position 7.

Using the correlation feature of the Real Statistics add-in for Microsoft Excel (Zaiontz, 2022), a Spearman correlation coefficient for ordinal data was conducted on the assessment results of the two items, *livability* and *housing affordability* (Table 6), showing a finding of .81.

Exploring further, Table 7 provides the results of assessments in response to a survey question about the extent to which factors such as *housing affordability* contribute to the livability of a smart city, where 25% of respondents selected position 4 (Neutral) on the scale, 25% selected position 6 on the scale, and 50% selected position 7. When asked to assess to what extent factors such as *useful community functions and purposes* contribute to increased value for data in smart cities, 25% of survey respondents selected position 6 (Sure), and 75% responded at the upper end position of 7.

Using the correlation feature of the Real Statistics add-in for Microsoft Excel (Zaiontz, 2022), a Spearman correlation coefficient for ordinal data was conducted on the assessment results of the two items, *housing affordability* and *useful community functions and purposes* (Table 7), showing a finding of .81.

Creswell (2018) advises that with correlations in the .66 to .85 range "good prediction can result from one variable to the other" and as such, "coefficients in this range would be considered very good." As

Table 7. Assessments for useful community functions and purposes and housing affordability

Items	Livability and Governance in SCs	Correlation
Housing affordability	25% (4); 25% (6); 50% (7)	.81
Useful community functions & purposes	25% (6); 75% (7)	

such, variables assessed in this chapter may be particularly important for considerations of health and well-being, quality of life, and resilience in smart cities.

Qualitatively, findings of interest pertaining to an awareness of smart cities are articulated by an educator who noted that "smart city means something totally different to different people but 'most' people who have heard the term think of big data and machine learning, often with relevance to sustainability" adding that "smart also needs to be synonymous with social, useful and having clear purpose." Regarding the making of a smart city, an individual involved in public administration commented on the importance of: "being aware of being a citizen of the city, having a say on governance, and learning via interconnections with other cities" adding that while social media and other aware technologies are greatly affecting the experience of the city, they are not yet affecting experiences "enough in terms of governance." A community leader claimed that "cities have grit" and this may speak to the resilience needed for greater adaptability (Pafka, 2020).

As depicted graphically in Fig. 2, the interconnectivity of governance, livability and sustainability is aided by understandings of health, well-being, and quality of life, giving way to the notion and potentials of ambient governance, ambient livability, and ambient sustainability and the enriching of urban life and the ambient in smart cities.

Figure 2. Urban life and the ambient – governance, livability, and sustainability in smart cities

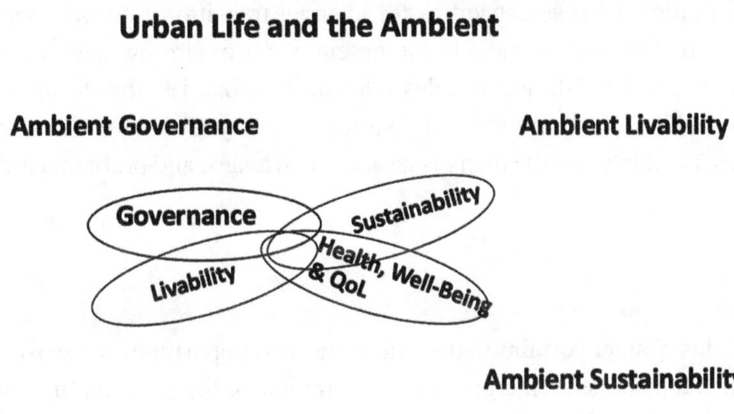

7. LIMITATIONS AND FUTURE DIRECTIONS

Possible limitations of this chapter are discussed in this section in contributing to challenges and opportunities going forward for research and practice.

Limitations

Limitations of this chapter pertain possibly to the extent of the literature review on urban life and the ambient in relation to governance, livability, and sustainability in smart cities and this is mitigated by the potential to extend the review going forward. Among other limitations of this chapter is that of the small sample size for the underlying study of this work, and as such, caution is encouraged in the interpretation of correlations calculated from the assessment of variables used as proxies for governance, livability, and sustainability in smart cities. Going forward, as variables for governance, livability, and sustainability evolve, opportunities may emerge for further explorations and for the identification and use of other variables, resulting in equally strong correlations with similar prediction potentials.

Future Research Directions and Opportunities

For research, future directions emerge in terms of the potential for conducting additional exploration, testing, and validation of the variables identified in this work relevant to governance, livability, and sustainability for urban life and the ambient in smart cities. Additionally, opportunities for explorations going forward exist for the identification of other and possibly more dynamic variables relevant to emergent notions of ambient governance, ambient livability, and ambient sustainability for urban life in the context of smart cities, learning cities, future cities, and the potential for flourishing.

Of note in Tables 6 and 7 is the presence of neutral responses for *housing affordability,* identifying a potential area of interest for further exploration.

Future Practice Directions and Opportunities

Future directions for practice include the potential for pursuing real world explorations with variables identified in this work relevant to urban life and the ambient, focusing on governance, livability, and sustainability in smart cities, such as capacities for change, transitions, or well-being (Table 4). Opportunities going forward for practice include the potential for exploring spaces with other, possibly more dynamic, emergent, and real-time variables relevant to urban life that foster notions of ambient governance, ambient livability, and ambient sustainability in everyday contexts in smart cities, learning cities, and future cities, in addressing the many issues, controversies, and problems presented in Table 5.

8. CONCLUSION

A key contribution of this chapter pertains to the challenges and opportunities identified associated with enriching urban life and the ambient emerging from the review of the research literature provided from multiple perspective involving urban governance, livability, and sustainability. Another contribution of this chapter is the formulation of a conceptual framework for enriching urban life and the ambient aided by urban governance, livability, and sustainability in smart cities based on the literature review where emerging issues, controversies, and problems contribute to operationalization of the framework for use in this chapter. Through an exploratory case study approach, together with an explanatory correlational design, the proposition under study in this chapter is explored. Based on survey data, a correlation is conducted for *housing affordability* and *livability,* in exploring livability and sustainability, with the finding of a Spearman correlation coefficient of .81, as "very good" with "good prediction" potential among variables. A correlation is also conducted for *housing affordability* and *useful community functions and purposes,* as proxy variables for livability and governance, with the finding of a Spearman correlation coefficient of .81, as "very good" with "good prediction" potential among variables. Limitations of the study underlying this chapter pertain to, among other things, the small sample size and caution is urged in the interpretation of findings, providing opportunities going forward for further testing and validation. Future directions guiding research and practice activities going forward are provided in terms of encouraging initiatives to identify other, additional variables relevant to urban life and the ambient in relation to the notions of ambient governance, ambient livability, and ambient sustainability, in the context of smart, learning, future cities, and the notion of flourishing.

Key insights and ideas emerging from this chapter are provided, along with exercises encouraging further exploration, as follows:

Insights emerging from this chapter include the following:

1. The importance of the relationship between *livability* and *housing affordability* in smart cities
2. The importance of factors pertaining to people and their needs in smart cities drawing on governance guided by *useful community functions and purposes*
3. The potential for many more factors to emerge for understanding *governance, livability,* and *sustainability* in smart cities in terms of the ambient, as in, more dynamic, emergent, and adaptable

Ideas emerging from this chapter include the following:

1. Potential opportunities for exploring urban life and the ambient in relation to emergent variables pertaining to ambient governance in the context of smart cities
2. The opportunities potential for the identification of other and additional variables for exploring urban life and the ambient in smart environments in relation to ambient livability
3. The use of more integrated approaches to the governance of urban life, allowing for increased collaboration across sectors in the making of smarter, more sustainable, more livable cities, and more flourishing cities, now and for future cities

Exercises for the reader to consider, encouraging further exploration include the following:

1. Identify one or more ways in your city that smart urban governance matters
2. Identify one or more ways in which some aspect of livability would enhance the quality of life in your city
3. Describe one or more ways in which a sustainability mindset would enhance the quality of life in your city

The audience for this chapter is intended to be interdisciplinary in nature, including but not limited to community members and leaders, educators, businesses, researchers, urban planners, urban designers, city staff and administrators and leaders, and anyone concerned with governance, livability, and sustainability in urban life, while being attentive to emergent potentials for the ambient and associated practice and research opportunities and challenges in smart cities, learning cities, and future cities.

REFERENCES

Allam, Z., Sharifi, A., Bibri, S. E., & Chabaud, D. (2022). Emerging trends and knowledge structures of smart urban governance. *Sustainability, 14*(9), 5275. doi:10.3390u14095275

AMPS. (2023). *New York – Livable Cities: A Conference on Issues Affecting Life in Cities.* Architecture, Media, Politics, Society. Retrieved 11 October 2022 from https://amps-research.com/conference/livable-cities-new-york/

Batty, M. (2018). *Inventing future cities.* MIT Press. doi:10.7551/mitpress/11923.001.0001

Bickley, S. J., Macintyre, A., & Torgler, B. (2021). *Safety in smart, livable cities: Acknowledging the human factor.* CREMA Working Paper, No. 2021-17, Center for Research in Economics, Management and the Arts (CREMA).

Braem, B., Panzica La Manna, V., Devis-Clavijo, J., Lefever, S., Lievens, B., & Marquez-Barja, J. (2019). Environmental sensing testbeds for livable smart cities. In *Proceedings of the 5th EAI International Conference on Smart Objects and Technologies for Social Good (GoodTechs '19).* Association for Computing Machinery. 10.1145/3342428.3342669

CGC. (2014). How can we make cities flourish? In *Challenges of Government Conference.* Blavatnik School of Government, University of Oxford. Retrieved 3 March 2022 from https://www.bsg.ox.ac.uk/news/how-can-we-make-cities-flourish

Cheshire, J., & Uberti, O. (2021). *Atlas of the invisible: Maps and graphics that will change how you see the world.* Norton.

Creswell, J. W. (2018). *Educational research: Planning, conducting, and evaluating quantitative and qualitative research* (6th ed.). Pearson.

Drapalova, E., & Wegrich, K. (2020). Who governs 4.0? Varieties of smart cities. *Public Management Review, 22*(5), 668–686. doi:10.1080/14719037.2020.1718191

Eneqvist, E., & Karvonen, A. (2021). Experimental governance and urban planning futures: Five strategic functions for municipalities in local innovation. *Urban Planning, 6*(1), 183-194. DOI: doi:10.17645/up.v6i1.3396

Evans, J., Karvonen, A., Luque-Ayala, A., Martin, C., McCormick, K., Raven, R., & Palgan, Y. V. (2019). Smart and sustainable cities? Pipedreams, practicalities and possibilities. *Local Environment, 24*(7), 557–564. doi:10.1080/13549839.2019.1624701

Fernandez, F., Sanchez, A., Velez, J. F., & Moreno, B. (2020). The augmented space of a smart city. *2020 International Conference on Systems, Signals and Image Processing (IWSSIP)*, 465-470. 10.1109/IWSSIP48289.2020.9145247

Garau, C., & Pavan, V. M. (2018). Evaluating urban quality: Indicators and assessment tools for smart sustainable cities. *Sustainability, 10*(3), 575. doi:10.3390u10030575

Gasco-Hernandez, M., Gil-Garcia, J. R., & Luna-Reyes, L. F. (2022). Unpacking the role of technology, leadership, governance and collaborative capacities in inter-agency collaborations. *Government Information Quarterly, 101710*(3). Advance online publication. doi:10.1016/j.giq.2022.101710

Giles-Corti, B., Vernez Moudon, A., Lowe, M., Cerin, E., Boeing, G., Frumkin, H., Salvo, D., Foster, S., Kleeman, A., Bekessy, S., Hérick de Sá, T., Nieuwenhuijsen, M., Higgs, C., Hinckson, E., Adlakha, D., Arundel, J., Liu, S., Oyeyemi, A. L., Nitvimol, K., & Sallis, J. F. (2022). What next? Expanding our view of city planning and global health, and implementing and monitoring evidence-informed policy. *The Lancet. Global Health, 10*(6), e919–e926. doi:10.1016/S2214-109X(22)00066-3 PMID:35561726

Giuffrida, I. (2021). Smart cities and sustainability: A new challenge to accountability? *William and Mary Environmental Law and Policy Review, 45*(3), 739. https://scholarship.law.wm.edu/wmelpr/vol45/iss3/5

Goodspeed, R. (2022). Leveraging the promise of smart cities to advance smart growth. In Handbook of Smart Growth: Promise, Principles, and Prospects for Planning. Edward Elgar Publishing.

Grossi, G., Meijer, A., & Sargiacomo, M. (2020). A public management perspective on smart cities: 'Urban auditing' for management, governance and accountability. *Public Management Review, 22*(5), 633–647. doi:10.1080/14719037.2020.1733056

Higgs, C., Badland, H., Simons, K., Knibbs, L. D., & Giles-Corti, B. (2019). The urban liveability index: Developing a policy-relevant urban liveability composite measure and evaluating associations with transport mode choice. *International Journal of Health Geographics, 18*(14), 14. Advance online publication. doi:10.118612942-019-0178-8 PMID:31185987

Jiang, H., Geertman, S., & Witte, P. (2020). Smart urban governance: An alternative to technocratic "smartness". *GeoJournal*, 1–17. Advance online publication. doi:10.100710708-020-10326-w PMID:33191972

Joss, S. (2018). Future cities: Asserting public governance. *Palgrave Communications, 4*(36). Advance online publication. doi:10.105741599-018-0087-7

Knaap, G.-J., Lewis, R., Chakaborty, A., & June-Friesen, K. (Eds.). (2022). *Handbook of Smart Growth: Promise, Principles, and Prospects for Planning*. Edward Elgar Publishing.

Lowe, M., Whitzman, C., Badland, H., Davern, M., Aye, L., Hes, D., Butterworth, I., & Giles-Corti, B. (2015). Planning healthy, liveable and sustainable cities: How can indicators inform policy? *Urban Policy and Research, 33*(2), 131–144. doi:10.1080/08111146.2014.1002606

Martino, N., Girling, C., & Lu, Y. (2021). Urban form and livability: Socioeconomic and built environment indicators. *Buildings and Cities, 2*(1), 220–243. doi:10.5334/bc.82

Norman, B. (2018). Are autonomous cities our urban future? *Nature Communications, 9*(2111). Advance online publication. doi:10.103841467-018-04505-0 PMID:29844312

Pafka, E. (2020). As coronavirus forces us to keep our distance, city density matters less than internal density. *The Conversation*. Retrieved 21 February 2022 from https://theconversation.com/as-coronavirus-forces-us-to-keep-our-distance-city-density-matters-less-than-internal-density-137790

Patorniti, N., & Stevens, N. (2017). Unlocking the secrets of street ambience. *The Conversation*. Retrieved 17 March 2022 from https://theconversation.com/unlocking-the-secrets-of-street-ambience-72662

Peris-Ortiz, M., Bennett, D. R., & Pérez-Bustamante Yábar, D. (Eds.). (2017). *Sustainable Smart Cities: Creating Spaces for Technological, Social and Business Development*. Springer. doi:10.1007/978-3-319-40895-8

Sengers, F., Späth, P., & Raven, R. (2018). Smart city construction: Towards an analytical framework for smart urban living labs. In *Urban Living Labs*. Routledge. doi:10.4324/9781315230641-5

Shadowen, N., Lodato, T., & Loi, D. (2020). Participatory governance in smart cities: Future scenarios and opportunities. In N. Streitz & S. Konomi (Eds.), Lecture Notes in Computer Science: Vol. 12203. *Distributed, Ambient and Pervasive Interactions. HCII 2020.* Springer. doi:10.1007/978-3-030-50344-4_32

Tan, K. G., Lim, T. O., Zhang, Y., & Tan, I. Y. E. (2020). *Global Liveable and Smart Cities Index: Ranking Analysis, Simulation and Policy Evaluation*. World Scientific. doi:10.1142/11587

Thomson Reuters. (2022). More people leaving Toronto, Montreal for smaller pastures as pandemic hastens urban exodus. *CBC News*. Retrieved 28 February 2022 from https://www.cbc.ca/news/canada/urban-exodus-canada-toronto-montreal-covid-19-1.6313911

Timeus, K., Vinaixa, J., & Pardo-Bosch, F. (2020). Creating business models for smart cities: A practical framework. *Public Management Review*, *22*(5), 726–745. doi:10.1080/14719037.2020.1718187

Timmer, V., & Seymoar, N.-K. (2005). *The Livable City.* The World Urban Forum 2006: Vancouver Working Group Discussion Paper, International Centre for Sustainable Cities.

Trencher, G., & Karvonen, A. (2019). Stretching "smart": Advancing health and well-being through the smart city agenda. *Local Environment*, *24*(7), 610–627. doi:10.1080/13549839.2017.1360264

UNECE. (2022). *Sustainable Smart Cities.* UNECE (United Nations Economic Commission for Europe). Retrieved 18 March 2022 from https://unece.org/housing/sustainable-smart-cities

Voices. (2020). Urban governance for a sustainable future. *One Earth, 2*(2), 117-119. doi:10.1016/j.oneear.2020.01.014

Yin, R. (2018). *Case study research and applications: Design and methods* (6th ed.). Sage Publications.

Zaiontz, C. (2022). *Real statistics using excel.* Retrieved from www.real-statistics.com

ADDITIONAL READING

Bickley, S. J., Macintyre, A., & Torgler, B. (2021). *Safety in Smart, Livable Cities: Acknowledging the Human Factor, CREMA Working Paper Series 2021-17*. Center for Research in Economics, Management and the Arts (CREMA).

CEE. (2020). *Shooting for the moon; Making smart cities ethical and sustainable.* News Stories. Carnegie Mellon University, Civil and Environmental Engineering. Retrieved 15 May 2022 from https://www.cmu.edu/cee/news/news-archive/2020/07-2020-shooting-for-the-moon-making-smart-cities-ethical-and-sustainable.html

Chatterjee, A., Vaidya, G., Paulose, N. K., & Das Gupta, P. (2020). Exploring livability in the era of smart city: A case of Bhopal. In T. Vinod Kumar (Ed.), *Smart Living for Smart Cities. Advances in 21st Century Human Settlements*. Springer. doi:10.1007/978-981-15-4603-7_3 de Hoop, E., Moss, T., Smith, A., & Emanuel Löffler, E. (2022). *Knowing and governing smart cities: Four cases of citizen engagement with digital urbanism.* Urban Governance. doi:10.1016/j.ugj.2021.12.008

Kim, H. M., Sabri, S., & Kent, A. (2020). Smart cities as a platform for technological and social innovation in productivity, sustainability, and livability: A conceptual framework. *Smart Cities for Technological and Social Innovation. Case Studies, Current Trends and Future Steps*, (2), 9-28.

Mirzahossein, H., & Mohghaddam, S. A. A. (2021). Increasing citizen's livability in the future city: Responsive city, a remarkable solution. *Theoretical and Empirical Researches in Urban Management, 16*(3), 23–41. https://www.jstor.org/stable/27035544

Naito, T. (2022). Redefining the smart city for sustainable development. In H. Kharas, J. W. McArthur, & I. Ohno (Eds.), *Breakthrough: The Promise of Frontier Technologies for Sustainable Development* (pp. 165–179). Brookings Institution Press.

Robbins, J. (2021). Why the luster on once-vaunted 'smart cities' is fading. *Yale Environment 360.* Retrieved 11 February 2022 from https://e360.yale.edu/features/why-the-luster-is-fading-on-once-vaunted-smart-cities

Sofeska, E. (2017). Understanding the livability in a city through smart solutions and urban planning toward developing sustainable livable future of the city of Skopje. *Procedia Environmental Sciences, 37*, 442–453. doi:10.1016/j.proenv.2017.03.014

UNECE. (2020). *People-Smart Sustainable Cities.* Geneva: United Nations Economic Commission for Europe (UNECE). Retrieved 15 May 2022 from https://unece.org/sites/default/files/2021-01/SSC%20nexus_web_opt_ENG_0.pdf

KEY TERMS AND DEFINITIONS

Ambient Governance: Ambient governance in the context of smart cities refers to more adaptive, dynamic, collaborative, and interactive ways of designing, managing, rethinking, implementing, and maintaining urban processes, practices, and services.

Ambient Livability: Ambient livability in the context of smart cities refers to smarter, more adaptive and dynamic approaches to the quality of life afforded to people, taking into consideration a range of factors such as cultural, social, economic, environmental, political, sensorial, technological, and so on.

Ambient Sustainability: Ambient sustainability in the context of smart cities, learning cities, and future cities refers to more adaptive, dynamic, collaborative, and interactive ways of ensuring the quality of life for all and for the planet, taking into consideration a range of factors such as cultural, social, economic, environmental, political, sensorial, technological, and so on.

Chapter 11
Human Flourishing as a Way Forward in Smart, Learning, and Future Cities:
Connection, Positivity, and Resilience

ABSTRACT

The purpose of this chapter is to explore the notion of human flourishing in the context of urban life and the ambient in navigating a way forward for smart, learning, and future cities. As such, a review of the research literature is provided for human flourishing and for components such as connection, positivity, and resilience. Issues, controversies, and problems emerging from the literature review are highlighted informing the formulation of a conceptual framework for human flourishing in support of urban life and the ambient in smart cities. Using an exploratory case study approach combined with an explanatory correlational design, variables pertaining to the exploration are identified, and the nature of their relationship is assessed. Through the lens of human flourishing, focusing on urban connection, positivity, and resilience, urban life and the ambient is explored in navigating a way forward for smart cities while informing directions for future research and practice.

1. INTRODUCTION

Briggs and Reiss (2021) identify dimensions of human flourishing as the material, relational, and transcendent with the pillars being truth, purpose, and meaning. Dahl, Wilson-Mendenhall, and Davidson (2020) address the notion of the plasticity of well-being in the context of human flourishing identifying the core dimensions of awareness, connection, insight, and purpose. Bickley, Macintyre, and Torgler (2021) refer to the notion of flourishing cities focusing on safety and the 'human factor' in relation to smart livable cities. Ekman and Simon-Thomas (2021) articulate the notion of human flourishing in terms of connection, positivity, and resilience (CPR) for the greater good, where, although conceptually akin to happiness, the term flourishing is used and described as "a broad quality of life." Stahl, Andreou, Brey,

DOI: 10.4018/978-1-6684-4096-4.ch011

Hatzakis, Kirichenko, Macnish, Laulhé Shaelou, Patel, Ryan, and Wright (2021) describe flourishing as "an ethical principle typically associated with virtue ethics" applied in relation to digital technologies, to explore "whether the influence of AI" as in, artificial intelligence, "on human flourishing is noticeable in the way AI is developed and deployed" where "what constitutes flourishing or how technology can contribute to it" is regarded as "AI for good." For Goldenberg (2022), being able to imagine flourishing futures requires a "radical curiosity" as in digging deeper, excavating, requiring "a wider definition of well-being." As such, this chapter is significant in that it seeks to explore the notion of human flourishing and flourishing more generally on the planet, in the context of urban life and the ambient in relation to smart cities, learning cities, and future cities giving rise to, and motivating, the explorations identified in the following objectives.

Objectives: The main objectives of this chapter are to: a) provide a review of the research literature for human flourishing and the associated dimensions of connection, positivity, and resilience in relation to urban life and the ambient in the context of smart cities; b) formulate a conceptual framework for human flourishing in support of urban life and the ambient in smart cities aided by connection, positivity, and resilience; and c) explore the relationship between variables in smart cities such as *creative opportunities* and *visualizations of data that inspire* as factors relevant to human flourishing that may provide insight into navigating the way forward in smart environments. The explorations in this chapter gives rise to the main research question, as follows: *Why are connection, positivity, and resilience important for urban life and the ambient in smart, learning, and future cities?*

2. BACKGROUND

Focusing on the theme of flourishing cities, the Challenges of Government conference (CGC, 2014) addressed the issue of how to make cities flourish, as in "more prosperous, sustainable, and inspiring" with four key factors emerging – people, sustainable planning, investment, and good governance." From the perspective of "designing future advanced smart cities" and concerned with "making society flourish", Sakamoto and Nakajima (2017) proposed the notion of "flourished crowdsourcing" in the form of "social infrastructure" in order "to increase the social awareness of citizens" enabling them to "see the necessity of their contribution. Anderson (2020) maintains that "to flourish in a 'smart city'" where "data technologies deployed in our cities grow smarter" what is critical is "that we continue to nurture the human capacity for learning, wonderment, experimentation, risk-taking, and creativity." Schooling, Enzer, and Broo (2020) argue the primary purpose of infrastructure should be to provide "a platform for human flourishing" in support of a vision that is "people focused" playing "a fundamental role in the social, economic and environmental outcomes that determine the quality of people's lives." According to Clements-Croome (2021), the notion of flourishing "refers to the experience of life going well" as in "a combination of feeling good and functioning effectively" consisting of "10 attributes of positive wellbeing" namely – *competence, emotional stability, engagement, meaning, optimism, positive emotion, positive relationships, resilience, self-esteem, and vitality,* as defined in a European Survey described by Huppert and So (2013). Levin (2021), in the context of preventive medicine, describes human flourishing as an "underdeveloped concept" that "may provide a conceptual template for collaboration" across disciplines. In the context of a "flourishing automation-enhanced world", Shneiderman (2022) envisions an integration of AI (artificial intelligence) with HCI (Human-computer interaction) approaches so as "to amplify, augment, and enhance human abilities" in order "to empower people, build their self-efficacy, support creativity, recognize responsibility, and promote social connections." As such, Shneiderman (2022) sees a world where human creativity flourishes. Vardi (2022) argues for the importance of the long-term goal of resilience, "enabled by redundancy and distributivity" as "a fundamental, but under-appreciated, societal need" instead of efficiency as a short-term goal for computing.

2.1 Definitions

Definitions from the research and practice literature are provided for key terms used in this chapter.

- **Human Flourishing:** Schooling, Enzer, and Broo (2020) describe human flourishing as "the ultimate good that people aim for, both as individuals and as society."
- **Connection:** Dahl et al. (2020) describe connection as "a subjective sense of care and kinship toward other people that promotes supportive relationships and caring interactions."
- **Resilience (smart cities):** Anderson (2020) proposes that "a resilient smart city is one where human as well as machine intelligence is maximized" and "is premised on a belief that preparing for unknown futures in a technology-infused society requires human agility and experimentation."

2.2 Overview

This chapter provides a review of the research and practice literature for human flourishing and the associated dimensions of connection, positivity, and resilience, in understanding evolving potentials for smart cities. This review enables the highlighting of key issues, controversies, and problems and the formulation of a conceptual framework for human flourishing in support of urban life and the ambient in smart cities aided by connection, positivity, and resilience that is then operationalized for use in this chapter. The main research question under exploration in this chapter is reformulated as a proposition for investigation and the methodology used is described, followed by the presentation of results and a discussion of findings. Key limitations of the study underlying this chapter are identified and future directions for research and practice are outlined. The conclusion provides an overview of findings, the identification of insights and ideas, development of several exercise questions encouraging further exploration, and identification of the target audience for this chapter.

3. PERSPECTIVES ON FLOURISHING IN SMART CITIES

This chapter provides a review of the research and practice literature for human flourishing in the context of urban life and the ambient in smart cities focusing on the dimensions of connection, positivity, and resilience.

3.1 Urban Life and Connection in Smart Cities

Komninos (2019) addresses the grand challenges going forward for smart cities in terms of "smart ecosystems" and "connected intelligence" focusing on smart growth, said to be "based on knowledge, innovation, and digital technologies"; safety and security as critical factors for quality of life; and sustainability on the planet. Anderson (2020) identifies connection as one of five keystone practices for an urban ecology. Dahl et al. (2020) claim that connection "may occur through positive social perceptions, such as gratitude and appreciation, as well as perspectives of shared humanity outside of one's immediate social circles." Citing the work of Ryff (2014), Dahl et al. (2020) note that "the capacity for caring relationships and positive social interactions figures prominently in scientific conceptions of well-being." Drawing on the work of Watson (2019), SmartCity (2020) highlights the "need to connect cities and live symbiotically with nature" and consider the use of "low-tech ecological solutions" for many challenges. From a practice perspective, Stoecker (2021) claims that connectivity is a key element for the success of smart cities, focusing on technology, applications, and buy-in from people where integrity, confidentiality, and accountability are key in sustainable network infrastructure, using the example of urban mobility. Bisson (2022) articulates the notion of the "ambient future" enabled through ambient computing and the smart connections afforded through the interactions we have with sensors, actuators, and the cloud "where the environment around us responds to our needs before we even know what we want." It is worth noting that the interdisciplinary International Conference on Connected Smart Cities (CSC, 2022) is now in its 8th year, inviting discussion on the social, behavioral, technical, economic, and environmental impacts of smart city applications and services. It is also worth noting the presence of the Smart Cities and Connected Communities (SCCC, 2022) research at the University of Maryland, College of Information Studies, focusing on "integrating people, data, devices, infrastructure, and institutions to realize the potential of communities.

3.2 Urban Life and Positivity in Smart Cities

Dunn (2014) reminds us that "there have always been future cities" and visions of such in fiction, architecture, and so on, claiming that "we may be moving toward a future more positive in outlook" with "a more people-based approach" concluding that "it is positive to see that many artists, architects and designers have remembered perhaps that the most important element of cities are their people." Dainow (2017) argues that frameworks "which seek to incorporate the human and the digital into integrated systems, offer our best hope of anticipating the future issues of smart cities." Rogers and Hunt (2019) describe an "aspirational futures methodology" and demonstrate application of the methodology to Bristol and Birmingham focusing on three visions – environment and resources; people and community; and work and economy, hypothesizing that "an intervention in a city that yields positive outcomes (or reduced negative outcomes) in all three extreme visions is likely to yield great value in terms of city enhancement." The approach (Rogers and Hunt, 2019) assesses an urban intervention in relation to sustainability, resilience, and livability while fostering "a transformation of the engineering of cities." Apanaviciene, Urbonas, and Fokaides (2020) argue that smart building promotes positivity for urban development and the national economy while improving quality of life and sustainability. Dahl et al. (2020) note the "positive social perceptions, such as gratitude and appreciation" that contribute to human connection while Shamshuddin (2020) indicates that the implementation process for urban resilience "can also have unintended positive effects on a city's capacity to prepare for and respond to shocks." Giles-Corti et al. (2022) describe a series of papers focusing on "urban design, transport, and health" identifying actions for city planning that "affect human and planetary health and amplify city vulnerabilities." Infrastructures in support of walking and biking are identified as positive while also acknowledging concerns with air and noise pollution exposure (Giles-Corti et al., 2022). Sareen et al. (2022) describe positivity in smart cities in terms of Positive Energy Districts (PEDs) that "constitute an emerging energy transition paradigm" involving "rapid upscaling to match the urgency of climate mitigation and adaptation." Sareen et al. (2022)

pose ten questions pertaining to PEDs and it is worth noting that elements addressed in response to the questions appear across the chapters of this current book, including: engagement, learning, governance, collaboration, sustainability, and urban futures. Schiavo and Magalhães (2022) speak of quality of life in smart sustainable cities in relation to Seligman's theory of well-being (2012) where positive emotions and positive relationships form two of five pillars, all of which are related to flourishing. Derkenbaeva, Halleck Vega, Hofstede, and van Leeuwen (2022) undertake to develop a comprehensive view of the positive energy districts (PEDs) concept using a complex adaptive systems approach enabling a focus on complexity and resilience.

3.3 Urban Life and Resilience in Smart Cities

Concerned with practices enabling smart cities to flourish, Anderson (2020) explores resilience in relation to creativity and uncertainty, identifying five keystone practices for an urban ecology – community, civility, communication, connection, and commitment contributing to adaptability, learning, and trust, among other factors. Regarding issues such as urban density, Pafka (2020) argues for the need to learn from the COVID-19 global pandemic "to rethink urban resilience from a broad and nuanced public health perspective" in figuring out how to "design a more adaptable city that maximises capacities for change in both its architecture and urban design" and in undoing the "rigidities of the 'master plan'." Shamshuddin (2020) speaks of the notion of "resilience resistance" described as "a new challenge for urban resilience" involving "the condition in which governance systems inherently develop barriers to change, flexibility, and adaptability through implementation" highlighting the aspects of "fatigue, complacency, and overconfidence." In the practice literature, Malewar (2021) describes work on heat and urbanization by Sarangi et al. (2021) where, among the concerns highlighted are those with health and well-being in cities associated with heat stress. Lim, Wilson, Grohs, and Pingel (2022) address the urban planning area of heat resilience and the implementation challenge of community engagement. It is worth noting that the Environmental Protection Agency (EPA) in the United States (2022) advances the notion of "cool pavements" as an approach to the reduction of the heat islands in cities however, Bloch (2019) describes work by Middel, finding that cool pavements "make people hot" under certain circumstances such as "a hot, dry day." Middel, Turner, Schneider, Zhang, and Stiller (2020) further advance the problem of cool pavements with the claim that this attempt to mitigate urban heat islands "can intensify pedestrian heat exposure." The OECD [2022] describes resilient cities as "cities that have the ability to absorb, recover and prepare for future shocks (economic, environmental, social & institutional)" and that "promote sustainable development, well-being and inclusive growth."

3.4 Literature Review Overview

An overview by author and date is provided in this section for the research and practice literature, in relation to smart cities, for human flourishing, connection, positivity, and resilience.

Table 1 provides an overview by author and date of the research and practice literature for urban life and the ambient pertaining to human flourishing, spanning the timeframe of 2014 to 2022.

The Challenges of Government conference (CGC, 2014) identified four key factors for building and developing flourishing cities – good governance, investment, people, and sustainable planning. Sakamoto and Nakajima (2017) propose flourished crowdsourcing as social infrastructure in support of a flourishing society. Anderson (2020) emphasizes the human capacities for learning, wonderment, experimenta-

Table 1. Overview of research for human flourishing in smart cities

Author(s)	Year	Human Flourishing
CGC	2014	Cities: good governance, investment, people, sustainable planning
Sakamoto & Nakajima	2017	Flourished crowdsourcing as social infrastructure for a flourishing society
Anderson	2020	Human capacity for learning, wonderment, experimentation, creativity, risk
Dahl et al.	2020	Well-being – awareness, connection, insight, purpose
Schooling et al.	2020	Infrastructure as a platform – economic, social, environment; QoL
Briggs & Reiss	2021	Material, relational, transcendent; Truth, purpose, meaning
Clements-Croome	2021	Model for multisensory human-centric design; Flourish theory
Ekman & Simon-Thomas	2021	Connection, positivity, resilience; Greater good, Quality of life
Levin	2021	Underdeveloped concept; Collaboration across disciplines potential
Stahl et al.	2021	Ethical principle, AI for good
Goldenberg	2022	Importance of curiosity and imagination for flourishing futures
Shneiderman	2022	Human-centered AI – amplify, augment, and enhance human abilities

tion, creativity, and risk-taking in support of human flourishing in smart cities. In the context of human flourishing, Dahl et al. (2020) identify key dimensions of well-being as awareness, connection, insight, and purpose. Schooling et al. (2020) identify infrastructure as a platform for human flourishing with economic, social, and environmental outcomes influencing the quality of life. Briggs and Reiss (2021) identify dimensions of human flourishing as material, relational, and transcendent, the key pillars of which are truth, purpose, and meaning. Clements-Croome (2021) advances a model for multisensory human-centric design drawing on flourish theory. Ekman and Simon-Thomas (2021) describe human flourishing in terms of connection, positivity, and resilience concerned with the greater good and quality of life. Human flourishing is described by Levin (2021) as an underdeveloped concept having the potential as a conceptual template for collaboration across disciplines. Stahl et al. (2021) describe human flourishing as an ethical principle, in relation to "the way AI is developed and deployed" as in "AI for good." Referring to a "flourishing automation-enhanced world", Goldenberg (2022) points to the importance of curiosity and imagination for developing flourishing futures. Shneiderman (2022) argues for human-centered AI to "amplify, augment, and enhance human capabilities" in support of a world where human creativity flourishes.

Table 2 provides an overview by author and date of the research and practice literature for urban life and the ambient pertaining to connection, spanning the timeframe of 2014 to 2022.

Komninos (2019) highlights the importance of smart ecosystems through connected intelligence to address the grand challenges pertaining to smart growth, safety and security, and planetary sustainability. Anderson (2020) identifies connection as one of five keystone practices for an urban ecology. Dahl et al. (2020) identify connection as one of several dimensions of well-being while citing the work of Ryff (2014) regarding caring relationships and positive social interactions where connection is aided by positive social interactions including gratitude and appreciation. SmartCity (2020) points to the work of Watson (2019) in emphasizing the importance of connecting cities and living symbiotically with nature in a move toward "low-tech ecological solutions" for wicked urban challenges. Stoecker (2021) highlights connectivity as key in terms of integrity, confidentiality, and accountability for sustainable network infra-

Table 2. Overview of research for connection in smart cities

Author(s)	Year	Connection
Ryff	2014	Caring relationships; Positive social interactions; Well-being
Komninos	2019	Smart ecosystems & connected intelligence: growth, safety, security, …
Watson	2019	Low tech solutions for connecting with nature & living on the planet
Anderson	2020	One of five keystone practices for an urban ecology
Dahl et al.	2020	Dimension of well-being; positive social perceptions, gratitude, appreciation
SmartCity	2020	Connecting cities & living symbiotically with nature
Stoecker	2021	Connectivity in sustainable network infrastructure – urban mobility
Bisson	2022	Ambient future with ambient computing – sensors, actuators, the cloud
CSC	2022	Behavioral, economic, environmental, social, technical impact of SC apps
SCCC	2022	Smart Cities & Connected Communities research projects

structure, such as urban mobility. Bisson (2022) speaks of the ambient future enabled through ambient computing and the smart connections made possible through our interactions with sensors, actuators, and the cloud. The Connected Smart Cities conference (CSC, 2022) invites discussion internationally on the behavioral, economic, environmental, social, technical impacts of smart city applications and services. The Smart Cities & Connected Communities (SCCC) research at the University of Maryland highlights the importance of "integrating people, data, devices, infrastructure, and institutions" in support of the potential of communities.

Table 3 provides an overview by author and date of the research and practice literature for urban life and the ambient pertaining to positivity, spanning the timeframe of 2014 to 2022.

Dunn (2014) claims that visions of future cities are "more positive in outlook" with "a more people-based approach" in design. Dainow (2017) advances the notion of integrated systems involving the human and the digital as the best hope for anticipating the issues of smart cities going forward. Rogers and Hunt (2019) describe and demonstrate use of an aspirational futures methodology as an approach

Table 3. Overview of research for positivity in smart cities

Author(s)	Year	Positivity
Dunn	2014	Future cities as more positive in outlook – people-based approach in design
Dainow	2017	Integrated systems & anticipating the future issues of smart cities
Rogers & Hunt	2019	Aspirational futures methodology – interventions transforming engineering
Apanaviciene et al.	2020	Smart buildings improving urban development, economy, QoL, sustainability
Dahl et al.	2020	Connection – positive social perceptions, gratitude, appreciation
Shamshuddin	2020	Urban resilience implementation and unintended positive effects
Derkenbaeva et al.	2022	Comprehensive view of PEDs – complexity, resilience
Giles-Corti et al.	2022	Urban design, transport, and health
Sareen et al.	2022	Positive Energy Districts (PEDs)
Schiavo & Magalhães	2022	QoL & well-being – PERMA – flourishing

for assessing an urban intervention in relation to sustainability, resilience, and livability in support of positive outcomes and transforming the engineering of cities. Apanaviciene et al. (2020) point to smart building design and development as promoting positivity in relation to urban development, the economy, quality of life, and sustainability. Dahl et al. (2020) highlight the importance of positive social perceptions such as gratitude and appreciation in contributing to well-being through connection. Shamshuddin (2020) points to the unintended positive effects associated with the implementation of urban resilience. Giles-Corti et al. (2022) refer to the positive health affects of urban design in support of walking and biking while also noting concerns with air and noise pollution. Derkenbaeva et al. (2022) provide a comprehensive view of positive energy districts (PEDs) while addressing complexity and resilience. Sareen et al. (2022) address positivity in smart cities in the form of Positive Energy Districts (PEDs) in response to "the urgency of climate mitigation and adaptation." Schiavo and Magalhães (2022) adapt Seligman's theory of well-being (2012) for smart sustainable cities involving PERMA – positive emotions, positive relationships, engagement, meaning, and accomplishment – as being important for flourishing.

Table 4 provides an overview by author and date of the research and practice literature for urban life and the ambient pertaining to resilience, spanning the timeframe of 2019 to 2022.

Anderson (2020) explores urban resilience in relation to creativity and uncertainty involving the keystone practices of community, civility, communication, connection, and commitment all which are important for trust, adaptability, and learning. Pafka (2020) encourages a rethinking of urban resilience to address urban density issues post-pandemic for adaptability and change in architecture and urban design. Shamshuddin (2020) articulates the notion of resilience resistance associated with governance systems where barriers to change, flexibility, and adaptability in implementation occur. Malewar (2021) describes work on heat and urbanization by Sarangi et al. (2021) giving rise to concerns with health and well-being in cities. The EPA in the US advances the notion of cool pavements to mitigate heat effects in cities although Bloch (2019) describes research work by Middel that finds cool pavements make people hot under certain circumstances with the potential to "intensify pedestrian heat exposure" (Middel et al., 2020). Lim et al. (2022) explore heat resilience in the context of urban planning and challenges associated with community engagement. The OECD [2022] describes resilient cities as being able to absorb, recover, and prepare for future shocks while promoting sustainable development, well-being, and inclusive growth. Vardi (2022) advances resilience as a long-term goal, important for redundancy and distributivity.

In terms of patterns and potential relationships, what emerges from this review of the research and practice literature is a comparative overview for human flourishing, connection, positivity, and resilience as presented in Table 5.

Elements that emerge across columns of Table 5, include, but are not limited to, appreciation, creativity, economic, environment, gratitude, learning, quality of life, social, sustainability, and well-being.

It is worth noting that the category of connection appears as an item within the positivity and resilience categories, while the category of resilience appears as an item within the human flourishing and positivity categories.

3.5 Issues, Controversies, and Problems

The review of the research and practice literature for urban life and the ambient pertaining to human flourishing, connection, positivity, and resilience highlights key issues, controversies, and problems as shown in Table 6.

Table 4. Overview of research for resilience in smart cities

Author(s)	Year	Resilience
Bloch	2019	Cool pavements make people hot in certain cases
Anderson	2020	Creativity, uncertainty; community, civility, communication, connection, commitment
Middel et al.	2020	Cool pavements and pedestrian heat exposure
Pafka	2020	Rethink urban resilience, urban density, adaptability in architecture urban design
Shamshuddin	2020	Resilience resistance in implementation; barriers to change, flexibility, adaptability
Malewar	2021	Urbanization and heat
Sarangi et al.	2021	Health and well-being; heat stress; urbanization
EPA	2022	Cool pavements to reduce heat islands in cities
Lim et al.	2022	Community-engaged heat resilience; urban planning
OECD	[2022]	Absorb, recover, prepare for future shocks; inclusive, sustainability, well-being
Vardi	2022	Long term goal supporting redundancy and distributivity

For human flourishing, issues, controversies, and problems pertain to AI for good – virtue ethics; greater good; inspiration; quality of life; flourished crowdsourcing; imagining flourishing futures; radical curiosity; re-thinking definitions of well-being; integration of AI and HCI; making cities more inspiring; prosperity; and sustainability.

Table 5. Patterns and potential relationships

Human Flourishing	Connection	Positivity	Resilience
AI for good	Ambient computing	Appreciation	Absorb
Awareness	Ambient future	Aspirational futures	Adaptability
Collaboration	Appreciation	Biking infrastructure	Change
Connection	Behavioral	Collaboration	Civility
Creativity	Caring relationships	Connection	Commitment
Economic	Connect cities	Engagement	Communication
Environment	Digital technologies	Governance	Community
Ethics	Economic	Gratitude	Connection
Experimentation	Environmental	Human connection	Creativity
Good governance	Intelligence	Human health	Distributivity
Infrastructure	Knowledge	Integrating human & digital	Economic
Insight	Gratitude	Learning	Environmental
Investment	Integrating people, data	Livability	Flexibility
Learning	Innovation	Planetary health	Inclusive
Material	Live symbiotically/nature	Positive energy districts	Institutional
Meaning	Low tech eco solutions	Positive outlook	Learning
People	Positive social interactions	Positive social perceptions	Prepare for shocks
Positivity	Positive social perceptions	Quality of life	Recover
Purpose	Quality of life	Resilience	Redundancy
Quality of life	Safety & security	Smart building	Resistance
Relational	Shared humanity	Sustainability	Rethink urban density
Resilience	Smart connections	Transformation in engineering	Social
Risk	Smart growth	Unintended positive effects	Sustainability
Social	Social	Urban futures	Trust
Sustainable planning	Sustainability	Urban resilience	Urban ecology
Transcendent	Technical	Walking infrastructure	Uncertainty
Well-being	Well-being	Well-being	Well-being

Table 6. Issues, controversies, and problems

	Issues	Controversies	Problems
Human Flourishing	AI for good/virtue ethics Greater good Inspiration Quality of life	Flourished crowdsourcing Flourishing futures Radical curiosity Rethinking well-being	Integration of AI & HCI Making cities inspiring Prosperity Sustainability
Connection	Connected intelligence Economic Environmental Safety & security	Appreciation Behavioral & Social Gratitude Shared humanity	Accountability Integrity Technical apps & services Sustainability
Positivity	Implementation People-based approach Positive future outlook Quality of life	Aspirational futures Human/digital integration Social interactions Transformation of engg	Capacity to handle shocks Infrastructures – bike, walk Smart buildings Well-being
Resilience	Adaptability Creativity Learning Uncertainty	Capacities for change Flexibility Health Resistance	Future shock ready Trust Urban density Well-being

For connection, issues, controversies, and problems pertain to connected intelligence; economic; environmental; safety and security; appreciation; behavioral and social; gratitude; shared humanity; accountability; integrity; technical applications and services in smart cities; and sustainability.

For positivity, issues, controversies, and problems pertain to implementation; people-based approach; positive future in outlook; quality of life; aspirational futures methodology; integration of human and digital frameworks; social interactions; transformation of the engineering of cities; capacity to handle shocks; infrastructures for biking and walking; smart buildings; and well-being.

For resilience, issues, controversies, and problems pertain to adaptability; creativity; learning; uncertainty; capacities for change; flexibility; health; resistance; ready to absorb recover and prepare for future shock; trust; urban density; and well-being.

It is worth noting that some elements emerge across this overview of categories for human flourishing, connection, positivity, and resilience such as quality of life, social aspects, sustainability, and well-being.

4. CONNECTION, POSITIVITY AND RESILIENCE IN SMART CITIES

Responding to the issues, controversies, and problems emerging from the literature review in the preceding section, a conceptual framework for human flourishing in support of urban life and the ambient in smart cities aided by connection, positivity, and resilience is formulated in Fig. 1, building upon elements in preceding chapters related to people and technology interactions on a past-present-future continuum while being attentive to learning cities and future cities.

The research question identified in Section 1 is reformulated as a proposition for exploration in this chapter, as follows:

P11: *Connection, positivity, and resilience are important for urban life and the ambient in smart, learning, and future cities in support of human flourishing generally, and more specifically, creativity, health and well-being, and quality of life*

Figure 1. Conceptual framework for human flourishing in support of urban life and the ambient in smart cities aided by connection, positivity, and resilience

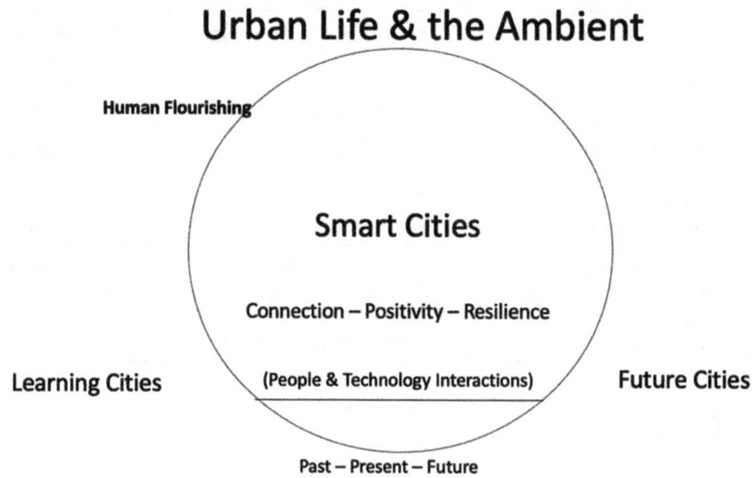

The research design used in exploring the proposition under study in this chapter is described in the Methodology section.

5. METHODOLOGY

Sections 5.1 to 5.3 describe the research design for this chapter in terms of the process used, sources of evidence for data collection, and data analysis.

5.1 Process

Use of an exploratory case study approach in this chapter is supported in view of the contemporary nature of the inquiry (Yin, 2018) and is combined with an explanatory correlational design since the two are said to be complementary (Yin, 2018) and as such, the former may be "needed to examine the underlying processes that might explain a correlation." An online space was used to describe the study and to invite people to participate. Interest emerged from individuals in many cities in multiple countries (Canada, United States, Greece, Ireland, Israel, etc.) reaching several continents.

5.2 Sources of Evidence

The online website enabled the gathering of demographic data from individuals during sign up for the study, which included age range, gender, and city. In addition, people could self-identify in one or more categories (e.g., business, community member, student, educator, etc.). A survey instrument, pre-tested prior to use, was used in this study and contained closed and open-ended questions. Survey questions pertained to a range of items related to smart cities pertaining to human flourishing and associated elements. Survey respondents were also invited to participate in an in-depth interview which was guided by an interview protocol that had been pre-tested prior to use in this study.

In addition, data were also gathered systematically through group and individual discussions, conducted in parallel with this study, with individuals in many sectors (e.g., students, faculty, and staff in higher education, business, community members and leaders, and city staff and leaders) in Canadian cities (e.g., Toronto, Vancouver, Victoria).

5.3 Data Analysis

Data analysis in this chapter is guided by the research question posed and the reformulated proposition for exploration, as follows:

P11: *Connection, positivity, and resilience are important for urban life and the ambient in smart, learning, and future cities in support of human flourishing generally, and more specifically, creativity, health and well-being, and quality of life.*

For the analysis of quantitative survey data, the use of descriptive statistics pertaining to connection, positivity, and resilience are explored in relation to elements such as *creative opportunities* and *visualizations of data that educate, inform and inspire in real time.* The correlation feature of the Real Statistics Resource Pack containing an-add in for Microsoft Excel (Zaiontz, 2022) enabled the conducting of a Spearman Correlation Coefficient for ordinal data in exploring the nature of relationships between items pertaining to *connection, creative opportunities,* and *visualizations of data that educate, inform and inspire in real time.*

Content analysis was used for qualitative data gathered through open-ended survey questions, interviews, and group and individual discussions so that inductive analysis from data collection was combined with deductive analysis using terminology from the literature review. Overall, an analysis was conducted for n=79 consisting of 42% females and 58% males for people ranging in age from their 20s to 70s.

6. RESULTS AND DISCUSSION

In exploring creativity and positivity, Table 7 provides the results of assessments in response to a survey question about the extent to which elements such as *creative opportunities* contribute to livability in smart cities. On a 7-point Likert type scale, where 1= Not at all and 7=Absolutely, for *creative opportunities,* 17% of survey respondents selected position 3 (Maybe), 50% selected position 6 (Sure), and 25% responded at the upper end position of 7. When asked to assess the extent to which factors such as *visualizations of data that educate, inform and inspire in real time* contribute to increased value for data in smart cities, 75% of respondents selected position 6 on the scale, and 25% selected position 7.

Using the correlation feature of the Real Statistics add-in for Microsoft Excel (Zaiontz, 2022), a Spearman correlation coefficient for ordinal data was conducted on the assessment results of the two items, *creative opportunities* and *visualizations of data that educate, inform and inspire in real time* (Table 7), showing a finding of .81.

Exploring further, Table 8 provides the results of assessments in response to a survey question about the extent to which factors such as *interactive public spaces* are associated with smart cities, where 17% of respondents selected position 5 (Sort of), 17% selected position 6 on the scale, and 66% selected position 7. When asked to assess to what extent factors such as *technology-driven services* are associ-

Table 7. Assessments for creativity and positivity in smart cities

Items	Creativity & Positivity in SCs	Correlation
Creative opportunities	25% (3); 50% (6); 25% (7)	.81
Visualizations of data that inspire	75% (6); 25% (7)	

Table 8. Assessments for connection (indirect) in smart cities

Items	Connection in SCs	Correlation
Interactive public spaces	17% (5); 17% (6); 66% (7)	.49
Technology-driven services	17% (5); 33% (6); 50% (7)	

ated with smart cities, 17% of survey respondents selected position 5 (Sort of), 33% selected position 6 (Sure), and 50% responded at the upper end position 7.

Using the correlation feature of the Real Statistics add-in for Microsoft Excel (Zaiontz, 2022), a Spearman correlation coefficient for ordinal data was conducted on the assessment results of the two items for connection, using the indirect proxy of *interactive public spaces* and *technology-driven services* (Table 8), showing a finding of .49.

Exploring further still, Table 9 provides the results of assessments in response to a survey question about the extent to which city-focused social media and other aware technologies give rise to possibilities such as *connecting*, where 17% of respondents selected position 5 (Sort of) on the scale and 83% selected position 7.

Using the correlation feature of the Real Statistics add-in for Microsoft Excel (Zaiontz, 2022), a Spearman correlation coefficient for ordinal data was conducted on the assessment results of the two items, *connecting* (a direct variable for connection) and *technology-driven services* (Table 9), showing a finding of .14.

Continuing to explore, Table 10 provides the results of assessments in response to a survey question about the extent to which city-focused social media and other aware technologies give rise to possibilities such as *collaborating* in smart cities, where 17% of respondents selected position 5 (Sort of), 17% selected position 6 on the scale, and 66% selected position 7. When asked to assess to what extent city-focused social media and other aware technologies give rise to possibilities such as *trust* in smart cities, 17% of survey respondents selected position 5 (Sort of), 33% selected position 6 (Sure), and 50% selected the upper end position of 7.

Using the correlation feature of the Real Statistics add-in for Microsoft Excel (Zaiontz, 2022), a Spearman correlation coefficient for ordinal data was conducted on the assessment results of the two

Table 9. Assessments for connection (direct) in smart cities

Items	Connection in SCs	Correlation
Connecting	17% (5); 83% (7)	.14
Technology-driven services	17% (5); 33% (6); 50% (7)	

Table 10. Assessments for connection (indirect) in smart cities

Items	Connection in SCs	Correlation
Collaborating	17% (5); 17% (6); 66% (7)	.23
Trust (aware technologies)	17% (5); 33% (6); 50% (7)	

items, *collaborating* (an indirect proxy for connection) and *trust (aware technologies)* (Table 10), showing a finding of .23.

Creswell (2018) advises that with correlations in the .66 to .85 range "good prediction can result from one variable to the other" and as such, "coefficients in this range would be considered very good." However, in the .35 to .65 range, correlations "above .35" are said to be "useful for limited prediction" (Creswell, 2018), while caution is encouraged with correlations in the .20 to .35 range where "there is only a slight relationship" and as such, "may be valuable to explore the interconnection of variables but of little value in prediction studies" (Creswell, 2018).

Qualitatively, among findings of interest is the comment by an Educator that "place is about living and thriving as human beings" adding that "post pandemic we are discovering that many office jobs can be done from home" rendering "office blocks" possibly "a thing of the past" and as such, "this will only emphasize place as being about people, homes and communities." The Educator also observed that "learning can be integrated into smart cities by thinking of cities as learning ecologies" where, for the human, integration with surroundings occurs through "discovery, sociability and connectedness." The Educator also provides "a word on creative thinking that must come into play to innovate smarter cities" stating that "people are the reason for cities (smart or otherwise)" and "ideas and people need to come first, then we invent the technology to support those ideas and purposes." Another Educator commented that "I believe social media has a huge influence" in that "it helps people to connect" while stating that what contributes to the making of a smart city is "citizen engagement and collaboration with the local authorities." According to a community leader, "cities have grit" a quality that may be important for the resilience needed for greater adaptability (Pafka, 2020).

During a virtual international conference (HCII 2022), when asked in a session question, *What can we do to make future cities flourishing spaces and places?* a session attendee responded with "Make technologies that promote walking and biking while fostering a disconnection technology in the midst of the city" supporting the claim by Giles-Corti et al. (2022) of the importance of infrastructure for walking and biking as positive elements for urban design.

A componential depiction of dimensions of human flourishing is presented in Fig. 2 showing the interrelationship of elements explored throughout the chapters in this book such as: adaptability (urbanizing as in adapting for urban uses and sharing, Chapter 2); collaboration (collaborating and privacy, etc., Chapters 3, 7, and 9); creativity (creative opportunities and collaboration, Chapter 7); meaning (meaningfulness and ICTs, etc. Chapter 6), and trust (trust-aware and connecting, Chapter 8).

As depicted graphically in Fig. 3, the interconnectivity of flourishing with connection, positivity, and resilience is aided in urban life by ambient computing giving way to the notion and potentials of ambient connection, ambient resilience, and an ambient future.

Figure 2. Urban life and the ambient in smart cities: components of human flourishing

Human Flourishing Components	
Adaptability	Urbanizing (adapting for urban uses) & Sharing (Chapter 2)
Collaboration	Collaborating & Privacy & Technology-driven services (Chapter 3) Collaborating & Creative opportunities (Chapter 7) Collaborating & Visualizations of data that educate, inform & inspire in real time (Chapter 9)
Creativity	Collaboration & Creative opportunities (Chapter 7)
Meaning	Meaningfulness & ICTs & Visual ways to show success in real time (Chapter 6)
Trust	Connecting & Trust involving aware technologies (Chapter 8)

7. LIMITATIONS AND FUTURE DIRECTIONS

Possible limitations of this work are discussed in this section along with future directions and opportunities for research and practice.

Limitations

Limitations of this chapter pertain possibly to the extent of the literature review on urban life and the ambient in relation to human flourishing, connection, positivity, and resilience in smart cities and this is mitigated by the potential to extend the review going forward. Among other limitations of this chapter is that of the small sample size for the underlying study of this work, and as such, caution is encouraged in the interpretation of correlations calculated from the assessment of variables used in this chapter such as *creative opportunities* and *visualizations of data that educate, inform, and inspire in real time* in smart cities. Going forward, as understandings of connection, positivity, and resilience in smart cities continue to develop, opportunities may emerge for the identification and use of other variables, resulting in correlations with good prediction potentials.

Figure 3. Urban life and the ambient – flourishing in smart cities

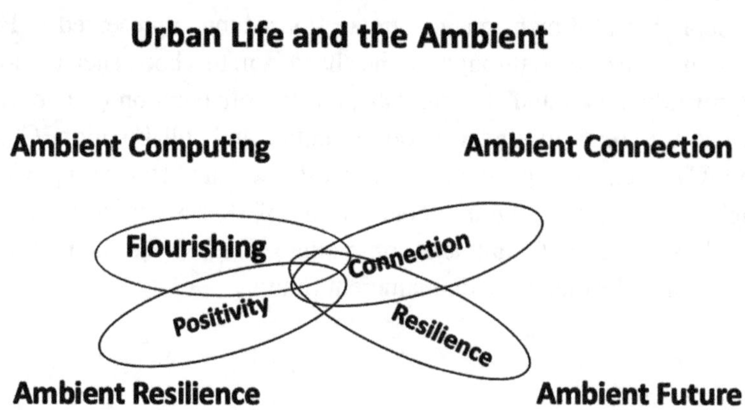

Future Research Directions and Opportunities

For research, future directions emerge in terms of the potential for conducting additional explorations on the variables identified in this work relevant to human flourishing in the context of urban life and the ambient in smart cities as identified in Table 5. As such, opportunities for explorations going forward emerge for the identification of other and possibly more dynamic variables relevant to emergent notions of ambient computing, ambient connection, ambient resilience, and an ambient future in learning more about urban life in the context of smart cities, learning cities, and future cities, in addressing the many issues, controversies, and problems presented in Table 6.

Future Practice Directions and Opportunities

Future directions for practice include the potential for pursuing explorations with variables identified in this work relevant to urban life and the ambient, focusing on the real world components of human flourishing such as connection, positivity, and resilience in smart cities, in terms of AI for good, smart connections, human connections, or human health (Table 5). Opportunities going forward for practice also include the potential for exploring spaces with other, possibly more dynamic, emergent, and real-time variables relevant to urban life that foster notions of ambient connection, ambient resilience, and human flourishing more generally in everyday contexts in smart cities, learning cities, and future cities.

8. CONCLUSION

A key contribution of this chapter is the identification of challenges and opportunities associated with human flourishing in smart cities in relation to urban life and the ambient that emerge from the review of the research literature provided from multiple perspective involving urban connection, positivity, and resilience. Another contribution of this chapter is the formulation of a conceptual framework for human flourishing in support of urban life and the ambient aided by connection, positivity, and resilience in smart cities based on the literature review where emerging issues, controversies, and problems contribute to operationalization of the framework for use in this chapter. Through an exploratory case study approach, together with an explanatory correlational design, the proposition under study in this chapter is explored. Based on survey data, exploring creativity and the dimension of positivity, among the correlations conducted is that for *creative opportunities* and *visualizations of data that educate, inform, and inspire in real time* with the finding of a Spearman correlation coefficient of .81, as "very good" with "good prediction" potential among variables. Limitations of the study underlying this chapter pertain to, among other things, the small sample size and caution is urged in the interpretation of findings, providing opportunities going forward for further testing and explorations. Future directions guiding research and practice activities going forward are provided in terms of further validating the findings in this chapter while also encouraging activities to identify other, additional variables relevant to human flourishing in the context of urban life and the ambient in relation to the notions of ambient connection, ambient resilience, and an ambient future, in the context of smart, learning, and future cities.

Key insights and ideas emerging from this chapter are provided, along with exercises encouraging further exploration, as follows:

Insights emerging from this chapter include the following:

1. The importance of the relationship between *creative opportunities* and *visualizations of data that educate, inform, and inspire in real time* in terms of positivity potentials in smart cities
2. The importance of factors pertaining to people and their needs in smart cities in terms of connection guided by variables such as *interactive public spaces and technology-driven services*
3. The potential for many more factors to emerge for understanding *connection, positivity,* and *resilience* in smart cities in relation to the notion of an ambient future

Ideas emerging from this chapter include the following:

1. Potential opportunities for exploring urban life and the ambient in relation to a variety of variables pertaining to connection, positivity, and resilience in the context of smart cities
2. The opportunities potential for the identification of other and additional variables for exploring human flourishing in the context of urban life and the ambient in smart environments in relation to ambient connection and ambient resilience
3. The use of more integrated approaches to human flourishing in urban life, allowing for increased collaboration across sectors in the making of smarter, more sustainable, and more livable cities, now and for future cities

Exercises for the reader to consider, encouraging further exploration include the following:

1. Identify one or more ways in your city that connection matters
2. Identify one or more ways in which some aspect of positivity would enhance the quality of life in your city
3. Describe one or more ways in which a resilience mindset would enhance the quality of life and human flourishing in your city

The audience for this chapter is intended to be interdisciplinary in nature, including but not limited to community members and leaders, educators, businesses, researchers, urban planners, urban designers, and anyone concerned with flourishing, connection, positivity, and resilience in urban life and the ambient, while being attentive to emergent potentials for creativity and associated practice and research opportunities and challenges in smart cities, learning cities, and future cities.

REFERENCES

Anderson, T. D. (2020). Keystone practices to enable smart cities to flourish. *Geography Research Forum, 40*, 171-192.

Apanaviciene, R., Urbonas, R., & Fokaides, P. A. (2020). Smart building integration into a smart city: Comparative study of real estate development. *Sustainability, 12*(22), 9376. doi:10.3390u12229376

Bickley, S. J., Macintyre, A., & Torgler, B. (2021). *Safety in smart, livable cities: Acknowledging the human factor*. CREMA Working Paper, No. 2021-17, Center for Research in Economics, Management and the Arts (CREMA).

Bisson, S. (2022). Ambient computing has arrive: Here's what it looks like, in my house. *ZDNET*. Retrieved 16 September 2022 from https://www.zdnet.com/article/ambient-computing-has-arrived-heres-what-it-looks-like-in-my-house/

Bloch, S. (2019). *The problem with 'cool pavements': They make people hot.* CityLab, Environment, Bloomberg. Retrieved 11 July 2022 from https://www.bloomberg.com/news/articles/2019-10-03/reflective-pavement-may-be-less-cool-than-it-seems

Briggs, A., & Reiss, M. J. (2021). *Human Flourishing: Scientific Insight and Spiritual Wisdom in Uncertain Times.* Oxford University Press online. DOI: doi:10.1093/oso/9780198850267.003.0010

CGC. (2014). How can we make cities flourish? In *Challenges of Government Conference.* Blavatnik School of Government, University of Oxford. Retrieved 3 March 2022 from https://www.bsg.ox.ac.uk/news/how-can-we-make-cities-flourish

Cheshire, J., & Uberti, O. (2021). *Atlas of the invisible: Maps and graphics that will change how you see the world.* Norton.

Clements-Croome, D. (2021). Flourish theory. In R. Appel-Meulenbroek & V. Danivska (Eds.), *A Handbook of Theories on Designing Alignment Between People and the Office Environment*. Routledge. doi:10.1201/9781003128830-14

Creswell, J. W. (2018). *Educational research: Planning, conducting, and evaluating quantitative and qualitative research* (6th ed.). Pearson.

CSC. (2022). *8th International Conference on Connected Smart Cities.* Retrieved 19 April 2022 from https://smartcities-conf.org

Dahl, C. J., Wilson-Mendenhall, C. D., & Davidson, R. J. (2020). The plasticity of well-being: A training-based framework for the cultivation of human flourishing. *Proceedings of the National Academy of Sciences (PNAS), 117*(51), 32197-32206. 10.1073/pnas.2014859117

Dainow, B. (2017). Smart city transcendent: Understanding the smart city by transcending ontology. *Orbit (Amsterdam, Netherlands), 1*, 1–15.

Derkenbaeva, E., Halleck Vega, S., Hofstede, G. J., & van Leeuwen, E. (2022). Positive energy districts: Mainstreaming energy transition in urban areas. *Renewable & Sustainable Energy Reviews, 153*, 111782. doi:10.1016/j.rser.2021.111782

Dunn, N. (2014). Past visions of future cities were monstrous, but now we imagine a brighter tomorrow. *The Conversation*. Retrieved 14 October 2022 from https://theconversation.com/past-visions-of-future-cities-were-monstrous-but-now-we-imagine-a-brighter-tomorrow-29188

Ekman, E., & Simon-Thomas, E. (2021). Teaching the science of human flourishing, unlocking connection, positivity, and resilience for the greater good. *Global Advances in Health and Medicine: Improving Healthcare Outcomes Worldwide, 10.* Advance online publication. doi:10.1177/21649561211023097 PMID:34123578

EPA. (2022). *Using cool pavements to reduce heat islands.* United States Environmental Protection Agency. Retrieved 11 July 2022 from https://www.epa.gov/heatislands/using-cool-pavements-reduce-heat-islands

Giles-Corti, B., Vernez Moudon, A., Lowe, M., Cerin, E., Boeing, G., Frumkin, H., Salvo, D., Foster, S., Kleeman, A., Bekessy, S., Hérick de Sá, T., Nieuwenhuijsen, M., Higgs, C., Hinckson, E., Adlakha, D., Arundel, J., Liu, S., Oyeyemi, A. L., Nitvimol, K., & Sallis, J. F. (2022). What next? Expanding our view of city planning and global health, and implementing and monitoring evidence-informed policy. *The Lancet. Global Health*, *10*(6), e919–e926. doi:10.1016/S2214-109X(22)00066-3 PMID:35561726

Goldenburg, S. (2022). *Radical Curiosity: Questioning Commonly Held Beliefs to Imagine Flourishing Futures*. Penguin Random House.

Huppert, F. A., & So, T. T. C. (2013). Flourishing across Europe: Application of a new conceptual framework for defining well-being. *Social Indicators Research*, *110*(3), 837–861. doi:10.100711205-011-9966-7 PMID:23329863

Komninos, N. (2019). *Smart Cities and Connected Intelligence: Platforms, Ecosystems and Network Effects*. Routledge., doi:10.4324/9780367823399

Levin, J. (2021). Human flourishing: A new concept for preventive medicine. *American Journal of Preventive Medicine*, *61*(5), 761–764. doi:10.1016/j.amepre.2021.04.018 PMID:34187711

Lim, T. C., Wilson, B., Grohs, J. R., & Pingel, T. J. (2022). Community-engaged heat resilience planning: Lessons from a youth smart city STEM program. *Landscape and Urban Planning*, *226*, 104497. doi:10.1016/j.landurbplan.2022.104497

Malewar, A. (2021). Urbanization causes uncomfortable heat time in cities. *TechExplorist*. Retrieved 14 June 2022 from https://www.techexplorist.com/urbanization-causes-uncomfortable-heat-time-cities/43325/

Middel, A., Turner, V. K., Schneider, F. A., Zhang, Y., & Stiller, M. (2020). Solar reflective pavement – a policy panacea to heat mitigation? *Environmental Research Letters*, *15*(6), 064016. doi:10.1088/1748-9326/ab87d4

OECD. (2022). *Resilient Cities*. Organisation for Economic Co-operation and Development (OECD). Retrieved 19 April 2022 from https://www.oecd.org/cfe/regionaldevelopment/resilient-cities.htm

Pafka, E. (2020). As coronavirus forces us to keep our distance, city density matters less than internal density. *The Conversation*. Retrieved 21 February 2022 from https://theconversation.com/as-coronavirus-forces-us-to-keep-our-distance-city-density-matters-less-than-internal-density-137790

Rogers, C. D. F., & Hunt, D. V. L. (2019). Realising visions for future cities: An aspirational futures methodology. *Proceedings of the Institution of Civil Engineers. Urban Design and Planning*, *172*(4), 125–140. doi:10.1680/jurdp.18.00010

Ryff, C. D. (2014). Psychological well-being revisited: Advances in the science and practice of eudaimonia. *Psychotherapy and Psychosomatics*, *83*(1), 10–28. doi:10.1159/000353263 PMID:24281296

Sakamoto, M., & Nakajima, T. (2017). Making citizens' activities flourish through a crowdsourcing-based social infrastructure. In S. Konomi & G. Roussos (Eds.), *Enriching Urban Spaces with Ambient Computing, the Internet of Things, and Smart City Design* (pp. 232–256). IGI Global. doi:10.4018/978-1-5225-0827-4.ch012

Sarangi, C., Qian, Y., Li, J., Leung, L. R., Chakraborty, T. C., & Liu, Y. (2021). Urbanization amplifies nighttime heat stress on warmer days over the US. *Geophysical Research Letters*, *49*(8). Advance online publication. doi:10.1029/2021GL095678

Sareen, S., Albert-Seifried, V., Aelenei, L., Reda, F., Etminan, G., Andreucci, M.-B., Kuzmic, M., Maas, N., Seco, O., Civiero, P., Gohari, S., Hukkalainen, M., & Neumann, H.-M. (2022). Ten questions concerning positive energy districts. *Building and Environment*, *216*, 109017. Advance online publication. doi:10.1016/j.buildenv.2022.109017

SCCC. (2022). *Smart Cities and Connected Communities*. University of Maryland, College of Information Studies. Retrieved 14 October 2022 from https://ischool.umd.edu/expertise-areas/smart-cities-and-connected-communities/

Schiavo, F. T., & Magalhães, C. F. (2022). Smart sustainable cities: The essentials for managers' and leaders' initiatives within the complex context of differing definitions and assessments. *Smart Cities*, *5*(3), 994–1024. doi:10.3390martcities5030050

Schooling, J., Enzer, M., & Broo, D. G. (2020). Flourishing systems: Re-envisioning infrastructure as a platform for human flourishing. *Proceedings of the Institution of Civil Engineers–Smart Infrastructure and Construction*, *173*(1), 166–174. doi:10.1680/jsmic.20.00023

Shamsuddin, S. (2020). Resilience resistance: The challenges and implications of urban resilience implementation. *Cities (London, England)*, *103*, 102763. doi:10.1016/j.cities.2020.102763 PMID:32501356

Shneiderman, B. (2022). *Human-centered AI: A new synthesis*. Keynote presentation at the Human-Computer Interaction International Conference (HCII2022).

SmartCity. (2020). *Ways to connect cities with nature – building truly sustainable future*. SmartCity Press. Retrieved 14 October 2022 from https://smartcity.press/connecting-smart-city-with-nature

Stahl, B. C., Andreou, A., Brey, P., Hatzakis, T., Kirichenko, A., Macnish, K., Laulhé Shaelou, S., Patel, A., Ryan, M., & Wright, D. (2021). Artificial intelligence for human flourishing – Beyond principles for machine learning. *Journal of Business Research*, *124*, 374–388. doi:10.1016/j.jbusres.2020.11.030

Stoecker, F. (2021). Connectivity: The fundamental ingredient of a successful smart city. *Open Access Government*. Retrieved 13 October 2022 from https://www.openaccessgovernment.org/smart-mobility/124429/

Thomson Reuters. (2022). More people leaving Toronto, Montreal for smaller pastures as pandemic hastens urban exodus. *CBC News*. Retrieved 28 February 2022 from https://www.cbc.ca/news/canada/urban-exodus-canada-toronto-montreal-covid-19-1.6313911

Vardi, M. Y. (2022). Efficiency vs. resilience: Lessons from COVID-19. In H. Werthner, E. Prem, E. A. Lee, & C. Ghezzi (Eds.), *Perspectives on Digital Humanism*. Springer. doi:10.1007/978-3-030-86144-5_38

Watson, J. (2019). *Lo-TEK. Design by Radical Indigenism*. TASCHEN.

Zaiontz, C. (2022). Real statistics using excel. Retrieved from www.real-statistics.com

ADDITIONAL READING

ALLEA. (2019). *Flourishing in a data-enabled society.* All European Academies (ALLEA) and The Royal Society. Retrieved 16 May 2022 from https://allea.org/portfolio-item/flourishing-in-a-data-enabled-society/

Crisp, N. (2021). Human flourishing in a health-creating society. *Lancet, 397*(10279), 1054–1055. doi:10.1016/S0140-6736(21)00585-7 PMID:33743858

Garrard, G., Williams, N., & Bekessy, S. (2018). Here's how to design cities where people and nature can both flourish. *The Conversation.* Retrieved 16 May 2022 from https://theconversation.com/heres-how-to-design-cities-where-people-and-nature-can-both-flourish-102849

IQSS. (2021). *Human flourishing program launches the global flourishing study with Baylor University researchers.* Harvard University, The Institute for Quantitative Social Science (IQSS). Retrieved 16 May 2022 from https://www.iq.harvard.edu/news/human-flourishing-program-launches-global-flourishing-study-baylor-university

Nelson, C. A. (2021). *Smart futures for a flourishing world: A paradigm shift for achieving global sustainability.* Changemakers Books.

Scanlon, B. R. (2022). Increasing resilience to climate extremes with emphasis on major urban areas. *CIRCLE Distinguished Lecture Series.* https://circle.cee.illinois.edu/circle-distinguished-lecture-series/

Schinkel, A. (Ed.). (2020). *Wonder, education, and human flourishing: Theoretical, empirical, and practical perspectives.* Uuitgeverij/VU University Press. doi:10.5040/9781350071926

Seligman, M. E. P. (2012). *Flourish: A Visionary New Understanding of Happiness and Well-Being.* Artia Books.

Smart Cities Dive Team. (2021). Reassessing the smart cities movement. *Smart Cities Dive.* Retrieved 11 March 2022 from https://www.smartcitiesdive.com/news/reassessing-past-future-smart-cities-movement/609694/

Townshend, T. G. (2021). *Urban Design and Human Flourishing: Creating Places that Enable People to Live Healthy and Fulfilling Lives.* Routledge. doi:10.4324/9781003154808

KEY TERMS AND DEFINITIONS

Ambient Connection: Ambient connection in a smart city context refers to more adaptive, dynamic, and in the moment approaches to activities, resources, or services in support of bridging, joining, and the fostering of relationships, collaborations, and trust.

Ambient Resilience: Ambient resilience in a smart city context refers to highly adaptive, responsive, and flexible capabilities in the face of uncertain, unexpected situations and circumstances.

Positivity: Positivity in a smart city context refers to a creative and solutions-oriented mindset in the face of complex, entangled, and wicked challenges and opportunities.

Chapter 12

A Componential View of Urban Life and the Ambient in Smart, Learning, and Future Cities: A Summary and Synthesis

ABSTRACT

This final chapter concludes with the purpose of providing a summary of urban life and the ambient in smart cities, learning cities, and future cities as presented across the chapters of this book. An integrated conceptual framework for urban life and the ambient in smart cities is formulated bringing together the conceptual frameworks provided in each chapter, resulting in a componential rendering. In summarizing the content across chapters, a tabular approach is used for literature review findings pertaining to issues, controversies, and problems as well as findings from the exploratory case study approach conducted in combination with an explanatory correlational design. Also, tabular presentations are provided for a variety of elements including variables explored and their correlations, and for insights, ideas, and exercises. The comparative glimpse across chapters enables a discussion of patterns and trends emerging across the dynamic, interdisciplinary space of urban life in smart cities.

1. INTRODUCTION

A summary of the preceding eleven chapters is provided in this chapter and as such, a key contribution is the formulation of an integrated conceptual framework for urban life and the ambient in smart cities. This work is significant in that it develops the lens of urban life and the ambient as a way of exploring and understanding smart cities, learning cities, and future cites. As a concluding chapter, this work undertakes to: a) provide a summary of urban life and the ambient from the perspectives presented in this book in the context of smart cities; b) provide an integrated conceptual framework for urban life and the ambient in smart cities to guide explorations going forward; and c) consider patterns and trends emerging in the relationship between variables explored that may provide insight into navigating the

DOI: 10.4018/978-1-6684-4096-4.ch012

Figure 1. Integrated conceptual framework for urban life and the ambient in smart cities

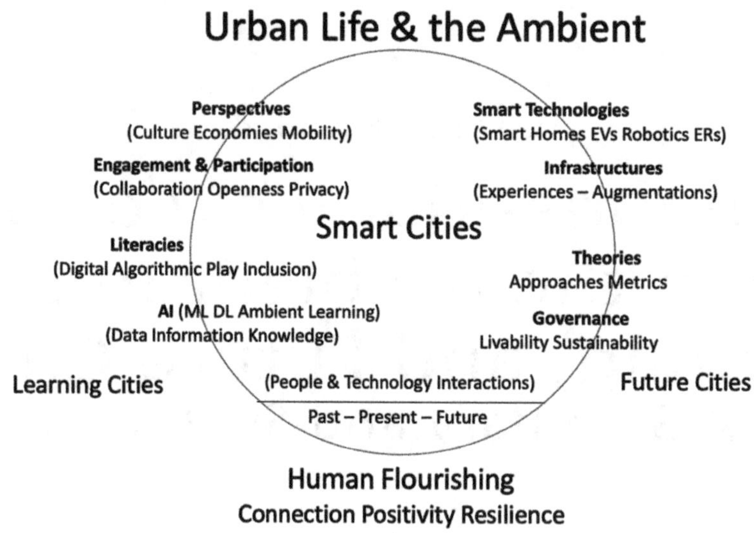

way forward in smart environments. These undertakings give rise to the main objective of this chapter as described below.

Objective: The main of objective of this chapter is to provide an integrated understanding of urban life and the ambient contributing a lens and broad framework for the study of smart cities, learning cities, and future cities while also providing guidance for researchers and practitioners on ways to experiment, explore, and approach practice, learning, and action in smart cities.

2. URBAN LIFE AND THE AMBIENT AS A LENS FOR EXPLORING SMART CITIES

This work develops the notion of urban life and the ambient as a lens for exploring smart cities, learning cities, and future cities.

2.1 Urban Life and the Ambient Integrated Conceptual Framework

This chapter revisits the conceptual framework for urban life and the ambient in smart cities presented in Chapter 1 together with the chapter – by – chapter conceptual frameworks presented in this book and provides an integrated conceptual framework for urban life and the ambient in smart, learning, and future cities as depicted in Fig. 1.

Broad components of the integrated conceptual framework for urban life and the ambient consist of smart cities, learning cities, and future cities contributing to the outcome of human flourishing focusing on connection, positivity, and resilience as encouraged by Ekman and Simon-Thomas (2021). More specific components consist of, but are not limited to, perspectives ranging from culture to economies to mobility; engagement and participation involving collaboration, openness, and privacy; literacies including the digital, the algorithmic, play, and inclusion; data, information, and knowledge; artificial

Table 1. Research questions for exploring urban life and the ambient in smart cities

Chapter	Research Questions
Chapter 1 – Q1	How does support for improving urban life emerge in the context of smart cities?
Chapter 2 – Q2	How does the ambient emerge in urban life in the context of smart cities?
Chapter 3 – Q3	Why do engagement and participation matter in urban life in the context of smart cities?
Chapter 4 – Q4	How are urban infrastructures evolving in smart cities?
Chapter 5 – Q5	Why are literacies for urban life in smart cities needed?
Chapter 6 – Q6	How are data, information, and knowledge related in urban life in smart cities?
Chapter 7 – Q7	How and why does artificial intelligence, including machine learning and deep learning, manifest in urban life and the ambient in smart cities?
Chapter 8 – Q8	How do smart homes, electric vehicles, extended realities, and robotics contribute to urban life and the ambient in smart cities?
Chapter 9 – Q9	Why are tools such as theory needed for understanding urban life and the ambient in smart, learning, and future cities?
Chapter 10 – Q10	How do governance, livability, and sustainability contribute to the enriching of urban life and the ambient in smart, learning, and future cities?
Chapter 11 – Q11	Why are connection, positivity, and resilience important for urban life and the ambient in smart, learning, and future cities?

intelligence (AI), machine learning (ML), deep learning (DL), and ambient learning; smart technologies including smart homes, electric vehicles (EVs), robotics, and extended realities (ERs); evolving infrastructures including experiences and augmentations; theories, approaches, and metrics; and governance, livability, and sustainability – all of which involve people and technology interactions across the past, present, and future continuum.

2.2 A Componential Approach to Understanding Urban Life in Smart Cities

Using a componential approach for exploring and understanding urban life and the ambient in smart cities enables an interdisciplinary space for engagement and interaction along with a wide range of perspectives from research questions to propositions explored to issues, controversies, and problems as discussed in the chapters of this book and as highlighted in Section 3 below.

3. PERSPECTIVES ON URBAN LIFE AND THE AMBIENT

Although presented in Chapter 1, this chapter again provides a listing in Table 1 of the research questions explored in each chapter of this book as a reminder of the extent of the explorations conducted for urban life and the ambient in smart cities, learning cities, and future cities.

Areas explored in these research questions include understandings of smart cities; the ambient in urban life in smart cities; engagement and participation; urban infrastructures; literacies; data, information, and knowledge; artificial intelligence (AI), deep learning (DL), and machine learning (ML); smart homes, electric vehicles (EVs), extended realities (ERs), and robotics; theory, approaches, and metrics; governance, livability, and sustainability; and human flourishing involving connection, positivity, resilience.

Again, although presented in Chapter 1, this chapter provides a listing in Table 2 of the propositions (reformulated from research questions) explored in each chapter of this book as a reminder of the nature of the explorations conducted for urban life and the ambient in smart cities, learning cities, and future cities.

Areas explored in these propositions pertain to understandings of smart cities involving funding and associated interactions and experiences, involving people and technologies; the ambient in urban life in smart cities; the evolving of urban infrastructures to include experiences, interactions, and augmentations; literacies such as the digital, the algorithmic, play, and inclusion; data, information, and knowledge including ways of knowing and computational intelligence (CI); artificial intelligence (AI), deep learning (DL), and machine learning (ML) in support of CI, learning, mobility, and other in-the-moment, real time activities; smart homes, electric vehicles (EVs), extended realities (ERs), and robotics in support of sustainability and livability on the planet; theory, approaches, and metrics involving the identification of variables to inform the work of researchers and practitioners; governance, livability, and sustainability in support of health and well-being, quality of life and urban life, and resilience; and human flourishing involving connection, positivity, resilience in support of creativity, health and well-being, and quality of life and urban life.

3.1 Chapter-by-Chapter Focus and Research Literature Explored

Each chapter of this book provides a review of the research literature for smart cities and key related concepts as shown in Table 3.

Chapter 1 focuses on urban life providing a glimpse of past, present, and future understandings, enriched with getting a sense of the city in relation to the concepts of smart cities, learning cities, and future cities.

Chapter 2 provides perspectives on the ambient and urban life in smart cities focusing on the concepts of culture, economy, and mobility.

Chapter 3 focuses on people and technologies in urban life and the ambient in relation to the concepts of engagement and participation, collaboration, openness, and privacy.

Chapter 4 focuses on urban life and infrastructures in relations to the concepts of augmentations, experiences, and interactions.

Chapter 5 focuses on urban life and the ambient in relation to literacies and the concepts of the algorithmic, the digital, inclusion, and play.

Chapter 6 focuses on data streams in urban life and the ambient and the related concepts of computational intelligence, information, knowledge, and ways of knowing.

Chapter 7 focuses on urban life and the ambient and artificial intelligence (AI) and the related concepts of deep learning (DL) and machine learning (ML), extending to ambient learning.

Chapter 8 focuses on smart technologies including smart homes, electric vehicles (EVs), and robotics, as well as extended realities (ERs) and the concepts and applications of augmented reality, mixed reality, and virtual reality.

Chapter 9 focuses on tools and resources for urban life and the ambient in smart cities in terms of theory, approaches, and metrics.

Chapter 10 focuses on urban life and the ambient in the context of the enrichment of smart cities aided by governance, livability, and sustainability.

Chapter 11 focuses on human flourishing as a way forward in smart cities in relation to the concepts of connection, positivity, and resilience.

Table 2. Propositions for exploring urban life and the ambient in smart cities

Chapter	Propositions
Chapter 1 – P1	Support for improving urban life in the context of smart cities emerges through questions pertaining to funding and associated interactions and experiences involving people and technologies
Chapter 2 – P2	The ambient emerges in urban life in the context of smart cities through interactions and awareness involving people and technologies
Chapter 3 – P3	Engagement and participation matter in urban life in the context of smart cities because they are influenced by the nature of collaboration, openness, and privacy
Chapter 4 – P4	Urban infrastructures are evolving in smart cities to include experiences, interactions, and augmentations
Chapter 5 – P5	Literacies for urban life in smart cities are needed to support understandings of elements such as the digital, the algorithmic, play, and inclusion
Chapter 6 – P6	Data, information, and knowledge are related in urban life in smart cities as shown through ways of knowing and computational intelligence
Chapter 7 – P7	Artificial intelligence, including machine learning and deep learning, manifest in urban life and the ambient in smart cities in increasingly pervasive ways in support of computational intelligence, learning, mobility, and many other in-the-moment, real time interactivities
Chapter 8 – P8	Smart homes, electric vehicles, extended realities, and robotics contribute to urban life and the ambient in smart cities in support of sustainability and livability on the planet
Chapter 9 – P9	Tools such as theory are needed for understanding urban life and the ambient in smart, learning, and future cities to identify variables important for the selection of approaches and metrics for researchers and practitioners
Chapter 10 – P10	Governance, livability, and sustainability contribute to the enriching of urban life and the ambient in smart, learning, and future cities influencing health and well-being, quality of life, and resilience
Chapter 11 – P11	Connection, positivity, and resilience are important for urban life and the ambient in smart, learning, and future cities, in support of human flourishing generally more specifically, creativity, health and well-being, and quality of life

What emerges from the literature reviews are a series of issues, controversies, and problems as shown in Tables 4-7 across the four sections of the book.

Table 3. Chapter-by-chapter literature review focus and key related concepts

Chapter and Focus	Related Concepts Explored		
1. Urban Life	Past – Present - Future	Sensing	Smart–Learning–Future
2. Perspectives	The Ambient	Culture – Economy	Mobility
3. People / Technologies	Engagement – Participation	Collaboration	Openness – Privacy
4. Infrastructures	Augmentations	Experiences	Interactions
5. Literacies	Algorithmic	Digital	Inclusion & Play
6. Data Streams	Computational costs	Information	Knowledge / Knowing
7. Artificial Intelligence	Deep Learning	Machine Learning	Ambient Learning
8. Homes/ER/EV/Robotics	Augmented Reality	Mixed Reality	Virtual Applications
9. Resources/Tools	Theory	Approaches	Metrics
10. Enrichment of SCs	Governance	Livability	Sustainability
11. Human Flourishing	Connection	Positivity	Resilience

Table 4. Section 1: issues, controversies, and problems

Chapter Topics	Issues	Controversies	Problems
Section 1: Introduction & Background to Urban Life & the Ambient			
Urban Life	Invigorate;Theory	Population density	Spiraling population, QoL, Environment
Smart Cities	Funding; Privacy	Security; Surveillance	Safety
Learning Cities	Digital; Literacies	Engagement funding	Funding data initiative for learning/ambient
Future Cities	Sustainability	Density; Livability	Technocratic governance
Ambient Culture	Meaningful	Multiple modes	Attention; Relevant; Understandable
Ambient Economies	Reliable	Ambient workforce	Economically viable ideas
Ambient Mobility	Environmental	Relevant meaningful	Interactions
Urban Life/Ambient	Environmental	Awareness	Air quality
Engagement	QoL	Happiness	Limitations of awareness; Sustainability
Participation	Role of citizens	Collaboration	Active/Meaningful, Data economics, Rights
Collaboration	Public sector	Authentic, Sustainable	Complexities
Openness & Privacy	Surveillance	Future	Side-channel attacks

Section 1, Table 4 identifies issues, controversies, and problems emerging from the introduction and background literature to urban life and the ambient, highlighting awareness, citizen involvement, collaboration, complexities, engagement, the environment, funding, literacies, meaningfulness, population density, quality of life (QoL), safety, sustainability, and theory.

Section 2, Table 5 identifies issues, controversies, and problems associated with evolving infrastructures, literacies, and data streams for urban life and the ambient, highlighting concerns with algorithms, collaboration, complexity, computational intelligence and costs, computer literacy, privacy, security, social media, and trust.

Section 3, Table 6 identifies issues, controversies, and problems emerging from AI to smart homes, e-vehicles, robotics, and extended realities from the research and practice literature as ambient urban living, highlighting accessibility, autonomy, control, creativity and AI, decision making, health and safety, human-machine interactions, infrastructure, machine learning, privacy, security, social, and trust.

Section 4, Table 7 identifies issues, controversies, and problems emerging from the theory, approaches, and metrics literature guiding understandings of urban life and the ambient, highlighting adaptability, collaboration, environmental, health and well-being, reimagining, quality of life (QoL), resilience, shocks, simulation, social interactions, sociotechnical, sustainability, and trust.

3.2 Overview of Variables Correlated for Urban Life and the Ambient

Variables identified from the research and practice literature for exploration of the propositions in each chapter are presented in Table 8 along with the resulting correlations.

In Chapter 1, variables explored include *success of smart city projects* and *livability* yielding what is considered to be a "very good" (Creswell, 2018) correlation of .81.

In Chapter 2, variables explored included *attuning to urban spaces* and *heightening urban sensibilities* yielding a Spearman correlation coefficient for ordinal data of .26; *sharing* and *urbanizing* (e.g., adapting

Table 5. Section 2: issues, controversies, and problems

Chapter Topics	Issues	Controversies	Problems
Section 2: Evolving Infrastructures, Literacies & Data Streams for Urban Life & the Ambient			
Augmentations	Social	Upgrade infrastructures	Tech-sentient infrastructures
Experiences	As a measure	Ambient media	Relevant, Meaningful
Interactions	Human aspects	Public spaces/openness	Social media + Common good
Urban life & Infrast	Multidimensional	Hyperconvergence/IT	Back-up plans, Diverseness
Algorithmic	Computer literacy	Better programming	Inspectability of code
Digital	Trust	Complexities	Collaboration
Inclusion	Prosperity	Action for inclusion	Policies & practices
Play	Playful practice	Players: smart planners	Complexity
Data	Technocentric	Technocratic	Need for thoughtful care
Information	Computational $	Privacy	Security
Knowledge	Further research	Map as reconstructing	Urban relevance
Knowing & CI	Algorithms	Reading data & effects	New urban science

for urban uses) yielding a correlation of .22; *openness* and *technology-driven services* yielding a correlation of .21; and *multiple modes of transport* and *walkability* yielding a correlation of .23. Correlations in the .20 to .35 range are said to have "only a slight relationship" such that, findings "may be valuable to explore the interconnection of variables but of little value in prediction studies" (Creswell, 2018).

In Chapter 3, variables explored included *collaborating* and *technology-driven services* yielding a correlation of .49; *collaborating* and *privacy* yielding a negative or inverse correlation of .57; and *community participation* and *privacy* yielding a correlation of -.57. According to Creswell (2018), correlations in the .35 to .65 range are "useful for limited prediction" and said to be "the typical values in prediction used to identify variable membership in the statistical procedure of factor analysis."

In Chapter 4, variables explored included *air quality* and *visible infrastructures* yielding a correlation of .57; and *air quality* and invisible infrastructures yielding a correlation of .81.

Table 6. Section 3: issues, controversies, and problems

Chapter Topics	Issues	Controversies	Problems
Section 3: From AI to Smart Homes, E-Vehicles, Robotics & Extended Realities as Ambient Urban Living			
Artificial Intelligence	Enhance life	Urban AI	Creativity & AI
Deep Learning	Accuracy/Precise	Decision making	Reassess policies/services
Machine Learning	AI/ML confusion	Un/Supervised/RL	Privacy
Ambient Learning	AutonomyControl	Ambient literacies	Making the invisible visible
Smart Homes	Support	Privacy & Control	Security
Electric Vehicles	Battery	Infrastructure	Materials
Robotics	Trust	Value	Human-machine interactions
Extended Reality	Health & Safety	Accessibility	Social concerns

Table 7. Section 4: issues, controversies, and problems

Chapter Topics	Issues	Controversies	Problems
Section 4: Theories, Approaches & Metrics in Guiding Understandings of Urban Life & the Ambient			
Theory	Collaboration	Reimagining	Understanding
Approaches	Simulation	Humanize urban design	Visualization
Metrics	QoUL	Sustainability	Prosperity
Governance	Local, day-to-day	Digital transformation	QoL
Livability	Data analysis	Social aspects	Health & Well-being
Sustainability	Adaptability	Resilience	Sociotechnical, Institutional
Human Flourishing	Inspiration	Prosperity	Sustainability
Connections	Environmental	Social, Behavioral	Technical
Positivity	Implementation	Social interactions	Shocks
Resilience	Adapt, Learning	Resistance	Trust

Table 8. Variables and their correlations for exploring urban life and the ambient in smart cities

Chap	Variables		Correlation
1	Success of Smart City projects	Livability	.81
2	Attuning to urban spaces Sharing Openness Multiple modes of transport	Heightening urban sensibilities Urbanizing Technology-driven services Walkability	.26 .22 .21 .23
3	Collaborating Collaborating Community participation	Technology-driven services Privacy Privacy	.49 .57 -.57
4	Air quality Air quality	Visible infrastructures Invisible infrastructures	.57 .81
5	Education Interactive public spaces Interactive public spaces Education Education	Placemaking Inclusiveness ICTs Interactive public spaces ICTs	.57 .36 .57 -.33 -.57
6	Access to public data Information ICTs Meaningfulness	Information ICTs Meaningfulness Ways to show success/real time	.63 .94 .57 .33
7	Collaborating	Creative opportunities	.81
8	Connecting Access to public data/success Access to public data/livability	Trust (aware technologies) Trust (data value) Trust (data value)	.56 .57 .77
9	Collaborating	Visualizations of data that inspire	.33
10	Livability Housing affordability	Housing affordability Useful community functions & purposes	.81 .81
11	Creative opportunities Interactive public spaces Collaboration Connecting	Visualizations of data to inspire Technology-driven services Trust (aware technologies) Technology-driven services	.81 .49 .23 .14

In Chapter 5, variables explored included *education* and *placemaking* yielding a correlation of .57; *interactive public spaces* and *inclusiveness* yielding a correlation of .36; *interactive public spaces* and *ICTs* (information and communication technologies) yielding a correlation of .57; *education* and *interactive public spaces* yielding an inverse or negative correlation of -.33; and *education* and *ICTs* yielding a negative correlation of -.57.

In Chapter 6, variables explored included *access to public data* and *information* yielding a correlation of .63; *information* and *ICTs* yielding a correlation of .94; *ICTs* and *meaningfulness* yielding a correlation of .57; and *meaningfulness* and *ways to show success in real time* yielding a correlation of .33.

In Chapter 7, variables explored included *collaborating* and *creative opportunities* yielding a correlation of .81.

In Chapter 8, variables explored included *connecting* and *trust* involving aware technologies yielding a correlation of .56; *access to public data* involving considerations of success and *trust* involving consideration of data value yielding a correlation of .57; *access to public data* involving livability and *trust* involving data value yielding a correlation of .77.

In Chapter 9, variables explored included *collaborating* and *visualizations of data that educate, inform, and inspire in real time* yielding a correlation of .33.

In Chapter 10, variables explored included *livability* and *housing affordability* yielding a correlation of .81; and *housing affordability* and *useful community functions and purposes* yielding a correlation again of .81.

In Chapter 11, variables explored included *creative opportunities* and *visualization of data that educate, inform, and inspire in real time* yielding a correlation of .81; *interactive public spaces* and *technology-driven services* yielding a correlation of .49; *collaboration* and *trust* involving aware technologies yielding a correlation of .23; and *connecting* and *technology-driven services* yielding a correlation of .14.

In summary, Table 8 shows a wide variety of correlation results for a range of factors explored pertaining to smart cities, learning cities, and future cities in support of insights, ideas, and exercises for further explorations going forward.

3.3 Overview of Insights, Ideas, and Exercises

Table 9 provides a listing by chapter of insights emerging from the explorations conducted.

In Chapter 1, key insights pertain to the importance of funding as a factor contributing to *success* in relation to smart city (SC) projects and, also to *livability* in relation to smart cities. Not to be overlooked is the potential for many more factors to emerge for understanding smart cities when emphasis is placed on the experience of the urban in everyday life.

In Chapter 2, key insights pertain to the importance of city-focused social media and other aware technologies giving rise to possibilities for the ambient in relation to *attuning to urban spaces* and *heightening urban sensibilities* and to ambient culture in the form of *sharing* and *urbanizing* (e.g., adapting for urban uses). Other factors for understanding the ambient include *people interacting with technologies* and *cultural events*.

In Chapter 3, key insights pertain to the importance of city-focused social media and other aware technologies giving rise to possibilities for *engagement* and *participation* in the form of *collaborating* as well as *technology-driven services*, and *privacy* and *community participation*.

In Chapter 4, key insights pertain to the importance of *invisible infrastructure* and experiences and mobile applications in relation to air *quality* for understanding through augmentations and interactions.

Table 9. Overview of insights for exploring urban life and the ambient in smart cities

Chapter	Insights
Chapter 1	- Importance of funding as a factor contributing to success in relation to SC projects - Importance of funding as a factor contributing to livability in relation to SCs - Potential for many more factors for understanding SCs experience of the urban
Chapter 2	- City-focused social media & other aware technologies giving rise to many possibilities: *the ambient* as *attuning to urban spaces & heightening urban sensibilities* - City-focused social media & other aware technologies as factors giving rise to *ambient culture* in the form of *sharing & urbanizing* (e.g., adapting for urban uses) - Potential for many more factors in understanding the ambient in SCs
Chapter 3	- City-focused social media & other aware technologies as factors giving rise to many possibilities for *urban engagement & participation* in the form of *collaborating* - Factors related to *urban engagement & participation* in SCs as in *technology-driven services* - Factors to explore *urban engagement & participation* in SCs *privacy & community participation*
Chapter 4	- Factors contributing to SCs – *invisible infrastructures & experiences* in the form of *air quality* - Importance of factors associated with *invisible infrastructures* in SCs: mobile apps *air quality* - Understanding *evolving urban infrastructures* in SCs - augmentations & interactions
Chapter 5	- Elements contributing to the success of a smart city project as factors giving rise to many possibilities for *urban literacies* in the form of *education* - Factors in SCs such as *placemaking & interactive public spaces & inclusiveness* - Understanding *urban literacies* in SCs using factors such as *play & the algorithmic*
Chapter 6	- The relationship between urban data & information in the form of *access to public data & information* as factors contributing to livability & the making of a SC - Importance of *ICTs & meaningfulness* associated with *data, info & knowledge* in SCs - Understanding *urban data, information & knowledge*; *visual ways to show success in real time*
Chapter 7	- City-focused social media & other aware technologies as factors giving rise to many possibilities for urban AI in support of more aware people in the form of *collaborating* - Importance of factors associated with urban livability & taking the needs of people into consideration in SCs in the form of *creative opportunities* - Understanding urban AI, creativity, and the ambient in SCs as ambient learning
Chapter 8	- The relationship between the value of urban data in the form of *access to public data* as a factor contributing to livability correlated with *trust* in the making of a SC - Importance of factors such as *trust* when associated with *connecting* in SCs - *Smart homes, EVs & robotics, extended reality* in SCs & associated capabilities
Chapter 9	- Relationship between *collaborating & visualizations of data that inspire* in SCs - Importance of factors pertaining to people & their needs in SCs - *Theory, approaches & metrics* in SCs as capabilities involving design thinking
Chapter 10	- Importance of the relationship between *livability & housing affordability* in SCs - Factors for people & needs in SCs: *useful community functions & purposes* - *Governance, livability & sustainability* in SCs as ambient - dynamic, adaptable
Chapter 11	- Relationship between *creative opportunities & visualizations of data that inspire* - People & their needs; connection guided by *interactive public spaces & tech-driven service* - *Connection, positivity & resilience* in SCs in relation to the notion of an ambient future

In Chapter 5, key insights pertain to the importance of factors related to the success of smart city projects giving rise to potentials for *urban literacies* as *education.* Also important are factors such as *placemaking, interactive public spaces,* and *inclusiveness* and the potential for other literacies such as *play* and *the algorithmic*.

In Chapter 6, key insights pertain to the importance of the relationship between urban data and information explored through *access to public data* and *information* as factors contributing to livability and the making of a smart city along with factors such as *ICTs* (information and communication technologies) and *meaningfulness* associated with data, information, and knowledge in smart cities and other potentials when used as *visual ways to show success in real time*.

In Chapter 7, key insights pertain to the importance of city-focused social media and other aware technologies as giving rise to possibilities for urban AI in support of more aware people through *collaborating* along with factors associated with urban livability attentive to people and their needs in the form of *creative opportunities* while opening the way for other factors to emerge such as ambient learning.

In Chapter 8, key insights pertain to the importance of the value of urban data as in *access to public data* as a factor contributing to livability in relation to *trust* and the importance of *trust* in relation to *connecting* in smart cities along with the potential for many factors to emerge for understanding *smart homes, EVs, robotics, extended reality,* and associated capabilities.

In Chapter 9, key insights pertain to the importance of the relationship between *collaborating* and *visualizations of data that educate, inform, and inspire in real time* in smart cities along with the importance of factors associated with people and their needs and factors for understanding *theory*, *approaches*, and *metrics* as in, capabilities related to the emergent, involving design thinking.

In Chapter 10, key insights pertain to the importance of the relationship between *livability* and *housing affordability* in smart cities as well as factors attentive to people and their needs such as *useful community functions* and *purposes* along with many other potential factors related to *governance*, *livability*, and *sustainability* in terms of the ambient as in, more dynamic, emergent, and adaptable.

In Chapter 11, key insights pertain to the importance of the relationship between *creative opportunities* and *visualizations of data that educate, inform, and inspire in real time* as well as factors such as *interactive public spaces* and *technology-driven services* with the potential to understand *connection*, *positivity*, and *resilience* in smart cities in relation to the notion of an ambient future.

Table 10 provides a listing by chapter of ideas emerging from the explorations conducted on urban life and the ambient in smart cities.

In Chapter 1, key ideas pertain to the opportunities potential for exploring funding from different perspective in urban life and smart cities research while on a practice level, exploring urban initiatives involving more creative funding arrangements. Also of note is the use of a componential approach for the study of urban life and the ambient in smart, learning, and future cities.

In Chapter 2, key ideas pertain to exploring the ambient from different perspectives in the context of urban life in smart cities while noting the opportunities potential of scenarios involving *ambient culture*, *ambient economies*, and *ambient mobilities* and the use of an interdisciplinary approach in the study and practice of urban life and the ambient in smart, learning, and future cities.

In Chapter 3, key ideas pertain to the exploration of *engagement* and *participation* in relation to variables such as *collaboration*, *openness*, and *privacy* in smart cities, among many other potential factors, and the use of interdisciplinarity and multi-methods as approaches to study and practice.

In Chapter 4, key ideas pertain to the opportunities and challenges of exploring evolving *urban infrastructures* in relation to *augmentations*, *experiences*, *interactions*, and other factors in smart cities and environments using interdisciplinary and multi-method approaches.

In Chapter 5, key ideas pertain to the opportunities potential of exploring *urban literacies* using variables such as the *algorithmic*, the *digital*, *inclusion*, *play*, and other factors in smart cities and environments using interdisciplinary and multi-method approaches.

In Chapter 6, key ideas pertain to the opportunities potential for explorations of urban life and the ambient in relation to *data*, *information*, *knowledge*, and other factors in smart cities and environments using an interdisciplinary and multi-method approach for both research and practice.

Table 10. Overview of ideas for exploring urban life and the ambient in smart cities

Chapter	Ideas
Chapter 1	- Potential for exploring funding from different perspective in urban life & SCs - Potential for urban initiatives involving more creative funding arrangements - Componential approach for the study of urban life & smart, learning & future cities
Chapter 2	- Exploring the ambient from different perspectives in urban life & SCs - Urban scenarios involving the ambient as in ambient culture, economies, mobilities - Interdisciplinary approach for study & practice of urban life & the ambient in SCs
Chapter 3	- Exploring urban engagement & participation in relation to variousvariables in SCs - Identify variables for exploring urban engagement & participation in smart spaces - Interdisciplinary methods for the study & practice of engagement in SCs
Chapter 4	- Exploring evolving urban infrastructures in relation to a variety of variables - Identify variables for exploring evolving urban infrastructures in smart environments - Interdisciplinary methods for the study & practice of infrastructures in SCs
Chapter 5	- Potential for exploring urban literacies in relation to a variety of variables in SCs - Potential for identifying variables for exploring urban literacies in smart spaces - Interdisciplinary methods for the study & practice of urban literacies
Chapter 6	- Exploring urban life & the ambient in relation to a variety of variables in SCs - Identify variables for exploring urban life & the ambient in smart environments - Interdisciplinary methods for the study and practice of urban life & data in SCs
Chapter 7	- Exploring urban AI & the ambient in relation to a variety of variables in SCs - Potential for identifying variables for exploring urban AI & ambient learning - Interdisciplinary methods for the study & practice of urban AI & the ambient in SCs
Chapter 8	- Potential opportunities for exploring urban life & the ambient in relation to a variety of variables pertaining to smart homes, EVs & robotics in the context of smart cities - Opportunities potential for the identification of other & additional variables for exploring methods smart environments in relation to mobility & extended reality - Use of an interdisciplinary & multi-method approach for the study & practice of urban life & access to public data, connecting & trust in smart, learning & future cities
Chapter 9	- Potential opportunities for exploring urban life & the ambient in relation to a variety of variables pertaining to theory, approaches & metrics in the context of SCs - Opportunities potential for the identification of other & additional variables for exploring smart environments in relation to ambient theory or ambient metrics - Use of an interdisciplinary & multi-method approach for the study & practice of urban life & data visualizations & collaborating in smart, learning & future cities
Chapter 10	- Potential opportunities for exploring urban life & the ambient in relation to a variety of variables pertaining to ambient governance in the context of SCs - Opportunities potential for the identification of additional variables for exploring urban life & the ambient in smart environments in relation to ambient livability - Use of more integrated approaches to governance, allowing for increased collaboration across sectors in the making of smarter, more sustainable & livable cities
Chapter 11	- Exploring urban life & the ambient in relation to connection, positivity & resilience - Opportunities for identifying variables in exploring human flourishing smart environments in relation to ambient connection & ambient resilience - Use of more integrated approaches to human flourishing with increased collaboration across sectors for smarter, more sustainable & livable cities, now & for the future

In Chapter 7, key ideas pertain to potential opportunities for exploring urban AI and the ambient with variables such as *collaboration*, *creative opportunities*, and others in smart cities and environments to understanding more about ambient learning using interdisciplinary and multi-method approaches.

In Chapter 8, key ideas pertain to potential opportunities for exploring urban life and the ambient in relation to smart homes, EVs, robotics, and extended reality through *access to public data*, *connecting*, and *trust* as well as augmented, virtual, and mixed realities using interdisciplinary and multi-method approaches in the context of smart, learning, and future cities.

In Chapter 9, key ideas pertain to opportunities for exploring urban life and the ambient in relation to theory, approaches, and metrics and more specifically ambient theory and ambient metrics in smart environments using interdisciplinary and multi-method approaches involving data visualizations and collaborating.

In Chapter 10, key ideas pertain to opportunities for exploring urban life and the ambient in relation to *governance*, *livability*, and *sustainability* extending to the potential for variables in smart environments in relation to ambient governance, ambient livability, and ambient sustainability involving a more integrated approach through cross-sector collaboration in smart, learning, and future cities.

In Chapter 11, key ideas pertain to opportunities for exploring urban life and the ambient in relation to connection, positivity, resilience, and other factors in learning more about human flourishing and possibly ambient connection and ambient resilience using more integrated approaches involving collaboration across sectors in smart, learning, and future cities.

Table 11 provides a listing by chapter of exercises encouraging further explorations on urban life and the ambient in smart cities.

In Chapter 1, suggested exercises pertain to identifying factors in your city contributing to livability; ways in which funding would enhance quality of life (QoL) in your city; and elements that contribute to making your city a learning city.

In Chapter 2, suggested exercises pertain to identifying factors in your city that give rise to the ambient; ways in which the ambient would enhance the quality of life (QoL) in your city; and elements that contribute to making ambient culture, ambient economies, or ambient mobility possible in your city.

In Chapter 3, suggested exercises pertain to identifying factors in your city that are giving rise to engagement and participation; ways in which collaboration would enhance the quality of life (QoL) in your city; and elements that contribute to making engagement and participation possible in your city.

In Chapter 4, suggested exercises pertain to identifying factors in your city that are giving rise to invisible infrastructures; ways in which the augmentations to urban infrastructures would enhance the quality of life (QoL) in your city; and elements that contribute to evolving infrastructures in your city.

In Chapter 5, suggested exercises pertain to identifying factors in your city that are giving rise to the need for urban literacies; ways in which inclusiveness would enhance the quality of urban life (QoUL) in your city; and elements that would support urban literacies in your city.

In Chapter 6, suggested exercises pertain to identifying one or more ways in your city that data and information matter; ways in which data and information would enhance the quality of urban life (QoUL) in your city; and elements that contribute to ways of knowing in your city.

In Chapter 7, suggested exercises pertain to identifying ways in which your city is using machine learning (ML) and/or deep learning (DL); ways in which artificial intelligence (AI) would enhance the quality of urban life (QoUL) in your city; and one or more ways that ambient learning contributes to making your city smarter.

In Chapter 8, suggested exercises pertain to identifying one or more ways in your city that smart homes or electric vehicles (EVs) matter; one or more ways in which extended reality would enhance the quality of life (QoL) in your city; and one or more ways in which robotics would enhance the quality of life in your city.

Table 11. Listing of exercises for exploring urban life and the ambient in smart cities

Chapter	Exercises
Chapter 1	- Identify one or more factors in your city that are contributing to livability - Identify one or more ways in which funding would enhance the QoUL in your city - Describe one or more elements that contribute to making your city a learning city
Chapter 2	- Identify one or more factors in your city that are giving rise to the ambient - Identify one or more ways in which the ambient would enhance the QoUL in your city - Describe one or more elements that contribute to making ambient culture, ambient economies, or ambient mobility possible in your city
Chapter 3	- Identify one or more factors in your city that are giving rise to engagement and participation - Identify one or more ways in which collaboration would enhance the QoUL in your city - Describe one or more elements that contribute to making engagement & participation possible in your city
Chapter 4	- Identify one or more factors in your city that are giving rise to invisible infrastructures - Identify one or more ways in which augmentations to urban infrastructures would enhance the quality of urban life in your city - Describe one or more elements that contribute to evolving urban infrastructures in your city
Chapter 5	- Identify one or more factors in your city that are giving rise to the need for urban literacies - Identify one or more ways in which inclusiveness would enhance the QoUL in your city - Describe one or more elements that would support urban literacies in your city
Chapter 6	- Identify one or more ways in your city that data and information matter - Identify one or more ways in which data & information would enhance the QoUL in your city - Describe one or more elements that contribute to ways of knowing in your city
Chapter 7	- Identify one or more ways in which your city is using ML and/or deep learning (DL) - Identify one or more ways in which AI would enhance the quality of urban life in your city - Describe one or more ways that ambient learning contributes to making your city smarter
Chapter 8	- Identify one or more ways in your city that smart homes or electric vehicles (EVs) matter - Identify one or more ways in which extended reality would enhance the QoL in your city - Describe one or more ways in which robotics would enhance the quality of life in your city
Chapter 9	- Identify one or more ways in your city that urban theory matters - Identify one or more ways in which new approaches would enhance the QoL in your city - Describe one or more ways in which metrics would enhance the quality of life in your city
Chapter 10	- Identify one or more ways in your city that smart urban governance matters - Identify one or more ways that some aspect of livability would enhance the QoL in your city - Describe one or more ways that a sustainability mindset would enhance the QoL in your city
Chapter 11	- Identify one or more ways in your city that connection matters - Identify one or more ways that some aspect of positivity would enhance the QoL in your city - Describe one or more ways that a resilience mindset would enhance the QoL & human flourishing in your city
Chapter 12	- Identify one or more elements key to SCs, LCs, FCs that seem to be missing from this book - Identify one or more ways that some aspect of flourishing would enhance the QoL in your city - Describe one or more ways that flourishing could benefit your city or the Planet

In Chapter 9, suggested exercises pertain to identifying one or more ways in your city that urban theory matters; one or more ways in which new urban approaches would enhance the quality of life (QoL) in your city; and one or more ways in which urban metrics would enhance the quality of life in your city.

In Chapter 10, suggested exercises pertain to identifying one or more ways in your city that smart urban governance matters; one or more ways in which some aspect of livability would enhance the quality of life (QoL) in your city; and one or more ways in which a sustainability mindset would enhance the quality of life in your city.

In Chapter 11, suggested exercises pertain to identifying one or more ways in your city that connection matters; one or more ways in which some aspect of positivity would enhance the quality of life (QoL) in your city; and one or more ways in which a resilience mindset would enhance the quality of life and human flourishing in your city.

In Chapter 12, suggested exercises, appearing near the end of this chapter (Section 5), pertain to identifying one or more elements critical to smart, learning, or future cities that seem to be missing in this book; one or more ways in which some aspect of flourishing would enhance the quality of life in your city; and describing one or more ways that flourishing could benefit your city and the Planet.

4. PATTERNS, TRENDS, AND FUTURE DIRECTIONS

Patterns and trends emerging from this work appear in a variety of ways in relation, for example, to the interdisciplinary literature reviews across chapters; the issues, controversies, and problems identified; and the correlations among variables. The components identified for smart cities in Fig. 1 pertain also to learning cities, future cities, and human flourishing. The research questions (Table 1) and propositions (Table 2) each pertain to a particular component of the framework (Fig. 1). Issues, controversies, and problems that recur across components (Tables 4-7) include, for example, collaboration, complexity, environment, livability, privacy, quality of life (QoL), security, sustainability, and trust.

Strongly correlated variables as identified in Table 8 such as those with a Spearman correlation coefficient for ordinal data of .81 pertain to *livability* and *housing affordability*, *livability* and *success of smart city projects*, *air quality* and *invisible infrastructures*, *housing affordability* and *useful community functions and purposes*, and *creative opportunities* and *visualization of data to educate, inform, and inspire in real time*. It is worth noting that *trust* in the context of awareness enabling technologies in relation to *collaboration* yields a lower correlation of .23, suggestive of challenges that require further exploration, and that trust and collaboration may be dynamic and fluid factors in smart environments. The lower correlation of .14 between connecting and technology-driven services may be suggesting that connectivity is not always available to everyone on the one hand and on the other hand, on a non-technical level, the human dimension of connecting may need to be considered.

Pointing to the "long and unsettled future of smart cities", Karvonen, Cugurullo, and Caprotti (2018) speak of the need for inspirations pertaining to interpretations for smart technologies and it is hoped that this book begins to respond to this need. It is worth noting that Puron-Cid and Gil-Garcia (2022) question whether smart cities are too expensive in the long term, finding financial sustainability (FS) "to be a useful perspective to better understand smart city projects over time" building toward a more integrative perspective using "other financial analytical tools and frameworks." Indeed, coming full circle from Chapter 1 when the importance of funding was explored in relation to livability and the success of smart city projects, it would seem that opportunities exist for the development of smarter, more sustainable approaches to funding and financing smart, learning, and future cities. Also worth noting is the claim by Jacobs (2022) that the city of Toronto "wants to kill the smart city forever" although perhaps what is needed is a broader more flourishing understanding of smart cities.

While limitations of this work may pertain to the extent of the literature review coverage in some chapters, this opens the way for further research opportunities going forward and ongoing dialogue. Indeed, the literature reviews presented throughout this book could be said to provide a particular strength for this work in terms of the unique collections of research and practice that are brought together for discussion, inspiration, and interpretation. Additionally, limitations of this work associated possibly with the variables explored provide opportunities going forward for reflection and debate on more effective variables for exploration in future research. Given the small sample size used in the study underlying this work, further exploration, testing, and validation of the variables used is encouraged.

5. CONCLUSION

An understanding of smart cities, learning cities, and future cities is offered in this work through the lens of urban life and the ambient, consisting of a componential approach incorporating multiple dimensions and an interdisciplinary perspective. Following an exploration across chapters of the many components of smart cities in relation to issues, controversies, and problems, this book seeks to navigate a path forward assisted by the notion of not just human flourishing (Ekman and Simon-Thomas, 2021; CGC, 2014), but also, flourishing on the Planet (Hedlund and Esbjörn-Hargens (eds.), 2022; Miller, Douglass, and Rigg, 2020) more generally, for all life forms and for the Earth, while having consideration for forays into outer space and the broader universe that surrounds us. Batty (2018) argues that "we can invent – but not predict – the future of cities" and this is encouraging, suggesting as it does the importance of our actions, creativity, and imagination. As such, the componential approach to smart, learning, and future cities advanced in this work is intended to provide a supporting, integrated, and expansive framework. And while caution is encouraged for the predictive capabilities of the correlation of variables conducted in this work for understanding smart cities, opportunities for identifying factors that matter in smart, learning, and future cities offer creative challenges and directions going forward.

A key item provoking reflection on this work is the statement by an educator who participated in the research study underlying this research, "place is about living and thriving as human beings." Also, worth noting and reflecting upon is the idea that human flourishing (Ekman and Simon-Thomas, 2021; CGC, 2014) involves enhancing human capabilities, which is encouraged throughout the chapters of this book.

Exercises for the reader to consider, encouraging further exploration include the following:

1. Identify one or more elements critical to smart, learning, or future cities that seem to be missing from this book
2. Identify one or more ways in which some aspect of flourishing would enhance the quality of life in your city
3. Describe one or more ways in which flourishing could benefit your city and the Planet more generally

Reflecting upon and engaging with these exercises is important for everyone so that the target audience for this book is intended to be unlimited.

REFERENCES

Batty, M. (2018). *Inventing future cities.* MIT Press. doi:10.7551/mitpress/11923.001.0001

CGC. (2014). How can we make cities flourish? In *Challenges of Government Conference*. Blavatnik School of Government, University of Oxford. Retrieved 3 March 2022 from https://www.bsg.ox.ac.uk/news/how-can-we-make-cities-flourish

Creswell, J. W. (2018). *Educational research: Planning, conducting, and evaluating quantitative and qualitative research* (6th ed.). Pearson.

Ekman, E., & Simon-Thomas, E. (2021). Teaching the science of human flourishing, unlocking connection, positivity, and resilience for the greater good. *Global Advances in Health and Medicine: Improving Healthcare Outcomes Worldwide, 10*. Advance online publication. doi:10.1177/21649561211023097 PMID:34123578

Hedlund, N., & Esbjörn-Hargens, S. (Eds.). (2022). *Big Picture Perspectives on Planetary Flourishing: Metatheory for the Anthropocene* (Vol. 1). Routledge. doi:10.4324/9781003140313

Jacobs, K. (2022). Toronto wants to kill the smart city forever: The city wants to get right what Sidewalk Labs got so wrong. *MIT Technology Review*. Retrieved 7 July 2022 from https://www.technologyreview.com/2022/06/29/1054005/toronto-kill-the-smart-city/

Karvonen, A., Cugurullo, F., & Caprotti, F. (2018). Conclusions: The long and unsettled future of smart cities. In A. Karvonen, F. Cugurullo, & F. Caprotti (Eds.), *Inside Smart Cities: Place, Politics and Urban Innovation*. Routledge. doi:10.4324/9781351166201-19

Miller, M. A., Douglass, M., & Rigg, J. (2020). Governing resilient cities for planetary flourishing in the Asia-Pacific. *Urban Studies (Edinburgh, Scotland), 57*(7), 1359–1371. doi:10.1177/0042098020903955

Puron-Cid, G., & Gil-Garcia, J. R. (2022). Are smart cities too expensive in the long term? Analyzing the effects of ICT infrastructure on municipal financial sustainability. *Sustainability, 14*(10), 6055. doi:10.3390u14106055

ADDITIONAL READING

Arroyo, K., Fakunle, D., Franco, P. I., Grant, K., & Verstraete, K. (2022). The creative power of the arts: Reimagining human and planetary flourishing. *Issue Lab*. Retrieved 19 May 2022 from https://search.issuelab.org/resource/the-creative-power-of-the-arts-reimagining-human-and-planetary-flourishing.html

Beatley, T. (2020). Biophilic Cities. In V. Loftness (Ed.), *Sustainable Built Environments. Encyclopedia of Sustainability Science and Technology Series*. Springer. doi:10.1007/978-1-4939-2493-6_1033-2

Beatley, T. ([2020]). *Biophilic cities: Design for planetary health and human flourishing*. University of Virginia School of Architecture Urban and Environmental Planning. Retrieved 19 May 2022 from https://olc.worldbank.org/system/files/1.%20Tim%20Beatley%20-%20Biophilic%20cities.pdf

Brown, J. (2021). How cities are going carbon neutral. *BBC, 2045: Memories of the Future | Cities*. Retrieved 17 November 2021 from https://www.bbc.com/future/article/20211115-how-cities-are-going-carbon-neutral

Burbano, L. (2022). Neom, the smart city built from scratch in the Arabian Desert. *Tomorrow City Blog*. Retrieved 4 August 2022 from https://tomorrow.city/a/neom-saudi-arabia

Cheshire, J., & Uberti, O. (2021). *Atlas of the invisible: Maps and graphics that will change how you see the world*. Norton.

Constable, H. (2020). How do you build a city for the pandemic? The new coronavirus has spread rapidly in cities around the globe. How might the virus make us think differently about urban design in the future? *BBC*. Retrieved 28 October 2021 from https://www.bbc.com/future/article/20200424-how-do-you-build-a-city-for-a-pandemic

Daly, J., Dovey, K., & Stevens, Q. (2020). We can't let coronavirus kill our cities: Here's how we can save urban life. *The Conversation*. Retrieved 14 September 2021 from https://theconversation.com/we-cant-let-coronavirus-kill-our-cities-heres-how-we-can-save-urban-life-137063

Dovey, K., Morley, M., & Stevens, Q. (2021). What next for parklets? It doesn't have to be a permanent switch back to parking. *The Conversation*. Retrieved September 14, 2021 from https://theconversation.com/what-next-for-parklets-it-doesnt-have-to-be-a-permanent-switch-back-to-parking-159534

Franck, K., & Stevens, Q. (2007). *Loose Space: Possibility and Diversity in Urban Life*. Routledge.

Mukherji, A. (2020). Human development is dependent on planetary well-being. *International Science Council Blog*. Retrieved 19 May 2022 from https://council.science/current/blog/human-development-is-about-flourishing-within-planetary-boundaries/

KEY TERMS AND DEFINITIONS

Planetary Flourishing: Planetary flourishing in the context of urban life and the ambient in smart, learning, and future cities refers to a seeking of the good, for all life forms and for the planet.

About the Author

H. Patricia McKenna is the founder and President of AmbientEase (Emergent Adaptive Solutions Everywhere) Inc., exploring the rapidly evolving information landscape afforded by new, emerging, and next generation technologies. McKenna collaborates in team efforts to set up international, national, regional, and local information services, research projects, startups, and other creative and future-oriented initiatives. As founder and Director of the Urbanities Lab, McKenna is concerned with the implications of new technologies for cities, communities, and learning environments. McKenna engages with educators, learners, business, government, and local communities around use experience and unexpected possibilities for leveraging and generating new relevancies and vibrancies in 21st century information spaces for learning, living, and working. With a focus on smart cities, learning cities, and future cities, McKenna is the author of some 50 publications and has evolved a practice and learning space consisting of workshops, courses, project initiatives, community collaborations, presentations, and works in progress for sharing at conferences and in other physical and digital spaces. Patricia holds an undergraduate arts degree from the University of New Brunswick, a Master's degree in library science from McGill University, and a Doctorate in information management (with a focus on emerging technologies) from Syracuse University, iSchool.

Index

D

E

F

H

I

K

L

M

O

P

Q

R

S

T

U

V

W

www.ingramcontent.com/pod-product-compliance
Lightning Source LLC
LaVergne TN
LVHW080536300926
849162LV00007B/1624